Second Edition

Applying
Genomic and Proteomic
Microarray Technology
in Drug Discovery

Second Edition

Applying Genomic and Proteomic Microarray Technology in Drug Discovery

Robert S. Matson

CRC Press
Taylor & Francis Group
Boca Raton London New York

CRC Press is an imprint of the
Taylor & Francis Group, an **informa** business

First published in paperback 2024

First published 2013
by CRC Press
2385 NW Executive Center Drive, Suite 320, Boca Raton FL 33431

and by CRC Press
4 Park Square, Milton Park, Abingdon, Oxon, OX14 4RN

First issued in hardback 2019

CRC Press is an imprint of Taylor & Francis Group, LLC

© 2013, 2019, 2024 Taylor & Francis Group, LLC

Library of Congress Cataloging-in-Publication Data

Matson, Robert S.
  Applying genomic and proteomic microarray technology in drug discovery / Robert S. Matson. -- 2nd ed.
    p. ; cm.
  Includes bibliographical references and index.
  ISBN 978-1-4398-5563-8 (hardcover : alk. paper)
  I. Title.
  [DNLM: 1. Drug Design. 2. Genomics--methods. 3. Oligonucleotide Array Sequence Analysis--methods. 4. Proteomics--methods. QV 745]

615.1'9--dc23   2012037826

ISBN: 978-1-4398-5563-8 (hbk)
ISBN: 978-1-03-292207-2 (pbk)
ISBN: 978-0-429-10783-2 (ebk)

DOI: 10.1201/b13876

Visit the Taylor & Francis Web site at
http://www.taylorandfrancis.com

and the CRC Press Web site at
http://www.crcpress.com

Dedication

To my family—Jeanne, Erik, and Jacqueline.

Contents

Preface

Microarrays remain an invaluable tool for omics-based research in the life sciences, for drug discovery, clinical research, and diagnostic applications worldwide. They are used for gene expression analysis, the assessment of genomic mutation, and multiplex immunoassays. We can create patterns of macromolecules such as enzymes and living cells with array-based technologies. They have a future in companion diagnostics and personalized medicine. The purpose of this book is to introduce microarrays into your future.

In this second edition I have sought to add new information about the construction and use of microarrays while preserving the important earlier works that demonstrated key concepts. The commercial landscape has dramatically changed over the past few years so it has been necessary to make an accounting of these changes and reflect upon the contributions that these companies provided in the advancement of array technology. There are also new surface chemistries and immobilization strategies that have emerged for tethering and patterning of nucleic acids, proteins, carbohydrates, cells, and other ligands. Protein microarrays are used extensively. Thus, new applications as well as the performance of antibody arrays have been reviewed. A new chapter, "Multiplex Assays," has been added to examine the development and application of arrays across diverse platforms. We discuss applications for qPCR, multiplex lateral flow, multiplex bead assays, as well as microarrays; and offer platform-to-platform comparisons.

There is little doubt that these *omics* tools have changed our approach to problem solving. We are said to be omics-driven rather than hypothesis-driven, which is a kind of "shoot first and ask questions later" approach to discovery. After all of these years, I still view microarrays as an enormous opportunity.

Robert S. Matson, PhD, FACB
Orange, California

The Author

Robert S. Matson is president and co-founder of QuantiScientifics, LLC. The company is involved in the commercialization of multiplexed assays for life science, clinical research, and in vitro diagnostics. Previously, Matson was involved in the research and development of microarray technologies, detection chemistries, as well as point-of-care devices for more than seventeen years while at Beckman Coulter, Inc. He participated in the National Institute of Standards and Technology's (NIST) Advanced Technology Program sponsored Genosensor Consortium and collaborated with Sir Edwin Southern on the development of an in situ oligonucleotide array synthesis platform for the corporation. Other work included development of the A^2 MicroArray System, a microplate-based array platform for multiplexed micro-ELISA, which QuantiScientifics recently licensed for commercialization.

Matson has been granted twelve U.S. patents and six European patents on nucleic acid and protein microarray technology. He was inducted into Beckman Coulter's Inventors Hall of Fame in 2006 and was elected a Fellow of the National Academy of Clinical Biochemistry in 2008. He previously served in several technical management roles including Research and Development (R&D) Director, BioProbe International; R&D Director, Costar-Nuclepore; and Chemistry R&D Group Leader at BioRad Laboratories. Matson earned his PhD in biochemistry from Wayne State University. Following postdoctoral studies at the UCLA Medical School, he served as a principal investigator with the Veterans Administration Medical Center and as adjunct professor of biological chemistry at the University of California-Davis Medical School. He has also held a faculty lectureship at USC's Department of Chemistry and was assistant professor of chemistry at the University of Southern Maine, Portland. He is the author of *Applying Genomic and Proteomic Microarray Technology in Drug Discovery* (CRC Press, 2005) and *Microarray Methods and Protocols* (CRC Press, 2009).

1 Omics and Microarrays Revisited

These biomarker patterns will provide more individualized information, which will then provide support to clinicians and selection of optimal therapies.

Thomas O. Joos[*]

INTRODUCTION

In the previous edition of this book (2005) we spoke of an "omics-driven" paradigm shift toward quantitative biology in which the microarray played a key role. Indeed, we have witnessed the continued advancement of DNA-array-based platforms as powerful gene expression analysis tools for biomarker discovery. Protein arrays continue to play a role in biomarker discovery and validation but are also in demand for both clinical research and clinical diagnostic applications (Yu et al., 2010). Clearly, much debate remains over the clinical role for biomarkers, as well as the validation and regulation of derived multianalyte tests and multivariate index assays (Amur et al., 2008; Tahara et al., 2009; Boja et al., 2011; Smith, 2011; Dimond, 2012; Wellhausen & Seitz, 2012).

Nevertheless, microarrays are now transitioning for use in companion diagnostics. For example, Merck and Roche began collaboration in 2011 to develop the Affymetrix's AmpliChip p53 for companion diagnostics. Roche Diagnostics will validate and standardize tests for Merck's use in qualifying patients in clinical oncology trials. Clinical proteomics has emerged as a discipline; the first protein-based multivariate index assay, the OVA1 test for ovarian cancer, received U.S. Food and Drug Administration (FDA) approval in 2009.

We have also witnessed the emergence of other omics-driven research employing microarray technologies. The *glycomics* field (see Consortium for Functional Glycomics [CFG] at www.functionalglycomics.org) has adapted microarrays for the screening of glycan-protein interactions. For example, Smith and co-workers at Emory University (Atlanta, Georgia) have developed "shotgun glycomics" as a means to identify and characterize the functionality of glycans (Song et al., 2011). Oligosaccharide (glycan) arrays have been used in identifying immunogenic carbohydrate moieties on the pathogen, *Bacillus anthracis* exosporium (Wong et al., 2007). The Stanford University group has been active in the development of photochemical micropatterning of carbohydrates (Carroll et al., 2006) and in the photogeneration of epitope-specific glycan arrays using phthalimide terminated

---

[*] Protein Microarrays for Personalized Medicine, *Clin Chem* 56(3), 376–387, 2010.

self-assembled monolayers (PAMs) on photo-reactive surfaces (Wang et al., 2002). Godula and Bertozzi (2010) describe a poly(acryloyl hydrazide) (PAH) scaffold for the rapid immobilization of reducing glycans. They used biotin terminated PAH conjugated glycans to assemble a microarray on streptavidin-coated glass slides. The utility of glycan microarrays for elucidating the glycome was recently reviewed in the *Annual Review in Biochemistry* (Rillahan & Paulson, 2011). For additional information regarding the preparation and use of carbohydrate arrays as well as lectin microarrays, you are encouraged to review *Microarray Methods and Protocols* (Matson, 2009).

Metabolomics seeks to understand the functionality (pathways, networks) and biochemical status of metabolites within the *metabolome*, an accounting of approximately 1800 biochemicals with molecular masses of under 1500 daltons. These would include small molecules and compounds such as peptides, amino acids, sugars, antioxidants, and lipids. Psychogios et al. (2011) applied a number of metabolomic analytical tools to begin characterization of the serum metabolome (see the Serum Metabolome Database [SMDB] Web site at www.serummetabolome.ca). The SMDB has assembled at least 4500 biochemicals present in serum and plasma with information regarding concentration levels in normal and disease states. They used NMR (nuclear magnetic resonance) spectroscopy, along with various hyphenated mass spectroscopy (MS) methods: GC-MS (gas chromatographic MS), LC-MS (liquid chromatographic MS), TLC (thin-layer chromatographic)/GC-MS, as well as tandem mass spectroscopy (MS/MS) such as UPLC- (ultra-high-performance LC) MS/MS, and direct flow injection (DFI) MS/MS in their quest to identify, quantify, and validate plasma and serum metabolites. Progress in microbial metabolomics has been reviewed by Jane Tang from the Center for National Security and Intelligence (Tang, 2011). The author points out that studies aimed at an understanding of the gut metabolome are particularly important because microbes are intimately involved in drug detoxification, xenobiotics, allergy, and human disease. What is not used to any appreciable extent in metabolomics research at this time is microarray technology. However, Borgan et al. (2010) applied both gene expression microarray (transcriptome) analysis and magnetic resonance spectroscopy (metabolome) for investigations that resulted in a more refined subclassification of breast cancers.

Transcriptomics involves characterizing the *transcriptome*, essentially determining the functionality and relationship of some 100,000 mRNA transcripts being processed within cells of an organism. DNA microarrays have been extensively used for gene expression studies and the temporal analysis of transcripts. In their review, Malone and Oliver (2011) compared the virtues of gene expression microarrays and deep sequencing of RNA based upon RNA-Seq analysis. mRNA-Seq involves the use of fragment libraries of RNA that are then converted to cDNA by reverse transcription. This is followed by an adaptor ligation step and sizing for use on next-generation sequencing platforms. While RNA-Seq may be more applicable for discovery, the authors found both microarrays and deep sequencing to be useful: "We conclude that microarrays remain useful and accurate tools for measuring expression levels, and RNA-Seq complements and extends microarray measurements" (p. 1).

THE MICROARRAY FORMAT

The microarray, in its most elementary form, is but a collection of small spots of biological capture agents (DNA, antibody, carbohydrate, etc.) organized on a planar substrate such as a glass microscope slide. With this tool we are free to move beyond our rather myopic views of the cell in favor of a more global assessment of the cellular process. Moreover, the microarray provides us with *digital* information that we can assemble and then use to quantify biological events and relationships on a scale that was unimaginable a few decades ago. Finally, the microarray is a *parallel processor* providing the researcher with a rapid response to the biological query. In fact, so much data can be obtained in such a short time that it can be overwhelming, most often requiring the aid of sophisticated *bioinformatics analysis* software. Thus, the microarray has become a formidable instrument by which to quantify biology. The purpose of this book is to assess progress on the utility of microarray technology to solve important biological problems.

TERMS AND DEFINITIONS

Array technology finds its origins within molecular biology, a scientific field of study notoriously irreverent to the rules governing systematic nomenclature (Figure 1.1). Moreover, microarray applications have crossed over into other scientific fields, each with its own unique terminology. Below are some general terms that apply to the microarray field. Other terms and further elaboration with illustration can be found, for example, in the concise A-to-Z guide, *Microchips: The Illustrated Hitchhiker's Guide to Analytical Microchips* (Kricka, 2002). A more recent review traces the evolution of arrays from muffin pans and test tube racks (Kricka et al., 2010).

The DNA array has been categorized into different formats based upon what is immobilized to the **surface** (also known as the **solid-phase**, **substrate**, or **chip**) and what is captured out of the sample solution. Definitions change depending upon the format. For the classic **Southern dot blot** the sample was first spotted down on the surface, cross-linked, and then bathed with a radiolabeled oligonucleotide under **hybridization** (complementary nucleic acid strand base-pairing) conditions to detect the presence of a particular sequence within the sample. This was called probing. The oligonucleotide was the **probe**, and the DNA in the sample that was immobilized onto the surface became known as the **target**. Subsequently the Southern dot blot hybridization process was classified as **Format I** (or the forward blot) and was associated primarily with membrane blots. The invention of the **reverse blot** (**Format II**) led to some confusion in the early days of microarrays. Here many different probes are immobilized onto the substrate and the sample (target) labeled for hybridization (Figure 1.2). There are additional formats that will be described later on. In most instances, however, we will be discussing Format II microarrays for both nucleic acid and protein capture. The generally accepted meaning of the probe is as the solid-phase capture agent (Figure 1.3). In drug discovery, the target is usually the protein (or protein complex) that chemically interacts with a compound (drug candidate).

There are also terms related to the anatomy of the microarray (Figure 1.4). The probes are immobilized on the substrate at discrete (x, y) locations or **spatially addressable sites**. The probe spots (measured in microns- diameters for circular

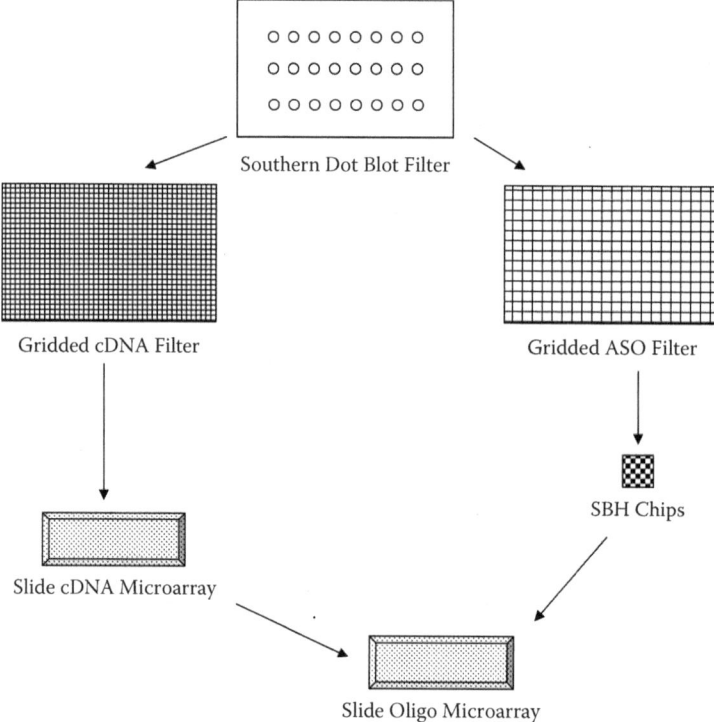

Southern Dot Blot Filter

Gridded cDNA Filter

Gridded ASO Filter

SBH Chips

Slide cDNA Microarray

Slide Oligo Microarray

FIGURE 1.1   Origin of the microarray.

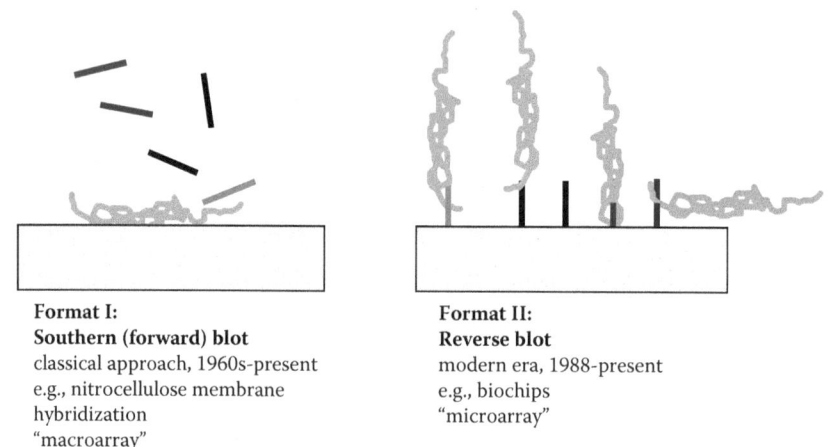

Format I:
Southern (forward) blot
classical approach, 1960s-present
e.g., nitrocellulose membrane
hybridization
"macroarray"

Format II:
Reverse blot
modern era, 1988-present
e.g., biochips
"microarray"

FIGURE 1.2   Hybridization blot formats.

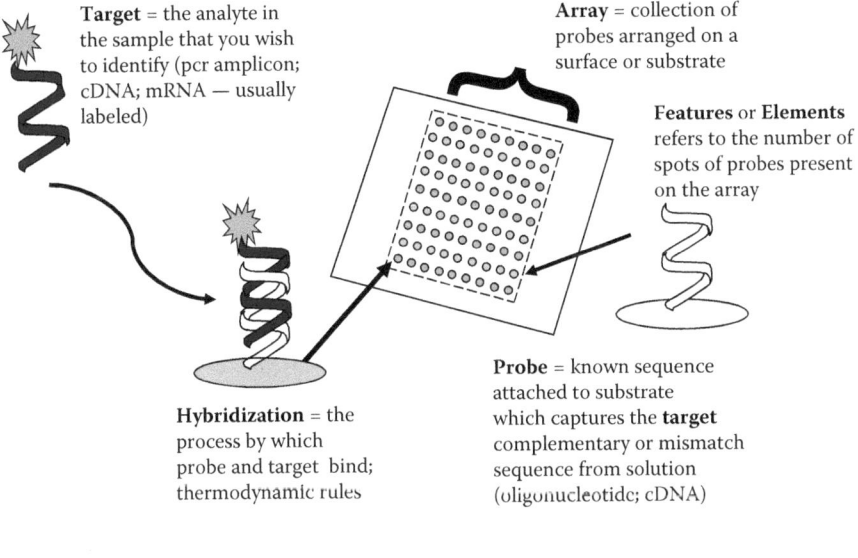

Target = the analyte in the sample that you wish to identify (pcr amplicon; cDNA; mRNA — usually labeled)

Array = collection of probes arranged on a surface or substrate

Features or **Elements** refers to the number of spots of probes present on the array

Probe = known sequence attached to substrate which captures the **target** complementary or mismatch sequence from solution (oligonucleotide; cDNA)

Hybridization = the process by which probe and target bind; thermodynamic rules

Substrate = the solid-phase surface (glass, silica, plastic, acrylamide, gold)

FIGURE 1.3 Microarray anatomy.

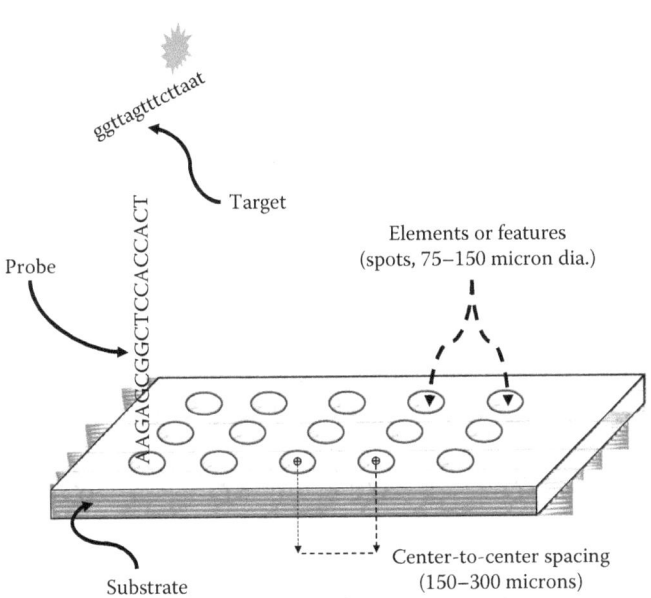

Target

Probe

Elements or features (spots, 75–150 micron dia.)

Center-to-center spacing (150–300 microns)

Substrate

FIGURE 1.4 Microarray terms.

spotted arrays; or the side of a square for in situ arrays) are often referred to as **features** or **elements** of the array. Thus, an array containing 10,000 features would have 10,000 probes arranged as an array on the substrate. Typical spotted arrays would have 100 to 150 micron (**μm**) diameter features, while the photo-lithographically prepared in situ arrays may have features on the order of 2 to 20 μm. The separation between elements is usually measured in terms of a **center-to-center distance**, **spacing**, or **pitch**. That is, for a printed array, two adjacent spots in the array (e.g., each at 100 micron spot diameter) may have a center-to-center distance of 150 microns. Or, the spots would be separated by 50 microns from their edge. The number of spots per square centimeter usually defines the **spot density**. As an example, Affymetix manufacturing arrays at >280,000 elements per 1.28 cm × 1.28 cm chip would have a spot density on a chip of 170,000 elements/cm². We can also classify arrays in terms of high, medium, and low probe density. There may be some argument regarding the precise boundary limits distinguishing microarrays on a density scale. Nevertheless, a high density array would contain >10,000 probes/cm²; medium density, 1000 to 10,000 probes/cm²; and low density, <1000 probes/cm². There are also definitions for arrays based upon whether or not they are **macro-** and **microarrays**. A macroarray could be regarded as having larger and fewer spots than a microarray. For example, the Southern blot on a standard sized membrane (~ 8 cm × 12 cm; nylon or nitrocellulose) with spot diameters of 500 microns would be considered a macroarray by most researchers. However, macroarrays can have thousands of printed spots per membrane and thus functionally perform at a level similar to that of the slide microarray.

The number of probe molecules per square millimeter defines the spot's **probe density**. Probes within an element or spot could have densities on the order of 10^9 to ~10^{12} molecules/mm² depending upon the molecular size of the nucleic acid (e.g., short oligonucleotide versus cDNA).

GENERAL UTILITY

As we will soon discover, microarray-based technologies have found utility in a number of fields. While DNA arrays are the most technically mature and have the broadest application portfolio, we have witnessed the ever-increasing generation of new kinds of probe arrays: antibody, antigen, enzyme, aptamer, carbohydrate, tissue, cell, and small molecule microarrays. The list undoubtedly will continue to expand. We can also describe microarrays in terms of prognostic, diagnostic, and predictive roles. A few examples that examine these applications are provided.

The Biomedical Testing Continuum

The primary focus for microarrays has been for biomedical-related analysis. Biomedical testing can be divided into four major areas: **discovery** testing, **clinical trial** testing, **specialized** testing, and **patient care** testing (Figure 1.5). Most emerging technologies are likely to have been first invented or developed during the discovery process. In the area of biomedical testing, new technologies will pass from research (academic, biomedical, pharmaceutical) into specialized testing arenas such as core facilities or outsourced (esoteric testing) laboratories, and hopefully

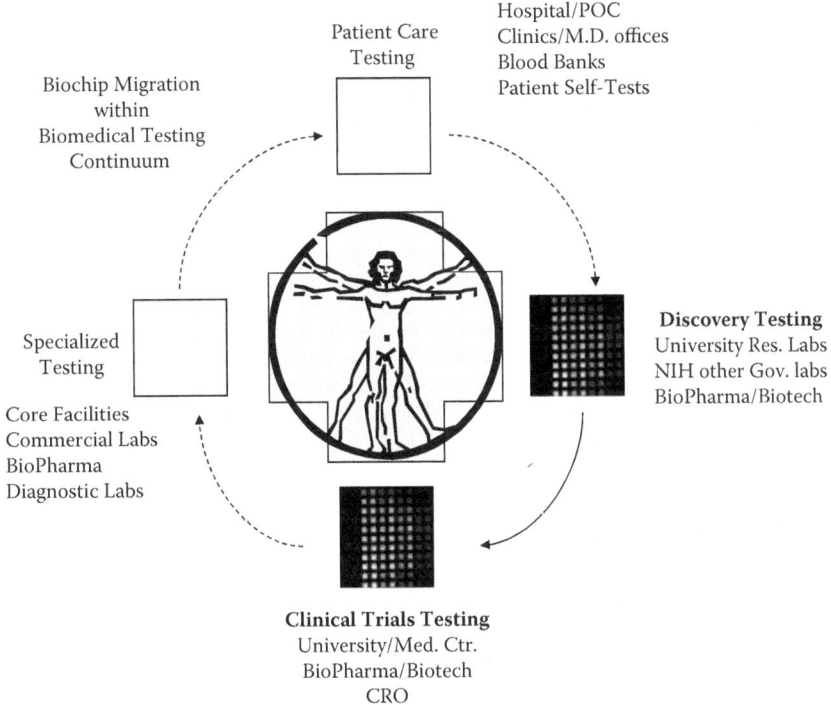

Patient Care
Testing

Hospital/POC
Clinics/M.D. offices
Blood Banks
Patient Self-Tests

Biochip Migration
within
Biomedical Testing
Continuum

Discovery Testing
University Res. Labs
NIH other Gov. labs
BioPharma/Biotech

Specialized
Testing

Core Facilities
Commercial Labs
BioPharma
Diagnostic Labs

Clinical Trials Testing
University/Med. Ctr.
BioPharma/Biotech
CRO

FIGURE 1.5 Biochip migration.

be adopted from these into the mainstream of diagnostics. Array technology finds its roots in academic circles as a means to sequence genomes. The reverse blot was originally constructed at a diagnostic research center as a new nucleic acid assay format for the simultaneous detection of infectious agents.

In the first edition (2005) we observed limited success at moving array technology directly into point of care (POC) or centralized laboratory testing. However, that has changed (see Chapter 7, "Multiplex Assays"), and you will now find microarray technology being adapted for POC in the form of biochips and lateral flow strips. Bead-based multiplex assays (or "fluidic" arrays) are in use for clinical research and are anticipated to migrate into in vitro diagnostics (IVD) and animal health diagnostics in the near future.

BIOTECH SECTOR TRENDS REVISITED

I previously characterized the application of array technology as riding upon the "omic" wave (Figure 1.6). There is no doubt that gene expression and genotyping microarray technology have evolved into mature products (see Chapter 2, "Commercial Microarrays"). These have benefited greatly from the availability of the human genome sequence database as well as other sequenced genomes (Genomics I). Whole genome chips are offered by numerous vendors. However, next generation

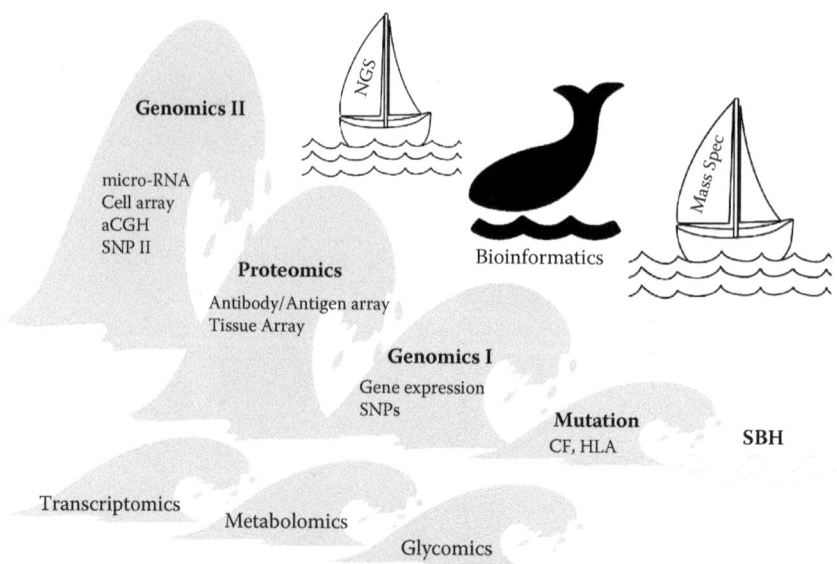

FIGURE 1.6   Microarray technology waves.

sequencing (NGS) has the potential to displace microarrays in certain fields. For a review of NGS technologies see Metzker (2010).

Protein microarrays, both planar and bead-based formats, have been widely adopted for biomarker discovery, biomarker validation, life science research, and clinical research. Multiplex immunoassays are being evaluated for use in animal health and human disease diagnostics. However, most proteomic studies continue to use two-dimensional (2D) gel electrophoresis and mass spectroscopy tools, especially matrix-assisted laser desorption-mass spectometry (MALD-MS). Improvement in mass spectroscopy for peptide and protein analysis (Whiteaker, 2010) may challenge microarrays for clinical use.

A second genomics wave (Genomics II) has emerged with growth in array-based genomic hybridization (aCGH) and single nucleotide polymorphism (SNP) analysis for genetic diagnosis. Micro-RNA (miRNA) gene expression profiling, especially in cancer research, is also underway using microarrays (see Chapter 5, "Gene Expression: Microarray-Based Applications").

Finally, microarray complexity has increased dramatically over the past decade requiring advanced statistical treatments, ever increasing our thirst for bioinformatics. Information technology (IT) plays a key role in microarray-based platform validation (Figure 1.7).

THE OMIC ERA

A major driving force in the advancement of microarrays has been the Human Genome Project and the development of the **genomics** field. Oddly enough, the origins of DNA array technology begin with efforts at sequencing by hybridization

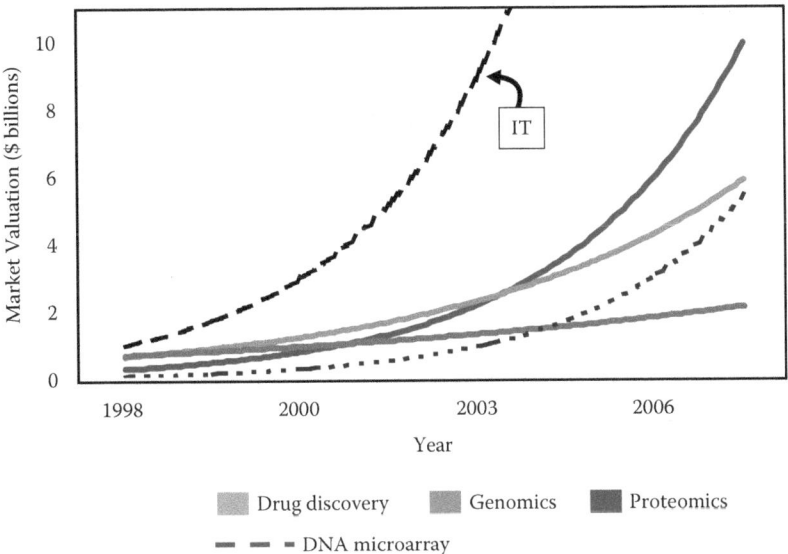

FIGURE 1.7 Biotech sector market growth.

(SBH) of the human genome. Except for a few private ventures such as the work at HySeq, the SBH approach was quickly supplanted by dideoxynucleotide dye-terminated gel-based sequencing. The other early use of DNA arrays was in mutation screening such as for cystic fibrosis (CF) mutations or in human leukocyte antigen (HLA) typing. The problem with screening for mutations associated with genetic disease remains a lack of therapy or cure. Without the availability of treatment there is very little incentive to produce a diagnostic test. The other issue with mutation detection is that the prognosis is often a statistical inference. For example, possessing the BRCA1 gene mutation only increases the probability risk or likelihood that a woman may develop familial breast cancer. The presence of the mutation alone cannot unequivocally determine whether or not someone will develop breast cancer. Other factors must be considered, especially in this case the family history regarding the occurrence of breast cancer. Thus, mutation detection based upon microarray technology has not grown as rapidly as once anticipated.

The breakout opportunity for widespread adoption of array technology came largely from gene expression studies in which the expression levels from two cell states (e.g., control versus drug-induced; normal versus diseased) were compared. So while sequencing efforts defined **structural genomics**, the gene expression microarray became the tool for **functional genomics**. In particular, gene expression microarrays were found to be well suited as a new kind of differential display technology applied to the drug discovery process. Here, **pharmacogenomics** examines the responsiveness of genes (cellular mRNA levels) to drug candidates. The microarray measuring the mRNA levels within the cell's genome permits identification of potential targets (albeit, mRNA serving as the surrogate of translated protein)

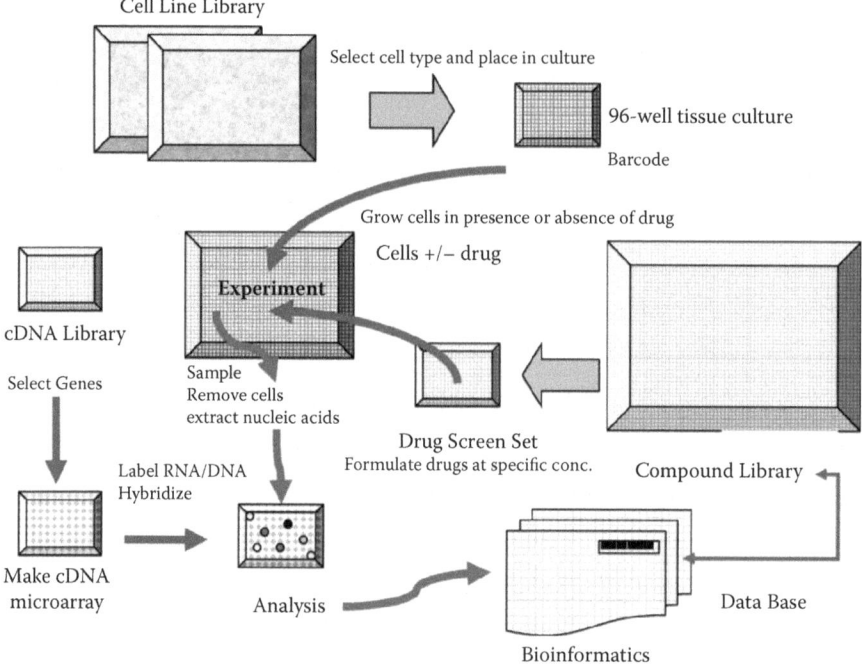

FIGURE 1.8 Off-target analysis.

or alterations to metabolic pathways, thereby implicating the participation of other targets. Metabolic alterations may also lead to "off-target effects" that are adverse or toxic. Therefore, microarrays have been found to be very useful in **toxicogenomic** applications. In essence, mRNA profiling using microarrays can potentially reduce the need for extensive animal model or tissue studies at the onset of the target discovery phase, as well as in the later preclinical phase looking at off-target toxicity (Figure 1.8).

Gene expression analysis has recognized limitations (Lillie, 1997; Clarke et al., 2001). The monitoring of transcriptional events serves as an indirect measure of protein (target) expression. Because proteins can undergo post-translational modification leading to subcellular localization and pooling, export, degradation, or complex formation with other proteins, the correlation between mRNA level and putative protein is often lost (Figure 1.9).

It is important to consider what both genome- and proteome-based arrays offer (Figure 1.10). The gene expression microarray monitors relative mRNA abundance between two cell populations, provided control and sample populations are processed in the same manner. This is often spoken of as a "snapshot" of gene activity. That is, unless we also sample these populations, often only a finite view of the changes in cellular activity between the two sets will be recorded. We may need at times many snapshots or time course studies. The great advantage of the gene expression array is that it permits a global analysis of the genome. Later I will be relating how genome-based approaches provide good information about cell cycles and pathways,

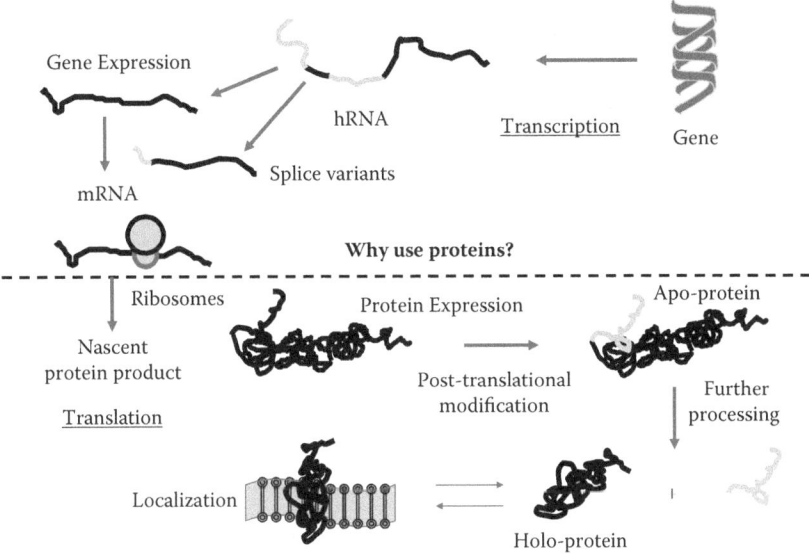

FIGURE 1.9 Why use proteins?

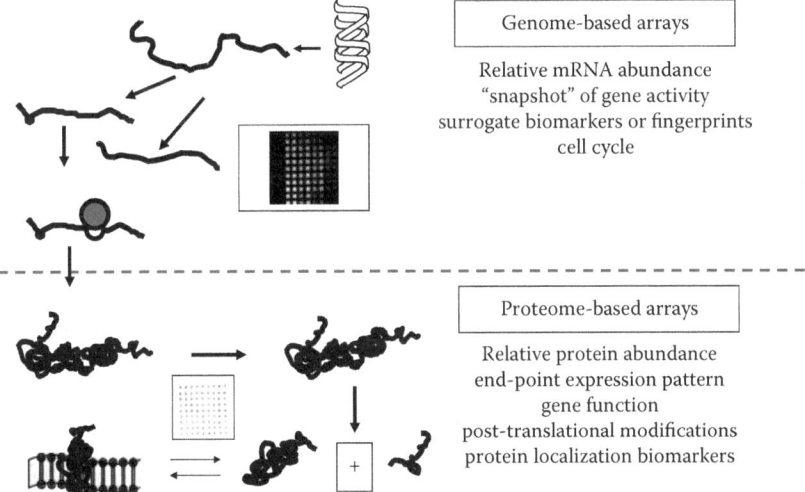

FIGURE 1.10 Genome- versus proteome-based arrays.

leading to the discovery of potential surrogate biomarkers. Gene expression microarrays remain an important tool for several reasons:

1. Availability of full genome representation
2. Ability to amplify for detection of lower abundant mRNAs
3. Simple capture process based upon hybridization thermodynamics and primary sequence information

Obviously for other reasons, such as RNA splice variants and the post-translational modification of proteins, one cannot rely on mRNA profiling alone. However, development of a complete "proteome chip" that would provide global inspection of relative protein abundance has been hindered. There is simply little means of obtaining a highly significant representation of a proteome (i.e., content) on any chip. Antibody arrays have been introduced but are of limited utility at this point for proteome-wide applications. Because microarrays are closed architecture technology platforms, they can only provide information based upon what is contained on the chip. So, we must first acquire enough protein content (antibody libraries or mimetic agents such as aptamers) necessary for proteome discovery work.

Some argue that it is a waste of time monitoring gene expression when ultimately we desire the end product, the protein target. More likely both are needed to further grasp at our understanding of cellular events (Clarke et al., 2001). We now have our list of genes in hand and a few good tools. Not surprisingly we have found biology to be even more complex than once thought. As John Weinstein suggests, "perhaps the most important (and least recognized) aspect of biology that makes it difficult to understand systematically is the fact that biological complexity was not produced by a watch maker or an engineer" (2002, p. 362).

So, while we can consider the strengths and weaknesses of various technologies and pursue "omic" approaches, we must also not lose sight that these are tools to aid in our study. Miller (2002) makes this point:

> These novel and significant tools, which are rapidly becoming indispensable, do not by themselves enlarge the bedrock of basic knowledge that underlies new discoveries. That foundation remains the detailed understanding of the biological basis of health and disease.

THE ROLE FOR GENE EXPRESSION MICROARRAYS IN DRUG DISCOVERY

Fundamental approaches in the drug discovery process have in themselves undergone a paradigm shift (Neamati & Barchi, 2002). The pharmaceutical industry has embraced molecular biology, adopted robotics for high-throughput screening, and supported those efforts aimed at genome sequencing and SNP identification as related to disease. Microarray technology, especially gene expression chips, has been well received in the drug discovery labs of Biopharma. According to Neamati and Barchi (2002) modern drug discovery can be arranged into five major areas (Figure 1.11):

1. Target Identification (discovery)
2. Target Validation
3. Lead Identification (hits)
4. Lead Optimization
5. Preclinical Pharmacology/Toxicology

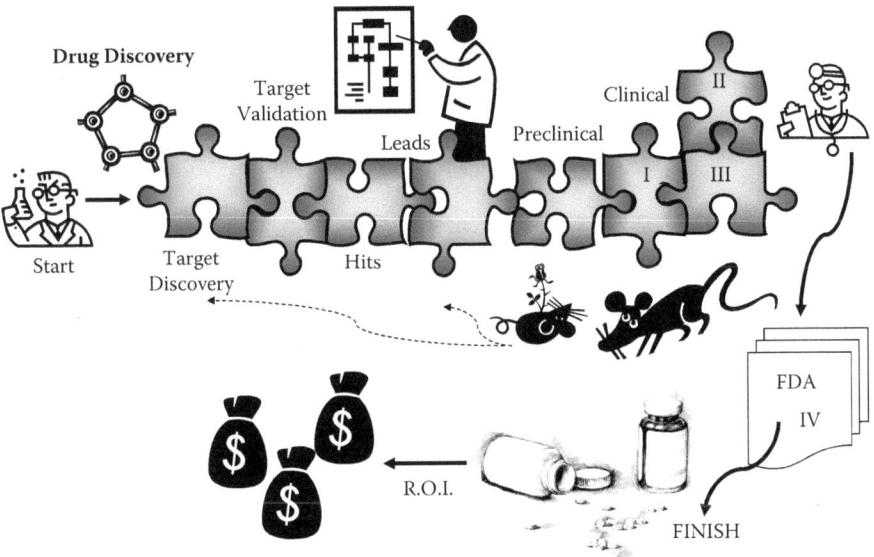

FIGURE 1.11 Drug discovery processes.

In the target discovery process we have witnessed an increased role for genomics. In fact, as much as 25% of new target identification efforts may now be based upon genomic approaches. Differential gene expression pattern analysis (control versus drug response phenotype; normal versus diseased state) developed using the microarray chip technology continues to play an important role. High-density microarrays such as Affymetix's GeneChip have been found to be very useful in target identification. Microarrays based upon normalized cDNA libraries have also been successfully used in the discovery of novel genes that are potential candidates for drug targeting (Katsuma, 2001). cDNA arrays have now given way to high-density oligonucleotide chips that represent whole genomes.

The human genome is "roughly" complete as are the sequences of useful model genomes (e.g., mouse, rat, yeast) nearly complete with other genomes being sequenced at a rapid pace. The actual number of fully completed genomes is rather elusive, but counting microbes there are most likely several hundred completed or nearly completed sequences. Gene expression arrays should continue to serve with increased capacity provided they are not displaced by next-generation sequencing technologies. Gene expression profiling of metabolic pathway enzymes can also play a pivotal role in guiding efforts toward identifying new targets.

While we may be remised in that differential gene expression is but a surrogate methodology, we do not yet have an equivalent proteomic tool. Albeit, the NAPPA (nucleic acid programmable protein array) approach (Ramachandran et al., 2008) could potentially be used for this purpose (Anderson et al., 2011).

Ringer et al. (2000) found pharmacogenomics was rapidly becoming an accepted route in the later stages of the drug discovery process involving both the preclinical and clinical phases. A key factor in the acceptance of the pharmacogenomic (and pharmacogenetic) approach has been that both drug efficacy and toxicity are well-correlated to changes in gene expression. Microarrays offer both high throughput and sensitivity. These attributes are particularly advantageous in reducing the time and cost in determining drug toxicity during the preclinical stage. During late stage clinical trials the profiling of an individual's genetic variation (SNP) correlated to drug response has been an important screening process. Several microarray-based SNP "calling" platform technologies (e.g., Affymetrix Genome-Wide Human SNP Array; Illumina's Golden Gate SNP Assay) can be used in defining specific polymorphisms associated with variable drug responses within individuals and various populations.

Without a doubt one of the most historic testimonies to the power of the microarray has been in the characterization of cancer cell lines and tumor gene expression. The National Cancer Institute's study of cDNA microarray gene expression patterns across 60 human cancer cell lines (with an activity pattern database on each cell line individually challenged by over 70,000 compounds) remains the "tour de force" in microarray-based profiling (Ross et al., 2000). Even more impressive has been the assembly of the integrated gene expression-molecular pharmacology database for the NCI60 cell lines (Scherf et al., 2000). And while the NCI60 study had some recognized limitations (limited activity assay data; limited number of arrayed genes; surrogate relationship of cell line to tumor cells), there are those encouraging examples where the gene expression approach has unraveled mechanisms of drug resistance.

Rew (2001) considered five areas in cancer research applicable for the use of microarray technologies:

1. Tumor classification—Drug treatment regimes often depend upon tumor type. The origin of metastatic tumor cells can be difficult to determine by conventional histopathology. Gene expression profiling may complement the more traditional methodology where tumors are difficult to classify.
2. Mutation detection—Genetic mutation leading to disease states such as in familial breast cancer can be detected by arrays. This is provided that sufficient gene probe sequence content is available in order to make a statistically significant prognosis.
3. Gene copy number—Tumors are known to contain variable numbers of genes relative to normal tissue. Comparison between the normal versus tumor gene populations is useful in identifying tumors and their differentiated state.
4. Cancer therapeutics—Different tumor types and the differentiated states of individual tumors are known to exhibit unique gene expression patterns, called molecular signatures or fingerprints. The molecular signature may be useful in tumor phenotype classification.
5. Drug sensitivity—Microarrays are useful in monitoring both on-target and off-target drug responses. Knowledge of which genes are up- or down-regulated leads to an understanding of the mechanisms of action and new treatments.

In the review by Zanders (2000), a good case is made for the use of gene expression microarrays to monitor changes in signaling pathway activity. The rationale is that under environmental stress such as in tumor growth or during an inflammatory response, signaling pathways are activated or repressed, and these events can be measured by gene activity by mRNA profiling. That is, "transcription of mRNAs could be exploited as a 'surrogate marker' of signaling pathway activation" (Zanders, 2000, p. 378). Cited examples include studies concerning the expression profiling of signaling pathway enzymes in yeast such as the mitogen-activated protein kinases (MAPKs) during times of growth or differentiation or when under stress (Roberts et al., 2000). Homologous mammalian MAPKs have been reported and are being investigated as potential drug targets. In the study by Iyer (1999), the use of gene expression microarrays revealed the induction of genes involved in wound healing during serum-stimulated human fibroblast growth.

TOXICOGENOMIC APPLICATIONS

Almost without exception, gene expression is altered during toxicity, as either a direct or indirect result of toxicant exposure (Nuwaysir et al., 1999).

This early cause-and-effect revelation most certainly promoted the early use of the gene expression microarray technology at the National Institutes of Health (NIH) resulting in the creation of the *ToxChip* cDNA microarray. A **toxicant signature** is derived from gene expression relative fold changes between control and treated cell populations. The ToxChip contained approximately 2000 human gene cDNAs arranged in various functional categories such as apoptosis, cell-cycle, cytochrome P-450s, and so on, to detect response to toxic insult. Hodges et al. (2003) were able to elucidate the mechanism of action for tamoxifen, a drug used in the treatment of breast cancer, using the ToxChip. Lobenhofer et al. (2002) used the ToxChip v1.0 to study the mechanism of estrogen-induced proliferation (mitogenesis) using a hormone-responsive human breast cancer cell line.

Even though toxicogenomic profiling is very useful in providing an adjunct to animal studies, there are certain precautions in interpreting microarray data because "the transcriptome profile is extremely sensitive to any subtle changes surrounding cells, tissues, or individual organisms" (Shoida, 2004, p. 22). There are several such factors that may influence the outcome of a microarray experiment and bias the toxicant signature. For example, circadian rhythms are often overlooked. A simple change out of culture media can have a dramatic effect on cells in culture.

In conclusion, gene expression array technologies have been largely accepted in the scientific community. They are well suited as tools for drug discovery and the elucidation of drug-target mechanism(s) of action (Cunningham, 2000; Clarke et al., 2001). In particular, microarrays provide insight into

1. The discovery and validation of new targets
2. Determination of drug efficacy, resistance, and toxicity
3. Identification of new diagnostic biomarkers

PROTEOMICS TODAY—THE GREAT CHALLENGE

Riding upon the success of the Human Genome Project in cataloging the 30,000 to 40,000 genes of the human genome, the Human Proteome Organization (HUPO; http://www.hupo.org) was established to catalog the entire set of expressed proteins comprising the human proteome(s), the Human Proteome Project (HPP). The organization has grown to address 13 separate initiatives for targeted research, for example, Plasma Proteome Project (PPP), Human Antibody Initiative (HAI), Disease Biomarkers Initiatives (DBI), and so forth.

As Tyers and Mann (2003, p. 193) related,

> Proteomics would not be possible without the previous achievements of genomics, which provided the "blueprint" of possible gene products that are the focal point of proteomics studies.

For the development of new therapeutics *the proteome* is an important quest given that proteins make up the majority of drug targets. Yet, the proteome is perhaps greater than an order of magnitude more complex than the genome (Figure 1.12). While the genome contains 10^4 to 10^5 genes (including splice variants), the proteome may contain over a million proteins if post-translational modifications and isotypes (isoforms) are included. Furthermore, proteins vary in wide abundance and occur in multi-protein complexes localized within cellular and sub-organelle membranes.

The real challenge could be in the high-throughput, highly parallel, micro-preparation of this structurally diverse class of biomolecules in their native states. The assessment of **low abundance** proteins has required the adoption of **enrichment** strategies prior to detection by mass spectrometry (Figure 1.13). In a number of cases, protein complexes require isolation with associated membrane components in order to preserve activity. It will also be important to be able to reassemble these multi-protein complexes in their native and active states.

Yet, there are those who believe that because of its excellent sensitivity, **mass spectrometry** (mass spec) may be able to detect proteins in biological fluids such as serum without the need for separation. Petricoin and Liotta (2004) review the mass

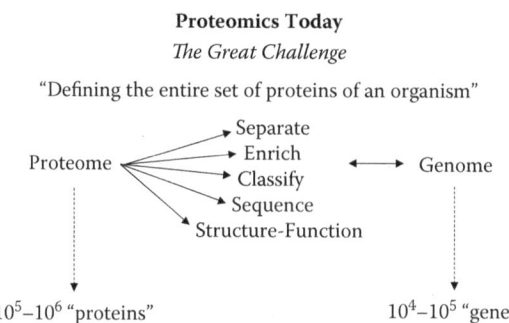

Proteomics Today

The Great Challenge

"Defining the entire set of proteins of an organism"

Proteome → Separate / Enrich / Classify / Sequence / Structure-Function ← → Genome

10^5–10^6 "proteins" 10^4–10^5 "genes"

FIGURE 1.12   Proteomics today—the great challenge.

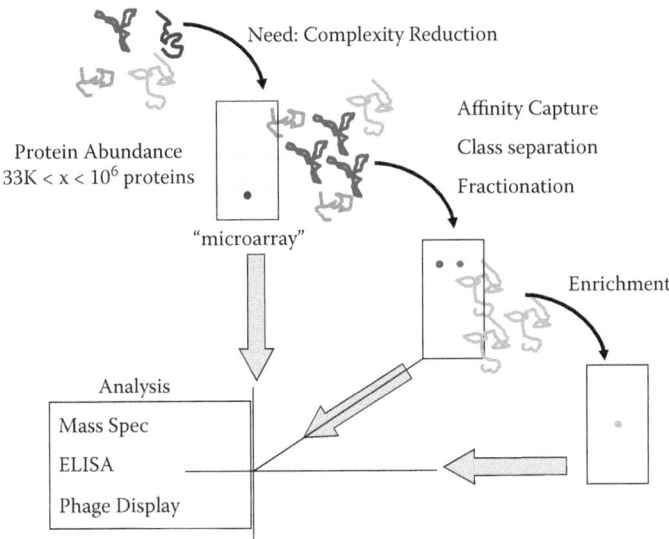

Need: Complexity Reduction

Affinity Capture

Class separation

Protein Abundance
33K < x < 10⁶ proteins

Fractionation

"microarray"

Enrichment

Analysis

Mass Spec

ELISA

Phage Display

FIGURE 1.13   The need for complexity reduction.

spec differential display pattern profiling of serum proteins and peptides associated with various cancers. Protein profiling offers an exciting opportunity as a noninvasive (or nearly so with a prick of blood) diagnostic tool similar in scope to that of the magnetic resonance imaging (MRI) scan for body tissues.

THE POTENTIAL ROLE FOR PROTEIN MICROARRAYS IN DRUG DISCOVERY

As we previously pointed out there are some very good reasons to consider proteomic approaches in drug discovery:

1. Most drug targets are proteins (largely receptors and enzymes).
2. Drug-mediated therapy is based upon less than 500 targets (a very small fraction of the proteome's estimated +100,000 to 10^6 proteins); therefore, finding many more effective targets is likely and highly desirable.
3. mRNA profiling is an indirect assessment of protein expression and cannot detect post-translational modification, an important signaling and regulatory process for drug targeting.

Because a modest number of well-characterized antibodies are readily available, it is no surprise that the first demonstrations in the use of protein microarrays have come from work on antibody-antigen arrays. Yet, the fundamental technology is not new. Roger Ekins (see reviews—Ekins et al., 1990; Ekins, 1998) introduced the *microspot* technology in the 1980s for clinical diagnostics. Later, MacBeath and Schriber (2000) and Haab et al. (2001) borrowing the tools and know-how from the cDNA microarray world promoted the use of slide-based antibody microarrays.

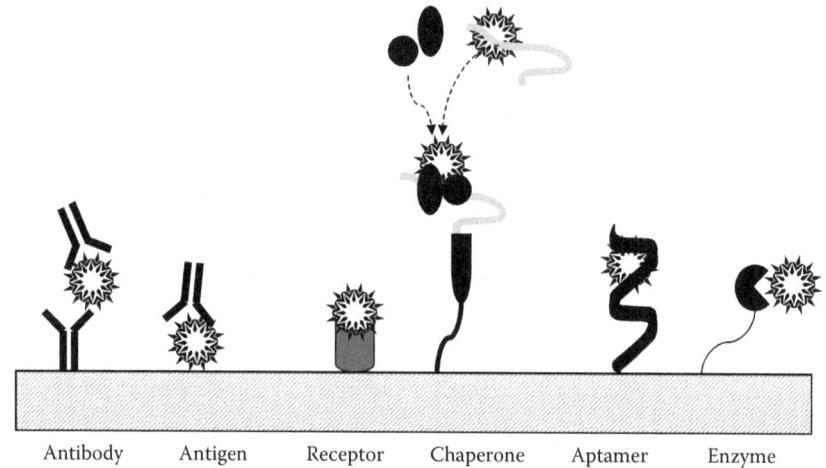

Antibody   Antigen   Receptor   Chaperone   Aptamer   Enzyme

FIGURE 1.14   Protein microarray formats.

Huels and co-workers (2003) have identified several areas within the drug development process where protein biochips have potential application. The antibody-based microarray ELISA format is now commonplace. Antigen arrays have been used to bait the capture and characterization of additional antibodies needed to fill in the proteome libraries. Several well-characterized antibodies are usually required in order to cover more than one epitope on each antigen. This is especially important if the sandwich assay is to be optimally employed. Antibody and antigen microarray formats are briefly described in more detail in Chapter 6, "Protein Microarray Applications," as well as in Chapter 7, "Multiplex Assays."

Future generation protein arrays will include in addition to antibody-antigen binding, other protein-protein or receptor-ligand interactions (Figure 1.14). These include protein-peptide, substrate-enzyme, aptamer, and protein–small molecule high-throughput assays. Proofs of principle on a variety of these formats have been the subject of numerous reviews. Essentially, protein microarrays are believed to be able to broadly cover the various phases of the drug discovery process similarly to what we have described for gene-expression microarrays (Figure 1.15). However, as with the DNA microarray, there remain a few hurdles ahead of us (Figure 1.16).

CRITICAL ISSUES WITH PROTEIN MICROARRAYS

Stability and Performance

While we can appreciate those first demonstrations of the utility for the protein microarray, in retrospect, they did not perform particularly well relative to a standard ELISA. As reported by Haab et al. (2001), only 50% of the antigens and only 20% of the antibodies performed well (i.e., quantitative detection of the cognate antigen/antibody) in the μg/mL range with some allowing detection into the sub-ng/mL range.

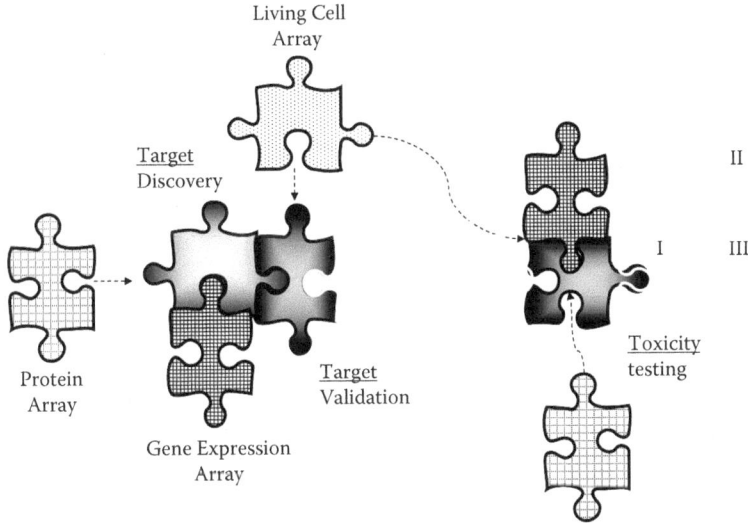

FIGURE 1.15 The role for microarrays in the drug discovery process.

Today, microarray-based immunoassays offer high sensitivity and large dynamic ranges at levels comparable to or better than conventional "singleplex" ELISAs.

As others have pointed out, assembling proteins on a substrate is a relatively easy task but much more of a challenge if all are to remain fully functional (Cahill, 2001; Sreekumar & Chinnaiyan, 2002; Valle and Jendoubi, 2003). Unlike the DNA array where the capture of cDNA or cRNA targets is relatively straightforward due to the structural similarity between probes and targets, proteins (even within the same

Critical Issues

DNA Microarrays	Protein Microarrays
Cross-platform standardization → MIAME	No PCR equivalent technology ← RCA
Improved printing technology[1]	Non-specific adsorption
	Cross-reactivity[2]
Guidelines on statistical methods of analysis	Detection sensitivity & dynamic range ← RLS
Continued access to "content"	Where to obtain "content" and at what cost

MIAME = Minimum Information About a Microarray Experiment

October 14, 2002   Adoption of MIAME for array standardization

Major scientific journals (including Cell, The Lancet, Nature and Science) endorse MIAME guidelines for publishing microarray gene expression data.

RCA = rolling circle amplification; see Figure 6.18

RLS = resonance light scattering; see Figure 6.23

[1] see Chapter 4. [2] see Chapter 6.

FIGURE 1.16 Critical issues facing DNA and protein microarrays.

class, e.g., antibodies) can vary considerably in their physical-chemical stability and binding character on surfaces. Another challenge remains in the selection of the optimal substrate(s), immobilization and presentation (orientation) of the ligand, as well as the development of sensitive assays for different protein classes.

Content

Antibody coverage of the human proteome is estimated to be about 5 to 10% of all human proteins and isoforms (Valle and Jendoubi, 2003). A major bottleneck in the use of protein expression arrays has been the lack of such a comprehensive set of these capture agents (Hanash, 2003). Since the equivalent of the polymerase chain reaction (PCR) process for mass amplification of low abundant proteins does not exist, the remaining library of proteome capture ligands may need to be generated by other means such as recombinant protein expression systems (Cahill, 2001). However, recombinant antibodies may be less stable and have lower binding affinities than monoclonal antibodies (Valle and Jendoubi, 2003). Therefore, in order to fully implement the microarray format a host of diverse capture agents could be required in addition to antibodies. These include peptides, small molecules, aptamers, ribozymes, or other molecular recognition probes yet to be discovered. However, it is also understandable because of the diverse nature of proteins that technologies besides microarrays will be used in proteomics research (Hanash, 2003).

Detection

As difficult as it is going to be to isolate and produce thousands of high affinity and specificity protein ligands, it may be even harder to come up with a good way to monitor binding of proteins to the chip (Kodadek, 2001).

In the review by Kodadek (2001) the following points are made regarding near-term approaches for detection on protein microarrays:

1. Dye labeling of proteins in cell extracts—A well-known problem is that different proteins have different labeling efficiencies; generating individual calibration curves for each protein within the array would be impractical, so quantification of protein arrays will be a challenge. Note: another approach would be to generate a labeled standard reference protein mixture or calibration set at different concentrations (Haab et al., 2001). These could then be mixed in with the cell extract providing some degree of normalization semi-quantitative information from the dye ratios. Schroder et al. (2010) utilized dual-color labeling in proteomic profiling experiments involving a microarray of 810 antibodies against differentially expressed cancer protein targets. Dual-color labeling was found to offer greater reproducibility over that of single-color labeling.
2. Chemical modification of proteins can lead to denaturation and aggregation. This can reduce both specificity (increased nonspecific binding) and sensitivity (decreased ligand affinity).
3. Sandwich assay—This format works well for ELISA. The success and potential shortcomings for microarrays are discussed in the next section.

4. Mass spectrometry—Excellent sensitivity but of low throughput and semi-quantitative. The ICAT (isotope-coded affinity tag) labeling method for proteins enables differential display analysis using mass spectrometry (Griffin and Abersold, 2001). This might be adaptable to work with protein microarrays using MALDI-TOF mass spectrometry similar to the original method developed by Sequenom for SNP detection. Pan et al. (2009) describe the use of internal standards of stable isotope labeled peptides to achieve targeted protein quantification by mass spectrometry.

Micro-ELISA Formats

Assays involving antibody or antigen arrays tend to be micro-ELISA formats. For instance, anti-cytokine monoclonal-monoclonal or monoclonal-polyclonal antibody pairs are readily available and well characterized for use in ELISA assays. Cytokines are important biological indicators used in drug discovery and toxicity testing. Therefore, it is no surprise that many of the early antibody microarray demonstrations involved the determination of cytokines from cell culture. They simply miniaturized and multiplexed the familiar sandwich immunoassay. The microarray-based cytokine assays generally claim sensitivities in the low pg/mL range with linear dynamic ranges from about 10 to 10,000 pg/mL depending upon the analyte. These are similar in performance to the standard ELISA.

Woodbury et al. (2002) developed a micro-ELISA assay for determination of hepatocyte growth factor (HGF) in human serum. In this case, a horseradish peroxidase (HRP)-catalyzed TSA-biotin amplification with streptavidin-Cy3 reporter was employed. They claimed sensitivity to 0.5 pg/mL or 6 fM HGF with a linear dynamic range of 12 to 4000 pg/mL in serum. This was good enough because *clinically relevant levels* varied 199 to 1640 pg/mL (breast cancer patients) and 153 to 998 pg/mL for age-matched normal controls. In addition, HGF and four other antigens found in serum were simultaneously quantified even though the analytes varied in physiological concentration from 20 to 60,000 pg/mL. The HGF microarray-ELISA correlated ($r^2 = 0.90$) with a standard 96-well-plate ELISA.

So, for the determination of a limited set of analytes, the antibody microarray works reasonably well and is comparable in performance if not better than that of a standard ELISA for individual analytes. Furthermore, we will discover that a suitable number of detection technologies are available for use on a variety of supports. The real problem may be in at what point antibody microarrays "**hit the wall**" for multiplexing. What is not generally appreciated is that each antibody pair needs to be matched up and cross-reactivity for all pairs determined. That involves not only the cross-reactivity of each capture antibody but also the cross-reactivity of all secondary antibodies, including the extent of secondary to secondary interactions. It is not uncommon to sort through several antibody pairs before finding compatible sets. Consider now having to do the same for an antibody array including hundreds to thousands of analyte-specific pairs. See Chapter 6 ("Protein Microarray Applications") for additional examples and discussion concerning the issue of cross-reactivity.

Protein Profiling Formats

Protein expression profiling (protein differential display) using microarrays is considered by most to be an important tool for proteomic discovery. It is similar in concept and approach to that of the gene expression microarray for mRNA profiling. Sreekumar and Chinnaiyan (2002) describe a general approach to using the microarray in order to monitor protein expression in cancer and normal tissues. Here are the steps:

1. Extract total protein from cancer and normal cells using a detergent (e.g., 1% NP40).
2. Remove excess detergent from lysate by adsorption onto a solid phase (e.g., beads).
3. Determine total protein concentration for each lysate.
4. Label equal amounts of protein from cancer and normal cell lysate (e.g., Cy5-NHS labeling of cancer lysate; Cy3-NHS labeling of normal lysate).
5. Remove free dye from dye-labeled proteins by gel filtration chromatography.
6. Mix Cy5-protein with Cy3-protein purified lysates.
7. Concentrate mixture.
8. Apply concentrate to antibody array.
9. Wash array free of unbound antigens, and then perform confocal scan.
10. Analyze data.

Here are some *potential issues* to consider: First, detergent extraction can be problematic. Not all proteins will extract or if so to the same extent. The amount of protein present can influence the efficiency and stability of detergent micelle formation. Inefficient removal of detergent as well as irreversible partitioning of proteins onto a solid support during purification is likely. If there were differences in protein abundance between test and control cell lysates before processing, there could be significant differential loss following processing. Individual proteins can have very different labeling efficiencies depending upon protein concentration, pH, ionic strength, and the number and accessibility of dye-reactive amino acid residues (Kodadek, 2001). As with the labeling of nucleic acids, Cy5 and Cy3 or other dyes may have different labeling efficiencies for the same protein. Dye-labeled proteins may differentially adsorb onto the solid phase used for purification. Concentrating may do more harm than good if proteins denature and aggregate forming protein complexes. If such complexes are applied to the antibody array both false positive and false negative associations are likely.

NEAR-TERM BIOMEDICAL APPLICATIONS

Cytokines

An abundance of vendors offer antibody-based microarrays for multiplexed cytokines analysis (see Chapter 2, "Commercial Microarrays," for commercial sources). Cytokines are biomarkers involved in cancer, cell injury, inflammation, and apoptosis. They are released by cells in culture in response to drug action (e.g., Turtinen et al., 2004) or are elevated in serum in various disease states. Moreover, there are numerous cytokines involved in cellular response with many serving as dual effectors (Asao and Fu, 2000). As a result, anti-cytokine microarrays are being evaluated in drug discovery for off-target toxicity testing to replace standard ELISA plate formats. An extensive listing of cytokines is available at COPE (Cytokines & Cells Online Pathfinder Encyclopedia) at http://www.copewithcytokines.org. Version 29.0 containing over 29,000 entries was released in spring/summer 2012.

Huang et al. (2002) prepared an antibody array for the simultaneous detection of 43 cytokines. They were able to verify the down-regulation of MCP-1 cytokine in transfected cells (human glioblastoma cells transfected with *cx43* expression vector) relative to control cells. The antibody array is an emerging technology. At least in one study based upon the use of a commercial membrane format, the cytokine microarray failed to accurately determine cytokine levels in bacterial and LPS-stimulated whole human blood (Copland, 2004).

Autoimmune Diseases and Allergy

Advancement in autoimmune and inflammatory disease treatment and diagnosis represents a critical worldwide need ranking only behind cardiovascular disease and cancer in importance to the medical practitioner. The list of related diseases is long but major categories include rheumatoid arthritis, asthma, diabetes type I, multiple sclerosis, and inflammatory bowel disease, such as celiac disease.

Antigen arrays also described as **"reverse-phase protein array"** (Paweletz et al., 2001) involve the immobilization of proteins to serve as bait for various protein-protein interactions (Sreekumar et al., 2002). For example, Joos et al. (2000) printed various auto-antigens present in sera with known association with various autoimmune diseases (e.g., Grave's disease, lupus, connective tissue disease, and others). The group then screened various sera for the presence of the auto-antibodies. By immobilizing on the array a serial dilution series for each antigen, the titers for these antibodies could be determined.

Feng et al. (2004) prepared an antigen microarray on polystyrene support including 15 auto-antigens useful for the detection of auto-antibodies involved in rheumatoid autoimmune diseases. In other studies, de Vegvar et al. (2003) used antigen microarrays to examine epitope-specific antiviral antibody responses in vaccine trials in an animal model for human immunodeficiency virus (HIV) infection. Huber et al. (2002) has reviewed different formats for antigen microarrays. The review by Robinson (2006) examines autoimmune disease, cancer, infectious diseases, and allergen profiling with antigen microarrays. Mezzasoma and co-workers (2002) used antigen microarrays to determine the levels of infectious agents in human sera. Antigens (*Toxoplasma gondii*, rubella virus, cytomegalovirus, herpes simplex virus, ToRCH antigens) were printed in an array format. Serum samples were applied and serodiagnosis determined using a sandwich assay employing fluorescently labeled secondary antibodies directed toward the primary sera antibodies. The Utz Lab has additional information regarding antigen arrays on their Stanford School of Medicine Web site (http://www.stanford.edu/group/utzlab/autoantibodies.htm).

The typing of various allergens using the antigen microarray has also met with success due in part to the availability of recombinant allergens (i.e., content). Jahn-Schmid and co-workers (2003) examined the analytical performance of an allergen (grass and tree pollens) microarray for the detection of allergen-specific serum IgG in 51 patient sera. While considerable variation was observed with intra- and inter-assay for some allergens, the sensitivity and specificity of the microarray were comparable to the conventional ELISA. Shreffler et al. (2004) constructed an antigen array including an overlapping series of peptide probes representing epitopes associated with the major peanut allergens. Examining 77 patient sera the group found considerable variation in patient IgE–epitope profiling suggesting such population heterogeneity might be of prognostic value.

Finally, Nishizuka et al. (2003) undertook the arduous task of proteomic profiling the NCI-60 cancer cell lines based upon high-density arraying of cell lysates. In this case, lysates are prepared in a urea denaturing buffer and maintained in a reduced state with dithiothretol (DTT). This allows opportunity for additional assessment from 2D-PAGE gels. Serial dilutions of each protein lysate were printed onto the substrate (FAST slides, nitrocellulose-coated glass, Schleicher & Schuell, Germany). Monoclonal antibodies screened by Western blotting to lysate were used for detection and SYPRO ruby protein stain (molecular probes) for determination of total protein. Fifty-two proteins (i.e., those with high specificity antibody recognition) were analyzed in lysates. Of these, 31 were matched to cDNA and GeneChip microarrays; 19 of those 31 showed significant correlation between the two gene-expression formats used earlier in characterizing the NCI-60 cell lines (e.g., Ross et al., 2000; Scherf et al., 2000). How do the expressions of these proteins correlate with the corresponding mRNA profiles? And the answer, please...? *Structure-related proteins are almost always better correlated with mRNA levels across the 60 cell lines.*

FUTURE MEDICINE—PHARMACOPROTEOMICS OR PHARMACOGENOMICS?

The end result for drug development is to successfully supply cost-effective drugs that provide for better patient treatment, offering cures from new therapies and improvements in disease management. The latter includes the advancement of diagnostics with tests that will more rapidly and more accurately determine disease state and monitor treatments. But, what road should we travel down in the future—pharmacogenomics or pharmacoproteomics (Figure 1.17)?

Jain (2004) defines *pharmacoproteomics* as the use of proteomic technologies in drug discovery and development. It is Jain's contention that pharmacoproteomics rather than genotyping will take the lead role in promoting the practice of personalized medicine. This remains to be seen. Key to that success will be the continued application of protein chips, enabling future discovery and development of drugs for personalized therapy, and entry into point-of-care diagnostics.

However, do not rule out pharmacogenomic approaches (Ginsburg et al., 2001). Sabatini's reverse-transfection method of creating "live cell" microarrays

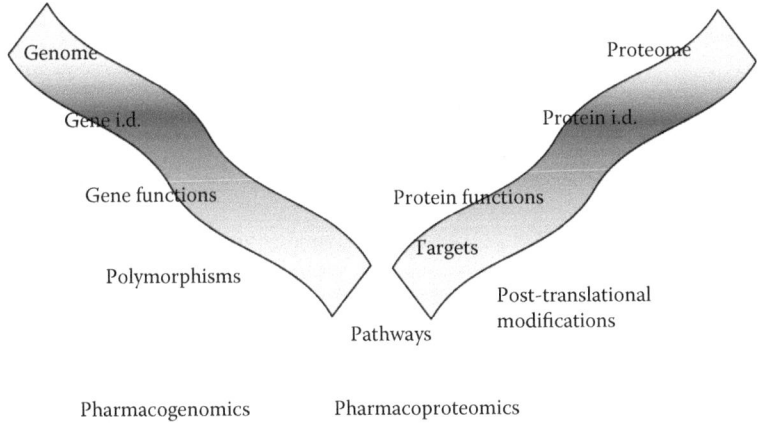

Genome

Gene i.d.

Gene functions

Polymorphisms

Proteome

Protein i.d.

Protein functions

Targets

Post-translational
modifications

Pathways

Pharmacogenomics   Pharmacoproteomics

Molecular Medicine

FIGURE 1.17 Where are we going from here?

offers even greater advantage for drug discovery and development (Ziauddin and Sabatini, 2001). The method, relying on arrayed cDNA expression vectors to transform adherent cells, provides localized, real-time gene expression analysis of the putative gene product. The live cell microarray could eventually displace the use of protein expression microarray for identifying drug targets. This is provided that more extensive libraries of full length cDNAs (needed to express the complete protein) become available. The approach could potentially be utilized for high-throughput screens to uncover genotype-phenotype relationships. Offering an alternative cellular approach, in 2011, the NIH and DARPA announced programs with the collaboration of the FDA to produce "body-on-a-chip" platforms of living cell types arrayed on microfluidic devices for detecting the toxicity of candidate drugs.

What appears to be certain is that microarrays in one omics-based form or another (DNA, protein, cell, tissue, or small molecule) are playing an increasingly important role in drug development and diagnostics (Figure 1.18). Perhaps **omics** will eventually evolve into an integrated "**systems biology**" approach (Ideker et al., 2001), one in which we monitor metabolic pathways, transport, compartmentalization, degradation, and so forth, and their inter-relationships for small molecules, cell surfaces (e.g., carbohydrate microarrays) (Wang et al., 2002), and biopolymers alike (Figure 1.19). We are just at the beginning of **molecular medicine** (personalized medicine) and **molecular diagnostics** (companion diagnostics). It remains an exciting era for science, technology, and personalized medicine.

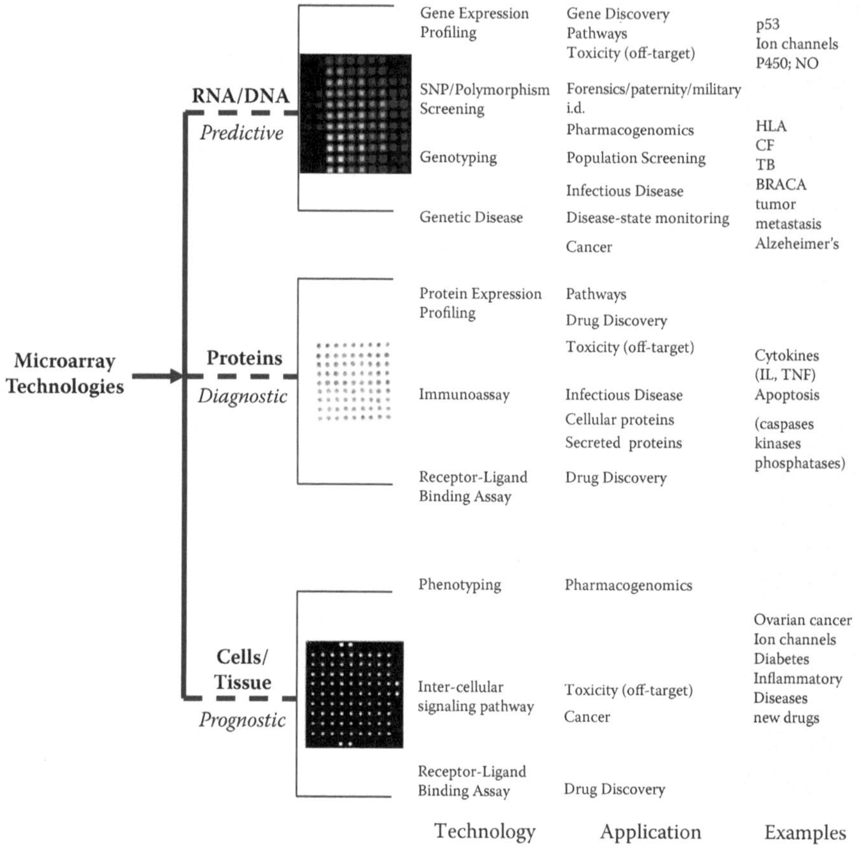

Technology	Application	Examples
RNA/DNA *Predictive*		
Gene Expression Profiling	Gene Discovery Pathways Toxicity (off-target)	p53 Ion channels P450; NO
SNP/Polymorphism Screening	Forensics/paternity/military i.d.	
	Pharmacogenomics	HLA
Genotyping	Population Screening	CF TB
	Infectious Disease	BRACA
Genetic Disease	Disease-state monitoring	tumor metastasis
	Cancer	Alzeheimer's
Proteins *Diagnostic*		
Protein Expression Profiling	Pathways Drug Discovery	
	Toxicity (off-target)	Cytokines (IL, TNF)
Immunoassay	Infectious Disease	Apoptosis
	Cellular proteins	(caspases
	Secreted proteins	kinases phosphatases)
Receptor-Ligand Binding Assay	Drug Discovery	
Cells/ Tissue *Prognostic*		
Phenotyping	Pharmacogenomics	
		Ovarian cancer Ion channels Diabetes
Inter-cellular signaling pathway	Toxicity (off-target) Cancer	Inflammatory Diseases new drugs
Receptor-Ligand Binding Assay	Drug Discovery	

Microarray Technologies

FIGURE 1.18   Microarray applications.

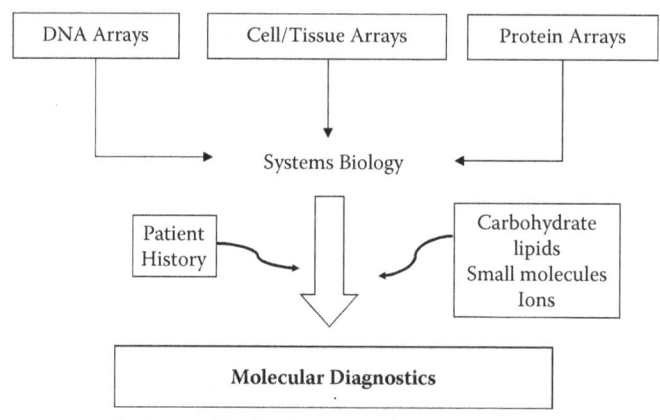

The future of the biochip rests within its ability to serve
as a fully integrated device for multiplexed analysis

FIGURE 1.19   Molecular diagnostics.

REFERENCES

Amur S, Frueh FW, Lesko LJ, Huang S-M. Integration and use of biomarkers in drug development, regulation and clinical practice: A U.S. regulatory perspective. *Biomarkers Med.* 2(3): 305–311, 2008.

Anderson KS, Sibani S, Wallstrom G., et al. A protein microarray signature of autoantibody biomarkers for the early detection of breast cancer. *J. Proteome Res.* 10(1): 85–96, 2011.

Asao H, Fu X-Y. Interferon-γ has duel potentials in inhibiting or promoting cell proliferation. *J. Biol. Chem.* 275(2): 867–874, 2000.

Boja ES, Jortani SA, Ritchie J, et al. The journey to regulation of protein-based multiplex quantitative assays. *Clin. Chem.* 57(4): 560–567, 2011.

Borgan E, Sitter B, Lingjaerde OC, et al. Merging transcriptomics and metabolomics—Advances in breast cancer profiling. *BMC Cancer* 10: 628, 2010.

Cahill DJ. Protein and antibody arrays and their medical applications. *J. Immunol. Methods* 250: 81–91, 2001.

Carroll GT, Wang D, Turro NJ, Koberstein JT. Photochemical micropatterning of carbohydrates on a surface. *Langmuir* 22: 2899–2905, 2006.

Clarke PA, te Poele R, Wooster R, Workman P. Gene expression microarray analysis in cancer biology, pharmacology, and drug development: Progress and potential. *Biochem. Pharmacol.* 62: 1311–1336, 2001.

Copeland S, Siddiqui J, Remick D. Direct comparison of traditional ELISAs and membrane protein arrays for detection and quantification of human cytokines. *J. Immunol. Methods* 284: 99–106, 2004.

Cunningham MJ. Genomics and proteomics—The new millennium of drug discovery and development. *J. Pharmacol. Toxicol. Methods* 44: 291–300, 2000.

de Vegvar HEN, Amara RR, Steinman L, et al. Microarray profiling of antibody responses against Simian-Human Immunodeficiency Virus: Post challenge convergence of reactivities independent of host histocompatibility type and vaccine regimen. *J. Virol.* 77(20): 11125–11138, 2003.

Dimond PF. Why biomarker discovery is hard. *Gen. Eng. Biotech. News,* June 25, 2012.

Ekins R, Chu F, Biggart E. Fluorescence spectroscopy and its application to a new generation of high sensitivity, multi-microspot, multianalyte, immunoassay. *Clin. Chim. Acta* 194: 91–114, 1990.

Ekins RP. Ligand assays: From electrophoresis to minaturized microarrays. *Clin. Chem.* 44(9): 2015–2030, 1998.

Feng Y, Ke X, Ma R, Chen Y, Hu G, Liu F. Parallel detection of autoantibodies with microarrays for rheumatoid diseases. *Clin. Chem.* 50(2): 416–422, 2004.

Ginsburg GS, McCarthy JJ. Personalized medicine: Revolutionizing drug discovery and patient care. *Trends in Biotechnol.* 19: 491–496, 2001.

Godula K, Bertozzi CR. Synthesis of glycopolymers for microarray applications via ligation of reducing sugars to a poly(acryloyl hydrazide) scaffold. *J. Amer. Chem. Soc.* 132: 9963–9965, 2010.

Griffin TJ, Abersold R. Advances in proteomic analysis by mass spectroscopy. *J. Biol. Chem.* 276(4): 45497–45500, 2001.

Haab BB, Dunham MJ, Brown PO. Protein microarrays for highly parallel detection and quantitation of specific proteins and antibodies in complex solutions. *Genome Biol.* 2(2): research0004.1–0004.13, 2001.

Hanash S. Disease proteomics. *Nature* 422: 226–232, 2003.

Hodges LC, Cook JD, Lobenhofer EK, et al. Tamoxifen functions as a molecular agonist inducing cell cycle-associated genes in breast cancer cells. *Mol. Cancer Res.* 1: 300–311, 2003.

Huang R, Lin Y, Wang CC, et al. Connexin 43 suppresses human glioblastoma cell growth by down-regulation of monocyte chemotactic protein 1, as discovered using protein array technology. *Cancer Res.* 62: 2806–2812, 2002.

Hueber W, Utz PJ, Steinman L, Robinson WH. Autoantibody profiling for the study and treatment of autoimmune disease. *Arthritis Res* 4(5): 290–295, 2002.

Huels C, Muellner S, Meyer HE, Cahill DJ. The impact of protein biochips and microarrays on the drug development process. *DDT* 7(18): S119–S124, 2003.

Ideker T, Thorsson V, Ranish JA, et al. Integrated genomic and proteomic analysis of a systematically perturbed metabolic network. *Science* 292: 929–934, 2001.

Iyer VR, Eisen MB, Ross DT, et al. The transcriptional program in the response of human fibroblasts to serum. *Science* 283: 83–87, 1999.

Jahn-Schmid B, Harwanegg C, Hiller R, Bohle B, Ebner C, Scheiner O, Muller MW. Allergen microarray: Comparison of microarray using recombinant allergens with conventional diagnostic methods to detect allergen-specific serum immunoglobulin. *E. Clin. Exp. Allergy* 33: 1443–1449, 2003.

Jain KK. Role of pharmacogenomics in the development of personalized medicine. *Pharmacogenomics* 5(3): 331–336, 2004.

Joos TO, Schrenk M, Hopfl P, et al. A microarray enzyme-linked immunosorbent assay for autoimmune diagnostics. *Electrophoresis* 21: 2641–2650, 2000.

Katsuma S, Tsujimoto G. Genome medicine promised by microarray technology. *Expert Rev. Mol. Diagn.* 1(4): 377–382, 2001.

Kodadek T. Protein microarrays: Prospects and problems. *Chem. Biol.* 8: 105–115, 2001.

Kricka LJ. 2002. *Microchips: The Illustrated Hitchhiker's Guide to Analytical Microchips.* AACC Press, Washington, DC.

Kricka LJ, Iami K, Fortina P. Analytical ancestry: Evolution of the array in analysis. *Clin. Chem.* 56(12), 1797–1803, 2010.

Lillie J. Probing the genome for new drugs and targets with DNA arrays. *Drug Development Res.* 41: 160–172, 1997.

Lobenhofer EK, Bennett L, Cable PL, Li L, Bushel PR, Afshari CA. Regulation of DNA replication fork genes by 17β-estradiol. *Mol. Endocrinol.* 16(6): 1215–1229, 2002.

MacBeath G, Schreiber SL. Printing proteins as microarrays for high-throughput function determination. *Science* 289: 1760–1763, 2000.

Malone JH, Oliver B. Microarrays, deep sequencing and the true measure of the transcriptome. *BMC Biol.* 9: 34, 2011.

Matson RS (ed.). *Microarray Methods and Protocols.* CRC Press, Boca Raton, FL, 2009.

Metzker ML. Sequencing technologies—The next generation. *Nature Rev.* 11: 31–46, 2010.

Mezzasoma L, Bacarese-Hamilton T, Di Cristina M, Rossi R, Bistoni R, Crisanti A. Antigen microarrays for serodiagnosis of infectious diseases. *Clin. Chem.* 48(1): 121–130, 2002.

Miller JP. Back to basics in the "omics" rush. *Targets* 1(1): 4–5, 2002.

Neamati N, Barchi, JJ Jr. New paradigms in drug design and discovery. *Curr. Top. Med. Chem.* 2(3): 211–227, 2002.

Nishizuka S, Charboneau L, Young L, et al. Proteomic profiling of the NCI-60 cancer cell lines using new high-density reverse-phase lysate microarrays. *PNAS* 100(24): 14229–14234, 2003.

Nuwaysir EF, Bittner M, Trent J, Barrett CJ, Afshari CA. Microarrays and toxicology: The advent of toxicogenomics. *Mol. Carcinogenesis* 24: 153–159, 1999.

Pan S, Aebersold R, Chen R, et al. Mass spectrometry based targeted protein quantification: Methods and applications. *J. Proteome Res.* 8(2): 787–797, 2009.

Paweletz CP, Charboneau L, Bichsel VE, et al. Reverse phase protein microarrays which capture disease progression show activation of pro-survival pathways at the cancer invasion front. *Oncogene* 20(16): 1981–1989, 2001.

Petricoin EF, Liotta LA. SELDI-TOF-based serum proteomic pattern diagnostics for early detection of cancer. *Curr. Opin. Biotechnol.* 5(1): 24–30, 2004.

Psychogios N, Hau DD, Peng J, et al. The human serum metabolome. *PLoS One.* 6(2): e16957, 2011.

Ramachandran N, Raphael JV, Hainsworth E, et al. Next generation high density self assembling functional protein arrays. *Nat. Methods* 5(6): 535–538, 2008.

Rew DA. DNA array technology in cancer research. *Eur. J. Surg. Oncol.* 27: 504–508, 2001.

Rillahan CD, Paulson JC. Glycan microarrays for decoding the glycome. *Annu. Rev. Biochem.* 80: 797–823, 2011.

Rininger JA, DiPippo VA, Gould-Rothberg BE. Differential gene expression technologies for identifying surrogate markers of drug efficacy and toxicity. *DDT* 5(12): 560–568, 2000.

Roberts CJ, Nelson B, Marton MJ. Signaling and circuitry of multiple MAKP pathways revealed by a matrix of global gene expression profiles. *Science* 287: 873–880, 2000.

Robinson WH. Antigen arrays for antibody profiling. *Curr. Opin. Chem. Biol.* 10: 67–72, 2006.

Ross DT, Scherf U, Eisen MB, et al. Systematic variation in gene expression patterns in human cancer cell lines. *Nat. Genet.* 24: 227–235, 2000.

Scherf U, Ross DT, Waltham M, et al. A gene expression database for the molecular pharmacology of cancer. *Nat. Genet.* 24: 236–244, 2000.

Shreffler W, Beyer K, Chu T, Burks A, Sampson H. Microarray immunoassay: Association of clinical history, in vitro IgG function, and heterogeneity of allergenic peanut epitopes. *J. Allergy Clin. Immunol.* 113(4): 776–82, 2004.

Schroder C, Jacob A, Tonack S, et al. Dual-color proteomic profiling of complex samples with a microarray of 810 cancer-related antibodies. *Mol. Cell. Proteomics* 9(6): 1271–1280, 2010.

Shoida T. Application of DNA microarray to toxicological research. *J. Environ. Pathol. Toxicol. Oncol.* 23(1): 13–31, 2004.

Smith KM. Exploring FDA-approved IVDMIAs. *IVD Technol.* 17(3): 1–7, 2011.

Song X, Lasanajak Y, Xia B, et al. Shotgun glycomics: A microarray strategy for functional glycomics. *Nat. Methods* 8(1), 85–92, 2011.

Sreekumar A, Chinnaiyan AM. Protein microarrays: A powerful tool to study cancer. *Curr. Opin. Mol. Ther.* 4(6): 587–593, 2002.

Tahara H, Sato M, Thurin M, et al. Emerging concepts in biomarker discovery; the US-Japan workshop on immunological molecular markers in oncology. *J. Transl. Med.* 7: 45, 2009. doi:10.1186/1479-5876-7-45.

Tang J. Microbial metabolomics. *Curr. Genomics* 12: 391–403, 2011.

Turtinen LW, Prall DN, Bremer LA, Nauss RE, Hartsel SC. Antibody array-generated profiles of cytokine release from THP-1 leukemic monocytes exposed to different amphotericin B formulations. *Antimicrob. Agents Chemother.* 48(2):396–403, 2004.

Tyers M, Mann M. From genomics to proteomics. *Nature* 422: 193–197, 2003.

Valle RPC, Jendoubi M. Antibody-based technologies for target discovery. *Curr. Opin. Drug Discov. Devel.* 6(2): 197–203, 2003.

Wang D, Carroll GT, Turro NJ, et al. Photogenerated glycan arrays identify immunogenic sugar moieties of *Bacillus anthracis* exosporium. *Proteomics* 7: 180–184, 2002.

Wang D, Liu S, Trummer BJ, Deng C, Wang A. Carbohydrate microarrays for the recognition of cross-reactive molecular markers of microbes and host cells. *Nat. Biotechnol.* 20: 275–281, 2002.

Weinstein JN. "Omic" and hypothesis-driven research in the molecular pharmacology of cancer. *Curr. Opin. Pharmacol.* 2: 361–365, 2002.

Wellhausen R, Sietz H. Facing current quantification challenges in protein microarrays. *J. Biomed. Biotechnol.* 1–8, 2012. doi: 10.1155/2012/831347.

Whiteaker JR. The increasing role of mass spectrometry in quantitative clinical proteomics. *Clin. Chem.* 56(9): 1373–1374, 2010.

Woodbury RL, Varnum SM, Zanger RC. Elevated HGF levels in sera from breast cancer patients detected using a protein microarray ELISA. *J. Proteome Res.* 1: 233–237, 2002.

Yu X, Schneiderhan-Marra N, Joos TO. Protein microarrays for personalized medicine. *Clin. Chem.* 56(3): 376–387, 2010.

Zander ED. Gene expression analysis as an aid to the identification of drug targets. *Pharmacogenomis* 1(4): 375–384, 2000.

Ziauddin J, Sabatini DM. Microarrays of cells expressing defined cDNAs. *Nature* 411: 107–110, 2001.

2 Commercial Microarrays

INTRODUCTION

The microarray marketplace has matured. Slide-based gene expression microarrays have become a commodity. As a result, there has been a concomitant decline in the number of manufacturers of array printing equipment for do-it-yourself consumers. This is especially true in the arena of genomic research with the commercial availability of complete genome array offerings by well-established, top-tier companies, principally, Affymetrix, Agilent, and Illumina. There has also been an increase in the use of contract research organizations (CROs) that offer printing services for custom microarrays as well as assay development and testing. Moreover, the inevitable shift toward next generation sequencing (NGS) platforms has commenced. Deep sequencing is reported to be displacing microarray-based sequencing, although array-based comparative genomic hybridization (aCGH) and single nucleotide polymorphism (SNP) analysis platforms continue to be adopted for gene-based diagnostics. While it is evident that the commercialization of "applied genomics" tools is undergoing transition, other microarray-based formats, especially protein microarrays, remain very much in demand for basic research as well as for biomarker discovery and diagnostics. Glycan and other carbohydrate-based microarrays have been adopted for glycomics research.

This chapter summarizes the commercialization of the microarray platform with a historical perspective aimed at recognizing key technological developments. Companies that may be regarded as no longer doing business (*nldb*) are discussed because of their contribution to the development of the microarray field.

Microarrays can be found in a number of different formats, all of which have been adopted in one form or another for commercial use. These can be classified as in situ, ex situ, and electronically active or addressable arrays. The fiber-optic array is an optically addressable, ex situ array (Figure 2.1). While historically speaking the origin of array-based technology can be traced back several decades (see review by Matson & Rampal, 2003), we examine here the major breakthroughs in commercialization that were realized beginning in the early 1990s and have continued to evolve over the past 20 years.

IN SITU DNA ARRAYS

Affymetrix is credited with the first commercial development of the oligonucleotide array based upon their pioneering work in the photolithographic masking process coupled with standard DNA chemical synthesis (Pease et al., 1994). The Affymetrix microarray, the *GeneChip* product line, is constructed by synthesis of oligonucleotides from the chip's (substrate) surface. Hence the term *in situ* applies to their synthesis process. This involves photolithographic masking of specific areas on the

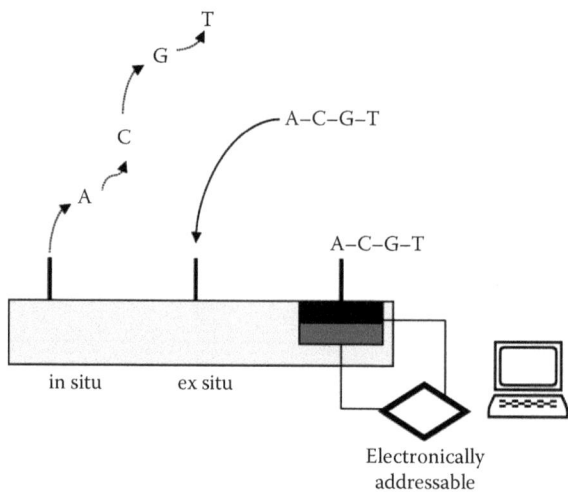

FIGURE 2.1   DNA array types.

substrate, followed by removal of the mask from a selected region, thereby exposing light sensitive deprotecting groups on the previously coupled monomers to ultraviolet (UV) light. With the base fully deprotected, and the light turned off, the chip is next flooded with bulk reagents containing the next monomer phosphoramidite which can now react. The growing oligonucleotide chain is thus extended by one base and the process repeated (Figure 2.2). GeneChips arrays are produced with 25mer oligonucleotides and contain over 500,000 features (20 µm² each) on a 1.28 cm × 1.28 cm chip. For example, the Human Genome U133A 2.0 chip contains over 22,000 probe sets for interrogation of 14,500 human genes (Figure 2.3). The new Human Mapping 500K Array includes two chips for SNP-based analysis, each containing about 250,000 SNPs. There are 6.5 million features on each array.

In order to commercialize the in situ array, Affymetrix needed access to certain intellectual properties, especially the Southern Array patents (see WO89/10977 patent family, **Oxford Gene Technologies** [OGT]). Affymetrix obtained license through a business relationship with **Beckman Coulter** that originally held the first and exclusive Southern license. Later, the company relinquished their exclusivity. Beckman Coulter and Affymetrix subsequently entered into an Array Automation, LLC, to automate the processing of Affymetrix chips. License to the Southern technology has been made available from OGT, so others are now enabled to commercialize in situ microarrays by alternative chemical synthesis approaches. Thus, Affymetrix's early entry into the DNA microarray market which had afforded them a formidable market position has been diminished of late by major competitors. They now share the stage with Agilent and Illumina. Yet, Affymetrix has expanded its product portfolio by the acquisition of **Panomics** (QuantiGene and Procarta product lines), **USB** (PCR reagents; enzymes and related biochemicals), and most recently **eBioscience** (2012), which positions the company for entry into immunology and oncology diagnostic markets. Affymetrix currently has several

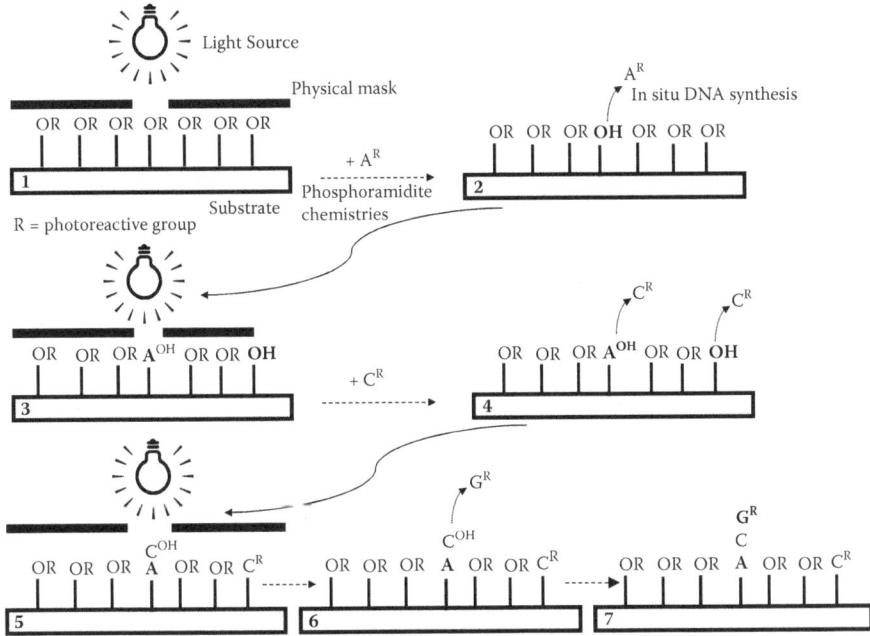

FIGURE 2.2 Photolithographic-based DNA synthesis. (From Pease AC, Solas D, Sullivan EJ, Cronin MT, Holmes CP, Fodor SPA. Light-generated oligonucleotide arrays for rapid DNA-sequence analysis. *PNAS* USA 91: 5022–5026, 1994. With permission.)

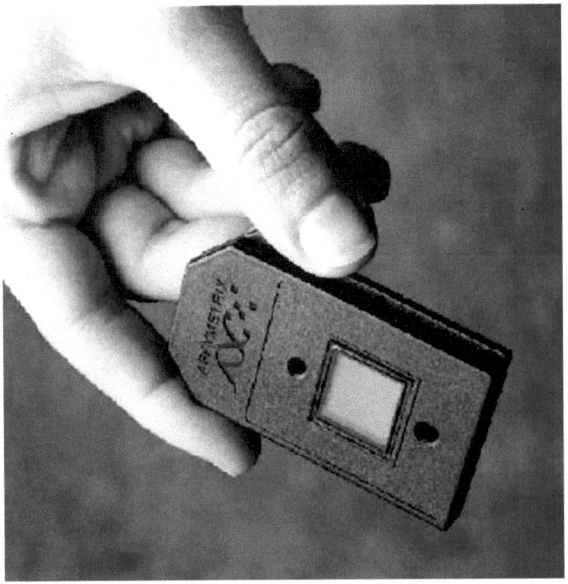

FIGURE 2.3 GeneChip. (Photo courtesty of Affymetrix, Inc., Santa Clara, CA.)

U.S. Food and Drug Administration (FDA)–approved diagnostic tests based upon the GeneChip.

Other earlier commercial efforts surrounding the development of in situ generated oligonucleotide arrays included that of **Rosetta Inpharmatics** (*nldb*—Merck acquired in 2001 but shuttered the Seattle research facility in 2008) and **Protogene** (*nldb*) whose technology relied upon the "sequential" dispensing of reactive phosphoramidite monomers by ink-jet printing to discrete locations on the substrate from which to grow the oligonucleotide sequence. Early work on the Rosetta-Agilent microarray (commercialized by Agilent) demonstrated the advantage of 60mer oligonucleotide probes for gene expression analysis (Hughes et al., 2001). Agilent's SurePrint Whole Human Gene microarray slides contain 41,000 elements (135 microns each) comprising 60mer probes that are in situ printed onto a standard 1" × 3" slide.

Another approach that attracted considerable attention was the introduction of the virtual (photolithographic) mask using the Digital Micromirror Device (DMD) (Texas Instruments). Singh-Gasson et al. (1999) from the University of Wisconsin built a maskless array synthesizer (MAS) incorporating the DMD processor (Figure 2.4). They were able to produce in situ oligonucleotide arrays of 76,800

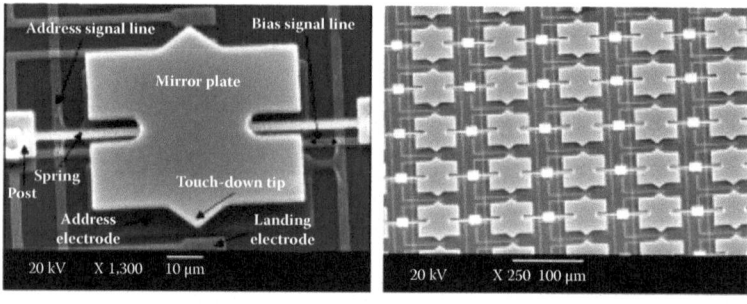

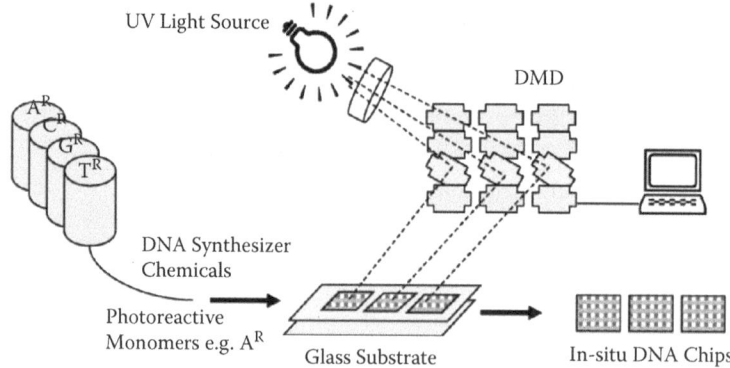

FIGURE 2.4   Digital Micromirror Device (DMD). (From Singh-Gasson S, Green RD, Yue Y, et al. Maskless fabrication of light-directed oligonucleotide microarrays using a digital micromirror array. *Nat. Biotechnol.* 17: 974–978, 1999; Lee, 2003. With permission.)

features (16 µm²) on standard microscope slides with a stepwise yield of about 95% for 18mers in 12 hours.

NimbleGen (Madison, WI—*nldb*, 2012) adopted the MAS approach to create arrays containing as many as 786,432 probe features in a 2.3 cm² area on a slide. Oligonucleotides in the range of 24mer to 90mer are synthesized with good step-wise yield. In order to accommodate current scanner resolution the densities have been reduced. Two formats are provided: 195,000 probes (1:4) or 390,000 probes (1:2) covering the area of a standard slide. A 24mer array can be synthesized within 2 to 4 hours. The average stepwise yield is 97.5%. NimbleGen was purchased by Roche in 2007. Roche plans to exit the DNA microarray marketplace and restructure NimbleGen at the end of 2012 to focus on sequencing technologies.

Febit introduced *geniom one* as a "benchtop microarray facility" which is a fully integrated instrument providing for DNA design, synthesis, hybridization, and anal-ysis. An eight-channel microfluidic device, *the DNA processor*, permits the simulta-neous construction of eight individual microarrays (Figure 2.5). Each microarray is produced with up to 6000 features (34 µm² each) or about 48,000 probes per dispos-able DNA processor. A complete experiment from design to hybridization analysis can be finished within 24 hours. Baum et al. (2003) used a four-channel device that permitted the synthesis of 25mers at $4 \times 12,880 = 51,520$ features. The detailed study describes results comparing the DNA processor to that of the Affymetrix GeneChip YG-S98 yeast genome chip. In side-by-side experiments the febit device performed in concordance to the Affymetrix chip (Baum et al., 2003) (Figure 2.5). Febit restructured in 2010 moving away from its base business to concentrate on micro-RNA discovery programs. At the end of 2011, febit GmbH changed its name to the **Comprehensive Biomarker Center GmbH** to continue their focus on genomic (micro-RNA) biomarker validation for cancer diagnostics.

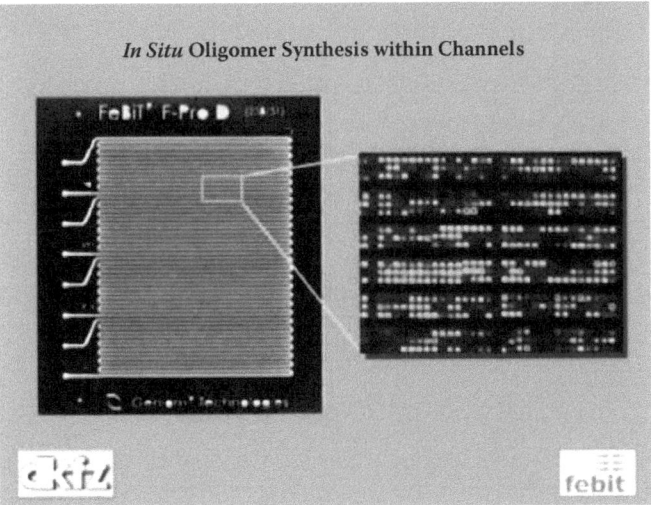

FIGURE 2.5   In situ DNA synthesis in channels. (Photo courtesy of febit AG, Mannheim, Germany.)

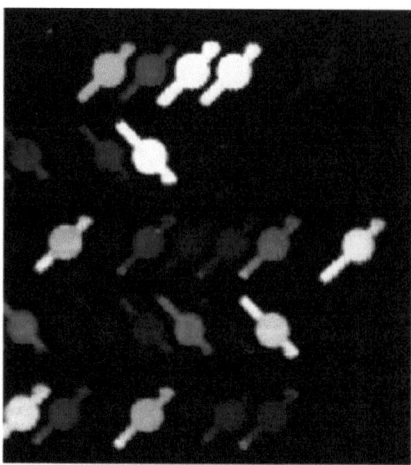

FIGURE 2.6   XeoChip. (Photo courtesy of Xeotron Corporation, Houston, TX.)

Xeotron (*nldb*, 2004) also used the DDM processor to produce arrays in micro-channels (Figure 2.6). In this case, however, standard oligonucleotide phosphor-amidite-based synthesis is performed. The virtual masking is directed instead toward deprotection of a photo-labile acid. The process is called *photo-generated acid* (PGA). Once the acid is liberated, the trityl group on the attached base can be removed to allow base extension to occur. Stepwise yields of 98.5% are reported for the XeoChip. In 2004, the company was acquired by Invitrogen which later became **Life Technologies** as a result of the merger with Applied Biosystems in 2008.

Arguably, a major disadvantage of the in situ format is that each coupling cycle is not 100% complete, thereby leading to reduced stepwise synthesis yields such that the final product at each site is a heterogeneous population containing truncated failure sequences. For the GeneChip process the oligonucleotide probe may only be 82 to 97% pure at each location. Investigations into the reasons for reduction in stepwise yield on glass revealed that coupling reactions are relatively inefficient when close to the surface (LeProust et al., 2001). As Graves (1999) points out, inefficient stepwise couplings reduce the number of probes at a particular site having an error-free sequence to somewhere between 1 and 36% of the total probe population. This would be acceptable for gene expression analysis or calling out a genotype but seemingly less so for SNP polymorphic scanning. However, match versus mismatch ratio metric analysis and the process of "tiling" of degenerate bases improves the performance of such arrays (Wang et al., 1998). In specific instances GeneChip microarrays have been shown to be comparable to those of gel-based sequencing. And, improvements in stepwise yield using the virtual masking approach and new coupling chemistries have been reported. In retrospect, probe purity is not as problematic as once thought.

However, Wu et al. (2007) assessed probes on Affymetrix Human Genome and SNP chips having strings of G nucleotide (G-stacking) and found abnormal binding and a poorer performance correlation rating compared to other probes. The longest G-stack in a probe on the HG-U133A array has nine (GGGGGGGGG) guanines. The authors conclude that "probes that contain G-stacks perform poorly on microarrays, because G-stacks tend to increase cross hybridization and reduce target-specific hybridization" (p. 2571).

McCall et al. (2011) examined the quality of GeneChip microarrays across more than 22,000 HGU133 arrays from data held in the Gene Expression Omnibus database. They observed that about 12% of the GeneChip HUG133a arrays reported were of poor quality. The study concluded that the most likely source of the quality issue is attributed to the array chip itself. It was also recognized that inter-/intra-lab microarray performance can vary greatly. Yet the investigators acknowledged that "this only explains a handful of these poor quality studies" (p. 8).

The other well-noted disadvantages of chip-based microarrays are cost and manufacture turnaround times. A 25mer in situ synthesis using the Affymetrix process theoretically requires $4 \times 25 = 100$ cycles. The number of cycles can be further reduced by applying algorithms to the design. However, there are limits even in software for what can be done to reduce the number of physical masks. Thus, the cost to manufacture by the physical photolithographic masking process requires larger-scale productions and has precluded significant offerings into the custom array market. On the other hand, the virtual masking (or maskless) approaches have demonstrated rapid in situ oligonucleotide synthesis making them well suited for lab scale with the capability to produce 24mer to 90mer probes (Nuwaysir et al., 2002). However, one issue regarding implementation of the DMD technology is control of stray light. Garland and Serafinowski (2002) considered a theoretical model to predict the impact of stray light at low photolithographic contrast ratios. This ratio essentially defines the spatial boundary required to reduce adjacent array element illumination.

At high contrast ratios (~105) (e.g., standard photolithography), very little stray light illumination is observed. At low to moderate contrast ratios (103 to 104) that are typical of the DMD technology, stray light may be problematic. For example, the model predicts that the synthesis of 20mers using DMD would result in the insertion of additional bases such that the major population of oligonucleotide probes would

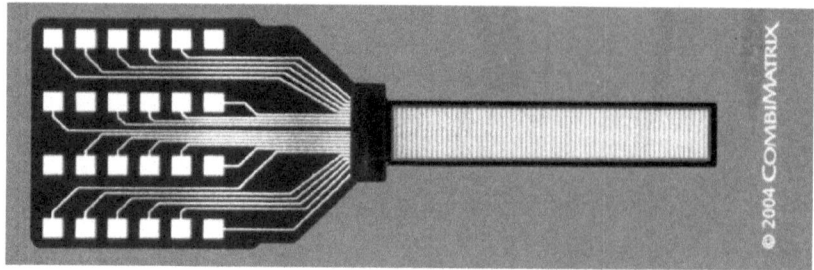

FIGURE 2.7 Early CombiMatrix chip. (Photo courtesy of CombiMatrix Corporation, Mukilteo, WA.)

be a mixture of 21mer and 22mers. This assumes the use of the direct 5′-photodeprotection process. At a contrast ratio of 400, the fraction of correct 20mer sequence would be 0.36, while at a ratio of 200 the fraction is reduced to 0.13. There is less of an issue with processes involving 5′-trityl deprotection of the base using a photogenerated acid (e.g., Xeotron's PGA). In this case, stray light generated acid can be buffered out. Garland (2002) demonstrated the stray light N+1 insertion for a T5 synthesis with 1% stray light. The occurrence of T6 that resulted from photoacid deprotection could be significantly reduced by the addition of n-octylamine that serves as an acid scavenger.

Finally, **CombiMatrix** emerged with an in situ process for generating oligonucleotide arrays on microchips with features of 1024 (94 micron diameter) or 13,416 (44 micron diameter). The features are actually "digitally addressable" electrode pads covered with a porous membrane reaction layer (Figure 2.7). Certain electrodes are turned on so as to generate protons over these electrodes. This permits controlled, site-specific deprotection of the attached base (removal of trityl) allowing the addition of the incoming phosphoramidite monomer. Trityl-phosphoramidite monomers and reagents are flooded over the chip and the process repeated through several cycles until the desired oligonucleotide is produced at that particular location. CombiMatrix launched the *CustomArray* in March 2004. The CustomArray product was designed for the custom synthesis of up to 40mer oligonucleotide probe arrays based upon 92 micron features with up to 902 user-defined probe elements. Sensitivity was reported to be 1.5 pM with 20% CV (intra- and inter-chip) and a shelf-life of 4 months. The system provided electrochemical detection. CombiMatrix also initiated some work on protein microarrays utilizing the porous reaction layer to link biotin using in situ chemistries generated from the electrode pads. Streptavidin is then used to bridge various antigens and antibodies to the surface through the covalently attached biotins (Dill et al., 2001).

In 2010, CombiMatrix closed its chip facility near Seattle and has successfully restructured as a cancer diagnostic service company, CombiMatrix Dx (Irvine, CA). The company provides array-based comparative genomic hybridization (aCGH) analysis based upon their proprietary *DNAarray* platform for genetic disorders and specific cancer screening.

EX SITU OR SPOTTED DNA ARRAYS

Ex situ or more often called *spotted* or *printed* arrays have become very popular formats especially for the building of custom arrays (noncommercial) used primarily by academic labs. (See the ABRF surveys on microarrays at www.abrf.org.) The printed cDNA microarray was largely developed out of the gene expression work originating from the Brown & Davis Labs at Stanford University (Schena et al., 1995). Plans for the construction of the microarrayer and split pin designs were available at the Brown lab Web site. This enabled the researcher to prepare his or her own microarrays useful for particular experiments. However, the demand for microarrayers and pin technologies was sufficiently large enough for the start of a cottage industry. This arose out of a number of precision engineering companies who had the expertise in fine tooling necessary to manufacture split pins. Others followed to produce microarrayers and develop slide scanners. Companies furnishing microscope slides turned their attention to the microarray community as well. Software companies were founded to offer packages for imaging and gene expression analysis. **Synteni** (founded by Stanford inventor Dari Shalon in 1994) was first to introduce custom microarray analysis based upon the new cDNA-based microarray slide format termed *gene expression microarray* (GEM) technology. The company was later acquired by Incyte Genomics in 1997 and became Incyte Microarray Systems until the unit was purchased by Quark Biotech (now Quark Pharmaceuticals, Fremont, CA) in 2002 for internal use in target discovery.

In 1999, an overview of the microarray world appeared as a supplement to Nature genetics, entitled "The Chipping Forecast." In that supplement David Bowtell (1999) highlighted what was then the industry leaders in equipment and services for microarrays. In the previous edition of this book (CRC Press, 2005) we noted a significant consolidation in the microarrayer business. That business has continued to erode, in particular for genomic arrays, as consumers opt for commercial microarray products rather than prepare homemade versions. There are only a few remaining manufacturers of microarrayer equipment. The major vendors include Arrayit, BioDot, DigiLab, Scienion (Germany), and Labcyte.

DigiLab (Marlborough, MA) acquired Genomic Solutions from Harvard Biosciences in 2007 which included the microarrayer portfolio of **Cartesian**, **Genomic Solutions,** and **BioRobotics** arrayer systems. **Genetic MicroSystems** was purchased by Affymetrix (Santa Clara, CA) who later abandoned the highly touted ring and pin microarrayer platform. **GeneXP Biosciences** (Woburn, MA) entered the marketplace offering custom microplate-based microarrays (BioGridArray™), as well as a high-throughput BioGridArrayer™. In 2004, **MetriGenix** (a spinoff of Gene Logic) who developed the flow-thru chip technology acquired GeneXP Biosciences and changed its name to **Xceed Molecular** (Toronto, Canada). **Axela Bioscience** (now Axela, Inc., Toronto, Canada) acquired Xceed Molecular in 2010 and currently offers the *Ziplex* flow-through chip microarray system for both nucleic acid and protein analysis.

Genetix (United Kingdom) expanded upon the success of their QBot colony pickers into the microarray world with a product line of QArray™ printers. However, in

2011 Genetix was sold to Molecular Devices (Sunnyvale, CA) and has discontinued their microarrayer product line.

TeleChem International, Inc. (Sunnyvale, CA), who originally introduced the first commercial quill printing pin, has continued to grow into one of the few full-service microarray companies. They have successfully brought out a line of microarrayers marketed under the NanoPrint™ and SpotBot™ product lines. The microarray products are offered by the company's **ArrayIt Corporation**.

About 80% of all microarray work involves the use of glass slides (ABRF Microarray Survey, 2001, www.abrf.org). The most common substrate is a standard glass microscope slide that has been surface treated with a coating such as poly-lysine that will bind nucleic acids (also proteins). As we learn in coming chapters there are a limited number of surface chemistries in use on commercial slides: aminosilane, epoxysilane, aldehyde modified, and thin-coated nitrocellulose are the most popular. Special graded slides for microarrays (ultra-flat; low intrinsic fluorescence) have been introduced. A set (library or panel) of capture probes (nucleic acid, protein, glycan, etc.) in print buffer (ink) are usually prepared in a source plate (e.g., 384-well microplate). The probes are spotted down on the surface using a precision x,y,z plotter (the microarrayer) to which is attached either a dispenser head (e.g., Scienion's piezoelectric dispenser, BioDot's solenoid-based dispenser, Labcyte's acoustic dispenser) for noncontact printing or pins (e.g., Telechem split pin) that touch down on the surface.

The most common nucleic acid spotted originally was cDNA prepared by reverse transcription polymerase chain reaction (RT-PCR) of cloned material, but now that the human genome has been fully sequenced, the chemically synthesized oligo probes are largely in use today. The main advantage of using cDNA had been that both known and unknown sequences could be arrayed and used for gene expression analysis. However, many additional genomes have now been sequenced so most have converted over to oligonucleotides probes. For gene expression, oligonucleotides in the range of 40mer to 70mer are used, while 15mer to 25mer are sufficient for resequencing protocols. Here again, the Southern and Affymetrix patents have prevailed forcing several companies to take license. Operon (Qiagen) and Genomic Solutions (DigiLab) discontinued sale of printed oligonucleotide arrays in view of Affymetrix's extensive patent portfolio.

CONTENT FOR DNA MICROARRAYS

The commercial success of the spotted microarray like that of the Affymetrix's GeneChip is content driven. That is, in order to provide the customer with a comprehensive gene expression microarray product, manufacturers must obtain gene-specific, annotated sequences covering a genome of major interest to the scientific community (e.g., the human genome, the yeast genome, or the mouse genome). Prior to completion of the sequences to these genomes, such content was largely derived from public databases including ESTs (expressed sequence tags). The National Center for Biotechnology Information (NCBI) maintains several databases of interest such as GenBank (Wheeler et al., 2003). GenBank's dbEST includes a database of submitted ESTs that are matched if possible to known genes using a process called UniGene (www.ncbi.nlm.nih.gov/UniGene/index.html). The UniGene clone collections are then available from government licensed vendors. The clones were primarily assembled from the I.M.A.G.E. consortium collections (Integrated Molecular Analysis of Genomes and their Expression) that maintained the libraries from 1993 to 2007. The clones were transferred to Open Biosystems (Thermo Scientific) in 2007.

The original U.S. distributors were the American Type Culture Collection (www. atcc.org), Open Biosystems (www.openbiosystems.com), and Research Genetics (now owned by Invitrogen, www.invitrogen.com). The European suppliers were MRC geneservice (now Source Biosciences) and RZPD German Resource Center for Genome Research, Berlin (renamed ImaGenes, acquired by Source Biosciences, 2010). A complete list of suppliers of full-length cDNA clones can be found at the NCBI Web site clone database or CloneDB at http://www.ncbi.nlm.nih.gov/clone/content/distributors.

Since completion of the human genome project other extensive clone libraries held in private were also made available including those of Incyte's LifeSeq sets including more than 13 million ESTs and 18,000 genes. These were available from Thermo Scientific's Open Biosystems but were discontinued as of June 2012. Invitrogen offers clone collections including those acquired from purchase of Research Genetics. In 2001, Invitrogen entered into a worldwide distribution agreement with the Institute for Genomic Research (TIGR) for their clones. Included were TIGR's extensive EST libraries with over 300,000 cDNA and genomic clones including 70K human, 10K rat, and 70K clones from microorganisms (http://clones.invitrogen.com).

Suppliers of DNA Microarrays

A list of suppliers of microarray products was published in *The Scientist* (2003) and *Nature Genetics* (2002). However, the number of companies offering spotted microarray products for gene expression analysis has greatly diminished with the availability of a growing portfolio of high-quality products from major vendors such as Affymetrix or Agilent. Arrayit (Sunnyvale, CA) and Phalanx Biotech Group (Belmont, CA) both offer a variety of DNA microarray slide products. Biocompare (www.biocompare.com) and SelectScience (www.selectscience.net), as well as BioArray News, which is part of the GenomeWeb group (www.genomeweb.com), are good places to begin sourcing microarray products.

COMPARISON OF COMMERCIAL DNA MICROARRAYS

Studies have compared the performance of in situ and ex situ spotted microarrays in gene expression analysis using commercial sources. The following studies illustrate some of the issues regarding the disparity between commercial microarray products and platforms. While these are now rather historical findings, we find similar persistence in intra-platform differences in the more recent miRNA microarray products (see Chapter 5, section entitled "The Nature of Platform-to-Platform Disparity").

Tan et al. (2003) evaluated Agilent, Amersham, and Affymetrix human genome gene expression microarray products. Amersham's CodeLink arrays (30mer probes) and Affymetrix's GeneChip (25mer probes) were assessed along with Agilent's cDNA array format using the same cRNA pools. A total of 2009 genes common to all three platforms were analyzed. While intra-platform correlation coefficients were high ($r > 0.9$), the Pearson's correlation coefficients were as follows: Affymetrix-Amersham (0.59), Agilent-Amersham (0.48), and Agilent-Affymetrix (0.50). In other words, in this particular study there was not significant agreement as to the gene expression profile relationships obtained using these platforms, further suggesting "that cross-platform differences arise from the intrinsic properties of the microarrays themselves." Amersham was acquired by GE in 2004. The CodeLink microarray products were discontinued in 2007.

In another study, Barczak et al. (2003) compared GeneChip arrays to "long" oligonucleotide arrays. In this study, 7344 genes from the human genome were analyzed using the Affymetrix U95 GeneChip along with two spotted arrays comprising 70mer probes (Operon Human Genome Oligo Set, versions 1 and 2). In this case, a good correlation for differential expression was obtained between the spotted 70mer arrays and the in situ 25mer arrays. However, the long oligonucleotide hybridization intensities were lower overall leading to less reliable calls such that when compared to the in situ arrays the correlation was significantly reduced ($r \sim 0.6$). Excluding 4467 genes with low intensity values improved the correlation ($r \sim 0.9$) for the remaining 2877 common genes. Obviously probe selection is critical. Another important point is made in this study: if possible, use reference samples with gene expression levels similar to those expected for the test samples. This should improve quantification at the low end. Moreover, these two studies also give cause to the need for the standardization of commercial microarray products and reference standards so that interlaboratory, as well as cross-platform performances can be properly assessed. While commercial microarrays from different vendors all claim substantial gene probe real estate, the number of probes in common is smaller. This makes it difficult to compare experiments across platforms, especially those from major suppliers such as Affymetrix, Agilent, and Amersham (Figure 2.8).

Recently, Llorens et al. (2011) compared expression data for the EGF-dependent signal transduction pathway obtained across Agilent, Operon, and Illumina microarrays. These platforms had considerable probe overlap providing a pool of 17,070 shared genes that could be used for cross-platform concordance studies. The authors summarized their findings that while there was overall agreement on probes common across all platforms, there were also "quite a large number of probes that give discrepant results" (p. 12).

The spotted array is also being used for SNP analysis based upon primer extension labeling of oligonucleotide or cDNA probes. The advantage of using oligonucleotides is that these may be synthesized in good quantity and highly purified (98 to 99%) prior to the attachment. For example, Sequenom first introduced chip-based high-throughput SNP screening with mass spectroscopy analysis. The original process involved the creation of arrays by immobilization of single-stranded PCR amplicons onto silicon chips (MALDI targets) that had been surface treated with N-succinimidyl (4-iodoacetyl) aminobenzoate. Disulfide terminated primers were used to generate amplicons and the reduced template (thiol-DNA) covalently coupled to the sulfide-reactive chip. A single-base primer extension reaction on the immobilized target was performed followed by the addition of 3-hydroxypicolinic acid (matrix). The extended primer sequence is then determined using matrix-assisted laser desorption

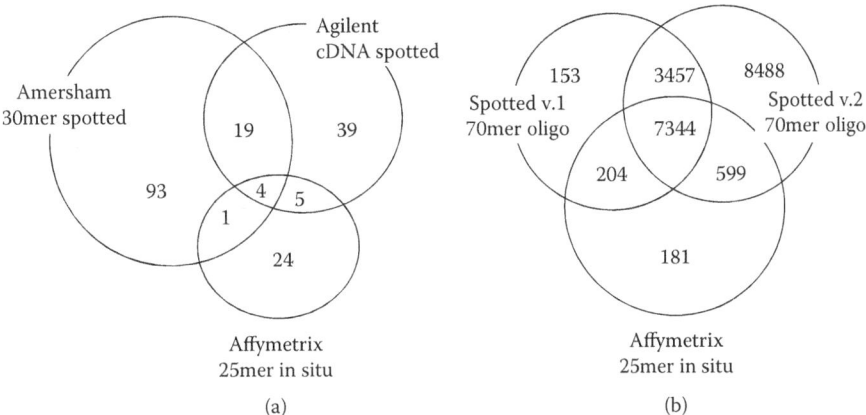

FIGURE 2.8 Microarray cross-platform showing differentially expressed gene clusters obtained from Amersham, Agilent, and Affymetrix products. (From Tan PK, Downcy TJ, Spitznagel Jr. EL, et al. Evaluation of gene expression measurements from commercial microarray platforms. *Nucleic Acid Res.* 31(19): 5676–5684, 2003; Barczak A, Rodrigues MW, Hanspers K, et al. Spotted long oligonucleotide arrays for human gene expression analysis. *Genome Res.* 13: 1775–1785, 2003. With permission.)

ionization-time of flight (MALDI-TOF) mass spectroscopy (MassArray™) to reveal the SNP (Tang et al., 1999). In later refinements, it was found to be unnecessary to use the APEX (anchored primer extension) approach with the analytical power of mass spectroscopy. Instead, solution-phase primer extension was employed followed by direct dispensing of the extended primer onto individual elements of the silicon chip that had been preloaded with matrix (Buetow et al., 2001).

Orchid introduced a zip code array approach to SNP analysis by printing capture oligonucleotides onto 384-gasket well, skirted glass plates. The complementary oligonucleotide was incorporated as a tag in the extending primer. Following a solution-based single base extension with labeled ddNTP the now labeled primer was captured and the SNP determined from the hybridized tag (Figure 2.9). This technology and business was purchased from Orchid by **Beckman Coulter** in 2002 and marketed as SNPstream™. Danaher Corporation acquired Beckman Coulter in 2011. The SNPstream product line is no longer listed with the company.

Low to medium density arrays have also entered the marketplace based upon the "array of arrays" format. **Genometrix** first commercialized using this format printing probes (nucleic acid or antibody) in a 96-well pattern on Teflon masked slides to perform micro-ELISA or genotyping (Mendoza et al., 1999; Weise et al., 2001). These were marketed as various services under the name GenoVista Partnership Program. The 96-well microarray platform was called the VistaArray. Unfortunately, the service provider model failed to generate or sustain sufficient revenues. Following the collapse of Genometrix, **High-Throughput Genomics** (now HTG molecular, Tucson, AZ) purchased Genometrix's intellectual property portfolio and has introduced an array within well microtiter plate (ArrayPlate™) for gene expression analysis. Assays are based upon HTG's proprietary multiplexed nuclease protection assay

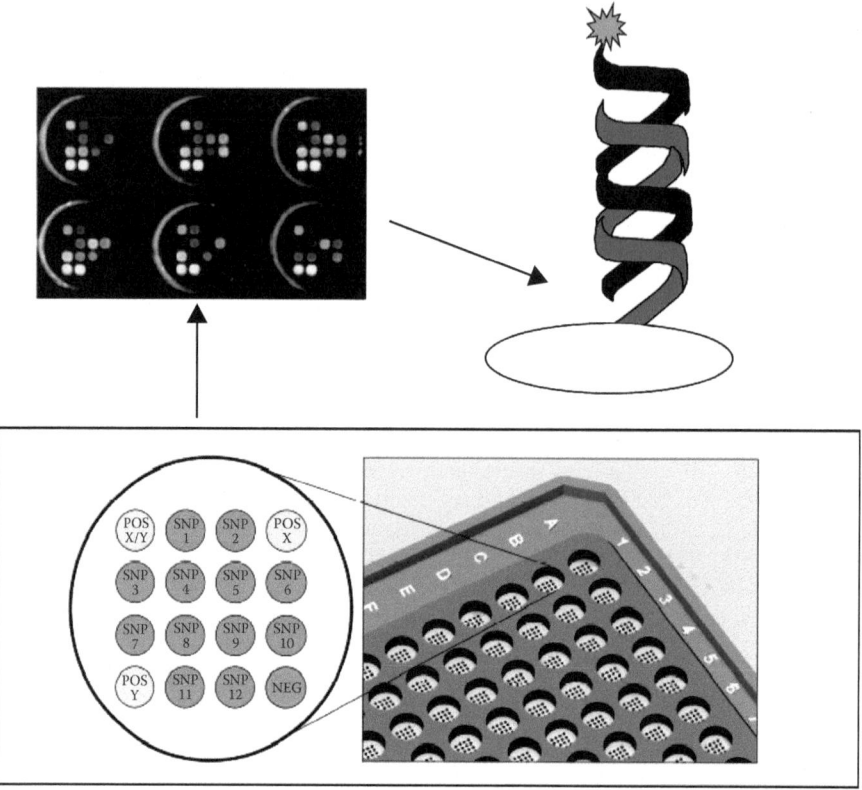

FIGURE 2.9 Single nucleotide polymorphism (SNP) detection using Orchid's SNPstream. (From Bell, 2002. With permission.)

(m-NPA). The primary advantage of m-NPA is that no amplification or sample processing of RNA need be conducted. Cell lysates may be added directly to the well and the released mRNA species captured and protected from nuclease digestion by sets of sequence-specific oligonucleotide probes. Once the RNA-probe hybrids are formed, other nucleic acids within the sample are digested with S1 nuclease. The nuclease protected RNA strands can then be destroyed by alkaline hydrolysis leaving behind the original probes that are now present in the sample in amounts quantitative to the protected RNA species. Probes are then captured within the array and their identities determined by the complementary sequence of the capture oligonucleotide. The amount of captured probe is estimated using a chemiluminescent reporter assay (Martel et al., 2002). This product is primarily focused on high-throughput drug-target gene profiling for drug discovery applications.

COMMERCIAL PROTEIN ARRAYS

The protein microarray marketplace (the majority of products are antibody arrays) has seen considerable growth over the past decade based upon a continued demand

and the readily available supply of antibody content. **Luminex** (Austin, TX) with its host of commercial vendor partners offers the *XMap* multiplex bead-based technology for use with Luminex 100/200/FlexMap 3D flow cytometers. The higher-throughput FlexMap 3D systems can analyze up to 500 analytes per microplate well. A lower-throughput CCD camera-based system, the MAGPIX, uses magnetic particles and can analyze up to 50 analytes. Luminex systems utilize laser-induced fluorescence for detection.

Mesoscale Discover (MSD, Gaithersburg, MD) offers the SECTOR series of imagers (CCD camera or photodiode array models) that read MULTI-ARRAY microplates (96/384-well) having spotted well arrays of antibodies in low density formats for detection of 4, 7, or 10 analytes by electrochemical chemiluminescence.

Aushon BioSystems (Billerica, MA) acquired Pierce's SearchLight product line from Thermo Fisher Scientific in 2009. The SearchLight format is a membrane-based microplate antibody array. Chemiluminescence is used for detection. In 2012, Aushon launched *Cira*, based upon a 12 spot circular designed microarray in a 96-well format. The circle array is reported to improve the performance of the Ciraplex™ multiplex assay by the precise location of the array within the optimal microfluidic path.

There are numerous vendors offering preprinted content (menu) consisting primarily of capture antibodies arrayed on glass slides or membranes. Cytokine arrays represent the most popular array product offering. A few companies also offer preprinted 96-well microplate microarray products. **R&D Systems** (Minneapolis, MN) a well-known supplier of antibodies and ELISA kits also offers microplate-based antibody array products. The *Mosaic* ELISA allows for the quantitative detection of up to 11 analytes per well based upon a sandwich immunoassay using a streptavidin HRP-generated chemiluminescent signal detected using a CCD camera imager. **Invitrogen Life Technologies** (Carlsbad, CA) offers the *ProtoArray* Human Protein Microarray of an insect expressed set of over 9000 full-length human proteins arrayed on a nitrocellulose coated glass slide. **Raybiotech** (Norcross, GA) supplies primarily cytokine and related inflammation panels arrayed on both membranes and glass slides. A quantitative microarray ELISA product, *Quantibody* Antibody Array, is based on a standard glass slide with 16 subarrays each spotted with an identical set of cytokine capture antibodies. Wells are formed with a 16-well removable gasket to process 16 samples. A slide holder allows for placement of four slides that permits processing of 64 samples. **Quansys Biosciences** (Logan, UT) supplies the *Q-Plex* multiplex ELISA products based on antibodies arrayed in a glass bottom 96-well or 384-well microplate primarily for the analysis of cytokines. Chemiluminescent or infrared signals are detected using a CCD camera system.

CONTENT PROVIDERS

There is an abundance of antibody content suppliers. However, not all antibodies and manufacturers of antibody products are created equal. It is prudent to test antibodies from at least two to three independent sources. For monoclonal antibodies pay careful attention to the clone identification and obtain as much background information as possible on your selection of either monoclonal or polyclonal antibodies—especially information regarding antigen source (i.e., native, recombinant, or synthetic peptide), host or protein expression system used, epitope mapping to the antigen, additives or stabilization agents such as BSA, purification, cross-reactivity, sensitivity, and intended use (e.g., ELISA, western blot, etc.). Do not assume that an antibody that works well in a standard ELISA will work well in a microarray format. If the capture antibody will be used in a sandwich assay format, it is very important to correctly identify matching reporter antibodies, especially matched pairs of monoclonals.

For sourcing of antibodies a good starting place would be **Lincott's Directory** of Immunological & Biological Reagents (www.linscottsdirectory.com, Mill Valley, CA) which has over 400,000 antibodies sourced (2012). **Abcam** (www.abcam. com, Cambridge, MA, and Cambridge, United Kingdom) also offers a large selection of primary antibodies, secondary antibodies, and antigens either prepared by the company or sourced from partners. **Biocompare** (www.biocompare.com, South San Francisco, CA) offers vendor sourcing as well. Other primary suppliers include **R&D Systems** (www.rndsystems.com, Minneapolis, MN); **Southern Biotech** (www.southernbiotech.com, Birmingham, AL); **Biolegend** (www.biolegend.com, San Diego, CA); **Santa Cruz Biotech** (www.scbt.com, Santa Cruz, CA); **BD Biosciences** (www.bdbiosciences.com, San Jose, CA); **Novus Biologicals** (www. novusbio.com, Littleton, CO); and **Abnova** (www.abnova.com, Taipei City, Taiwan, and Walnut Creek, CA). This is not by any means an all-inclusive list. Moreover, the intent here is to provide a place to start and not necessarily an endorsement of their services. *Nature* published an extensive table of antibody suppliers in their June 2007 issue (see *Nature* 447: 745–746, 2007).

AN OPEN PLATFORM APPROACH

As a commercial platform, the microarray has been used largely as a *closed platform*. That is, the data or information obtained would be limited by whatever probes (the content) are affixed to the microarray by the manufacturer. So, the consumer is restricted in their experimental design to examine biological content dictated by what content menus are commercially available. An *open platform* permits the user to design microarrays with content specific for their needs. The A^2 (A-squared) technology to be discussed here is an open platform for multiplex assay development based upon a unique microplate platform and proprietary self-assembly chemistry that enables the user to define assays based upon their own content. Because the A^2 approach removes the need to print arrays, assay development is accelerated.

A^2 Multiplex ELISA System

A-squared ($A^{2®}$) which stands for "array of arrays" is a microplate-based multiplex immunoassay platform originally developed by Beckman Coulter, Inc. Capture oligonucleotide "zip codes" are covalently tethered to the bottom of a specially designed 96-well polypropylene plate (the A^2 Plate). Complementary oligonucleotides may then be conjugated to selected antibodies, with the resulting oligo-antibody conjugates mixed together and delivered to the capture oligo plate. The antibody array is self-assembled under mild hybridization conditions at ambient temperature in 1 hour ready for assay development. The plate with its rounded square-wells was designed

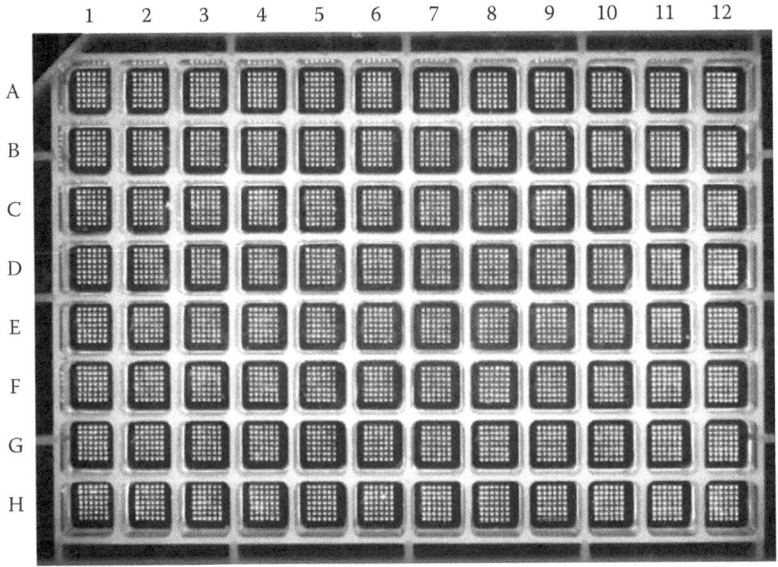

FIGURE 2.10   A^2 plate microarray.

for optimal mixing to improve assay performance and prevent evaporative wicking (Figure 2.10).

QuantiScientifics (Irvine, CA) who acquired the business from Beckman Coulter in 2009 now offers the *A^2 MicroArray System* that incorporates *Oligo-Link-it* technology to create protein microarrays based upon oligo-protein conjugates. Besides antibodies, other ligands such as antigens, peptides, aptamers, and small molecules may also be immobilized. The system includes a reader, fluorescent label detection reagents, and fully integrated software to perform multiplex quantitative μELISAs (Song et al., 2004; Matson and Rampal, 2010).

The capture oligos are arranged in triplicate throughout each of the 6 × 7 well arrays. In its present format 13 different capture oligos are represented along with a reference oligo that is used to register the automated software grid tool. Thus, each A^2 plate generates 4032 data points, resulting in 1248 analyte values (Figure 2.11). A developed A^2 plate is scanned using the A^2 MicroArray System reader. The reader is composed of a CCD camera system that incorporates up to two Zeiss Axio optical filter cube assemblies for dual wavelength epi-fluorescent imaging of each well bottom. An external fiber-optic tungsten halogen illuminator serves as the light source. The reader system includes an automated *x,y* translation stage for plate positioning, optical and positional calibration, image acquisition, data analysis, and reporting. If desired, the raw image files (.tiff) can be exported and analyzed in compatible image analysis software, and data reports may be exported into Excel spreadsheets.

In 2011, QuantiScientifics and automation's partner, BioDot, Inc. (Irvine, CA), introduced the **A^2 DxA System** (Diagnostic Assay Development System) for automated assay development. The system includes BioDot's MDx, a customized AD1500 series noncontact dispenser that has been configured for use with the A^2

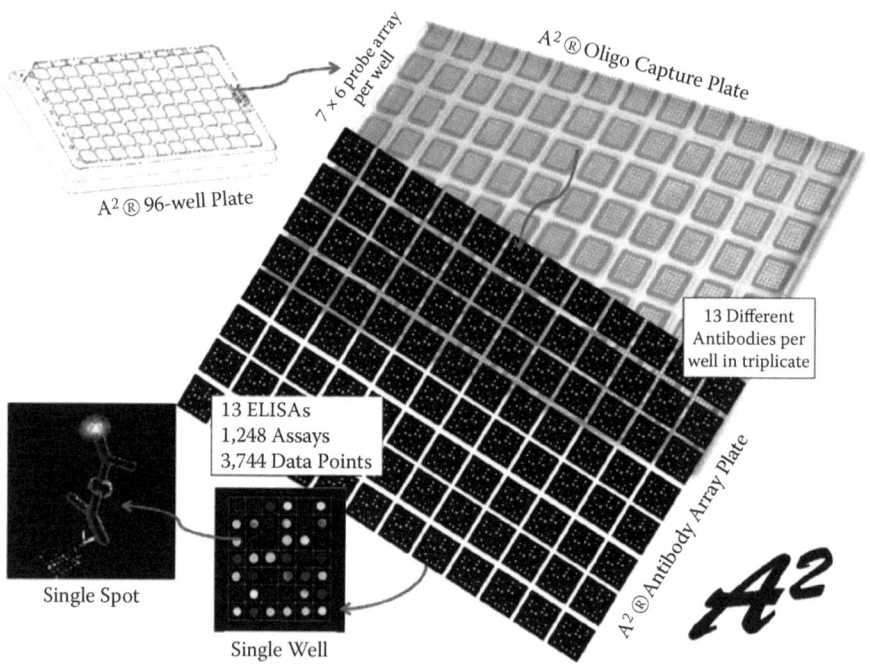

FIGURE 2.11   **(See color insert.)** Anatomy of the A^2 plate assay.

plate providing both print function and automated assay processing. The A^2 DxA reader is configured for quality control of printed plates as well as imaging and assay data analysis. The automated system can process two A^2 plates for the analysis of 160 samples, each for up to 13 analytes.

THREE-DIMENSIONAL (3D) AND FOUR-DIMENSIONAL (4D) CHIPS

The late Andrei D. Mirzabekov from the Engelhardt Institiute (Moscow) began development of hydrogel arrays in the late 1980s. Subsequently, collaboration with the Argonne National Laboratory (Argonne, IL) began in 1994 toward commercialization of hydrogel microarrays. In 1998, ANL entered into a research partnership that began the commercialization forays by Motorola and Packard (*nldb*, acquired by Perkin Elmer, 2001). Motorola Life Sciences abandoned the Mirzabekov gel in favor of the Surmodic's surface which they then brought to the market as CodeLink™. Amersham Biosciences acquired CodeLink from Motorola in 2002. As mentioned previously, following the GE acquisition the CodeLink product was abandoned.

Besides CodeLink several other 3D microarrays had entered the marketplace. These include Perkin Elmer's HydroGel™ polyacrylamide gel microarray (obtained from the acquisition of Packard Biosciences) for proteins. Biocept printed down droplets of a polyethylene glycol hydrogel onto glass slides offering a DNA microarray product called 3D HydroArray™ (Gurevitch et al., 2001). Biocept abandoned the microarray field about 2005, entered the microfluidics field, and successfully reinvented themselves as a cancer diagnostics company. Currently, at least two companies offer hydrogel microarray substrates. Schott North America (www.us.schott. com, Louisville, KY) provides the *Nexterion Slide P*, a glass microarray slide coated with a hydrophilic polymer that is N-hydroxy succinimide (NHS) activated for immobilization of proteins. The *Nexterion Slide H* provides a thin-film polymer coating with NHS groups useful for protein, oligonucleotide, and glycan immobilizations. Xantec (www.xantec.com, Dusseldorf, Germany) offers hydrogel microarray products for immobilization of proteins and DNA. Arrayit, Inc. (www.arrayit. com, Sunnyvale, CA) has partnered with Xantec to also offer these products.

Membranes cast upon glass slides also fall into the 3D surface category. Most notably has been the nitrocellulose coated FAST™ slides originally developed by Schleicher & Schuell BioScience (S&S). A cytokine micro-ELISA product under the trade name ProVision™ had been introduced as a single slide or offered in a 96-well spacing format (64 usable wells) called FASTQuant (Harvey, 2003). Historically, nitrocellulose membrane has been used for the sequestering of both proteins and nucleic acids. The adaptation by S&S of this microporous (0.2 micron pores) material cast into a microarray format has been relatively straightforward. The disadvantage of the NC slide is one of resolution because biomolecules more easily diffuse from the surface than seen with planar arrays. Also, membranes suffer from considerable light scatter and higher intrinsic fluorescence which are problematic for increased sensitivity. Membranes are better suited for detection of colorimetric or chemiluminescence reporters. S&S was acquired by Whatman in 2004 which in turn was acquired by GE Healthcare in 2008. In 2012, GE Healthcare granted Maine Manufacturing (Sanford, ME) the exclusive right to commercialization of the FAST® Protein Array product line.

Schott offered an open pore coated nitrocellulose slide, called the Nexterion® Slide NC, but this product has been removed from their Web site and is reported to have been discontinued (2012). Pall Life Sciences which offered the Vivid™ Gene Array slides based upon a coated nylon polymer has also exited the "nitrocellulose" microarray business.

FLOW-THRU BIOCHIPS

Beattie et al. (1995) described the fabrication and use of a flow-thru porous silicon genosensor prepared from channel glass. An array of square wells was prepared by etching into a silicon wafer. The wafer was then bonded to a second wafer with the alignment of the wells to square batches of acid-etched pores. For example, a packed array of 10 micron diameter pores are oriented perpendicular to the wafer face. Light could be transmitted through the porous silicon array thereby increasing resolution and detection sensitivity. GeneLogic licensed Beattie's invention but later abandoned commercialization. Instead, they formed a spin-off company, MetriGenix, which later introduced the 4-D Flow-Thru Chip (Figure 2.12). As previously discussed, Axela now offers the product as the Ziplex System (2012).

PamGene, a spin-off of Organon Teknika (Akzo Nobel, The Netherlands), founded in 1999, introduced a similar flow-thru technology based upon porous

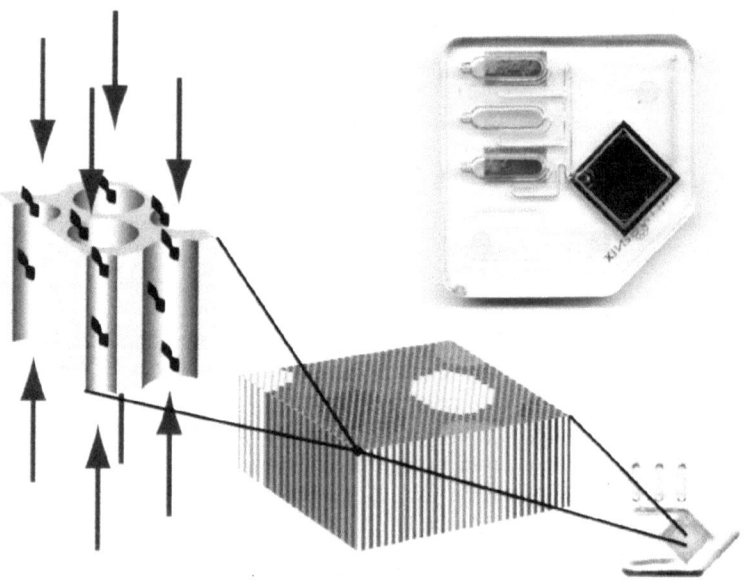

FIGURE 2.12 Early MetriGenix Flow-Thru Chip platform. (From Iyer M, Philip R, Matthai HE, Eastman E, O'Beirne AJ. Using 4-D diagnostic tools for genetic analysis. *IVD Technology*, July/August Issue, pp. 47–53, 2003. With permission.)

aluminum oxide. Their product is still marketed under the PamChip™ trade name (Figure 2.13). The capillary porous structure (200 nm pore diameter; 107 per mm^2) provides for a large surface area and small volume flow-path. Because of the capillary pore geometry, the printing process leaves sharp, well-defined probe spots as liquid is rapidly transported into the capillaries rather than diffusing across the substrate surface (Chan, 2002). The company currently offers PamChip peptide microarrays. The *PamChip 4* product contains four separate peptide arrays housed

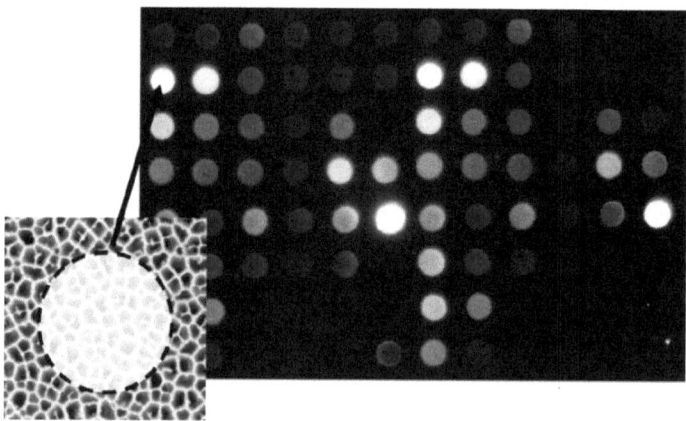

FIGURE 2.13 PamChip. (Courtesy of PamGene International BV.)

in a single device and is processed on the PamStation12. A phosphotyrosine kinase (PTK) array of 144 peptides and a serine-threonin kinase (STK) array of 144 peptides are available to assess cellular kinase activity.

ELECTRONIC BIOCHIPS

Electronically active chips (e.g., Nanogen's *NanoChip*) are true microchips in which microelectrodes (pads) become elements of the array (Figure 2.14). The microelectrodes are covered with materials that allow immobilization of probes. Each electrode is individually addressable so that specific probes can be attached to different electrodes. In the case of Nanogen, hybridization is accelerated by electromotive forces (emf) on the target. Enhanced stringency is also achieved by modulation of the emf (Heller et al., 2000). Nanogen (*nldb*) discontinued operations as a microarray diagnostic company in 2008.

Xanthon (*nldb*) was first to introduce a disposable 96-well microplate (X2A plate) microelectrode system (*X2AS* for Xanthon Xpression Analysis System) that could be used in high-throughput nucleic acid assays using a robotic workstation. The X2A plate allowed analysis of 96 samples for five specific sequences. They projected the ability to analyze 27,000 samples per day. The analysis was based upon cyclic voltammetry mediated by ruthenium ion redox in the presence of DNA. Unfortunately, the company went out of business failing to achieve mass production.

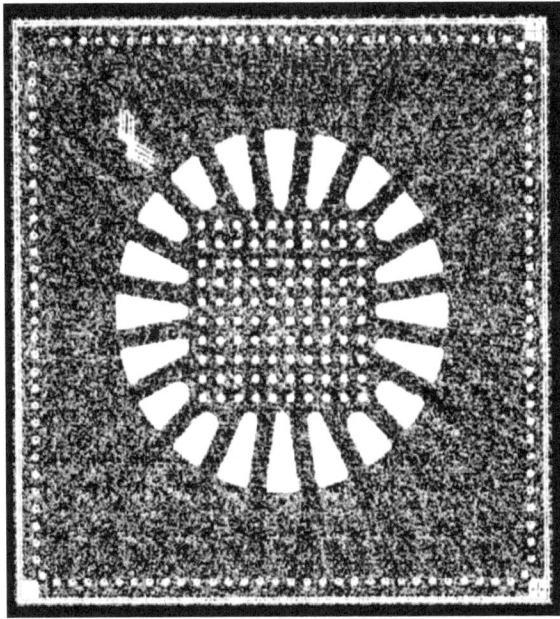

FIGURE 2.14 NanoChip electronic microarray. (From Heller MJ, Forster AH, Tu E. Active microelectronic chip devices which utilize controlled electrophoretic fields for multiplexed DNA hybridization and other genomic applications. *Electrophoresis* 21: 157–164, 2000. With permission.)

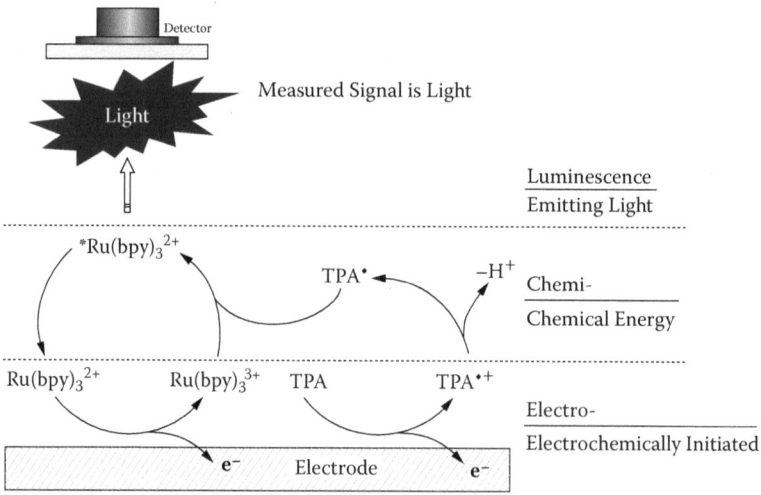

FIGURE 2.15   Electrochemiluminescence detection. (Courtesy of MesoScale Discovery.)

On the other hand, MesoScale Discovery (MSD) has succeeded in introducing product with a similar technology approach based upon ruthenium redox mediated electrochemical detection (Figure 2.15). MSD began as a joint venture between its parent company MesoScale and IGEN who pioneered much of the work on electrochemical detection based upon the ruthedium redox system. MSD's Multi-Spot™ and Multi-Array™ Plates have antibodies immobilized on multiple working electrode pads within each well allowing each spot within the well to serve as an individual assay. Multiplexed cytokine immunoassays can be performed in 96-well, 4-, 7-, or 10-spots per well patterns with detection limits of pg/mL with a broad linear dynamic range. Both 24- and 384-well electrode systems are also available.

Clinical MicroSensors (CMS) developed technology that involved DNA capture probes attached to a microelectrode pad through molecular wires of phenylacetylene attached to a gold substrate. Target (unlabeled) is allowed to hybridize under passive conditions. A signaling probe labeled with ferrocene serves as an electron donor that interacts with the gold electrode to produce an electronic signal (Umek et al., 2001). CMS technology may be adapted for point-of-care (POC) handheld devices (Figure 2.16) or microtiter plate formats. The major advantage of the electronically active or addressable chips is rapid analysis time. Nanogen had the added advantage of emf driven hybridization and stringency modulation. Clinical MicroSensors (CMS) was acquired by Motorola Life Sciences in 1999 to commercialize the *eSensor*. Motorola sold CMS to Osmetech in 2005, which became GenMark Diagnostics in 2010. GenMark offers a second-generation eSensor platform for molecular diagnostic applications, called the eSensor® XT-8 (2012).

Illumina first produced fiber-optic random bead arrays based upon the inventions of David Walt (Tufts University). In this case, latex beads are encoded using different fluorescent dye mixtures that are either adsorbed into the particle or attached to

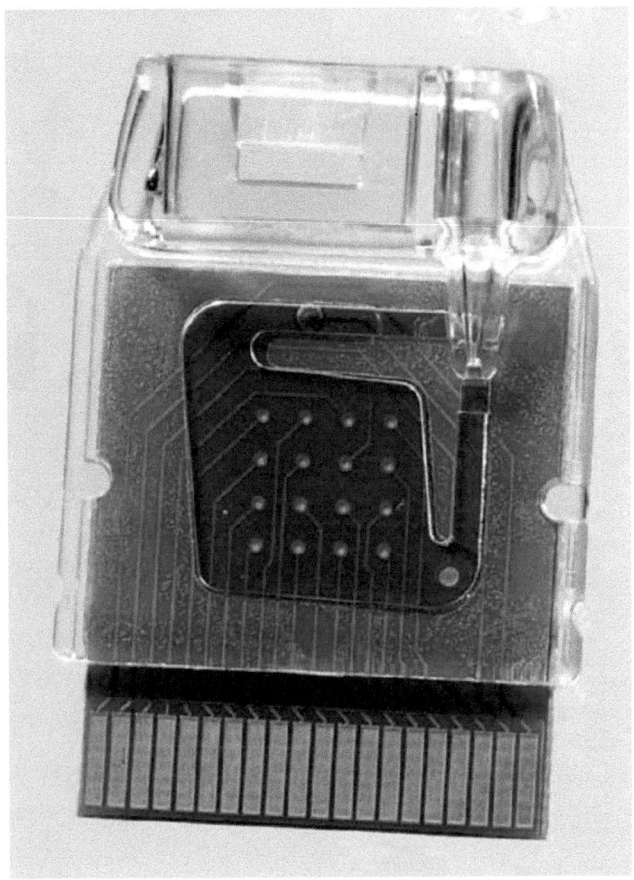

FIGURE 2.16   Early version of the eSensor biochip.

the surface. Presynthesized oligonucleotides are attached to selected bead populations such that a single dye or dye/dye ratio identifies the attached oligonucleotide. Populations are mixed in bulk and then loaded onto the tips of a fiber-optic in which one end has been acid etched to form microscopic nanowells. The nanowells are filled at random with the mixed bead population to create a BeadArray™. Such bundles can be arranged in an Array of Array™ format to match the well of 96-, 384-, or 1536-well plates (Figure 2.17). These products were initially offered under the trade name of Sentrix™ Array Matrices (Oliphant et al., 2002). Labeled targets are hybridized and located by imaging the fiber that has been coupled to a light source and CCD camera. Once signal and bead position have been mapped, each bead having a positive signal is decoded for the dye ratio and the sequenced determined. A major advantage of this technology is that a high array density can be achieved without printing. Density is determined by the number of etched wells. Very small volumes of sample can be addressed at high sensitivity. The major application for Illumina's products has been for high-throughput SNP genotyping. With densities

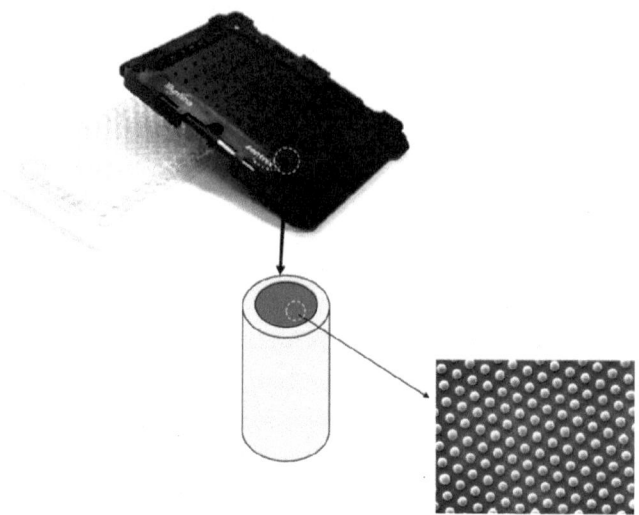

FIGURE 2.17   Sentrix fiber-optic chip. (From Oliphant A, Barker DL, Stuelpnagel JR, Chee MS. BeadArray™ technology: Enabling an accurate cost-effective approach to high-throughput genotyping. *BioTechniques* 32: S56–S61, 2002. With permission.)

up to 50,000 beads per optical fiber, a 96-well system can process 150,000 SNPs in parallel providing the potential to call greater than 1 million genotypes per day. A Sentrix BeadChip slide product was also introduced for gene expression analysis. Wells are etched in glass slides and in turn filled with 3 micron beads with each bead containing 50mer oligonucleotide gene-specific probes. The slide array can screen for 700 genes starting with 50 to 200 ng total RNA. Detection limits to 0.15 pM are claimed with fold-changes achievable at <1.3, dynamic range 2.8-fold, and array to array imprecision at <10% CV. The BeadArray/BeadChip platform has been incorporated into the *iScan* System for genotyping and gene expression analysis (2012).

Illumina's VeraCode technology (2012) uses digital holographic image coding to track microbeads. These are actually micro-cylinders (240 µm × 28 µm, diameter) with embedded 24-bit holographic elements. The *BeadXpress* System has dual-color laser excitation for generation of the holographic code image associated with each bead as well as the development of the assay signal. The micro-cylinder beads are surface activated for the immobilization of nucleic acids or proteins, thereby creating high-density microarrays for multiplex assays. These include genotyping, CpG methylation analysis, and protein screening (48-plex) by sandwich immunoassay.

REFERENCES

Barczak A, Rodrigues MW, Hanspers K, et al. Spotted long oligonucleotide arrays for human gene expression analysis. *Genome Res.* 13: 1775–1785, 2003.

Baum M, Bielau S, Rittner N, et al. Validation of a novel, fully integrated and flexible microarray benchtop facility for gene expression profiling. *Nucleic Acids Res.* 31(23) e151: 1–13, 2003.

Beattie KL, Beattie WG, Meng L, et al. Advances in genosensor research. *Clin. Chem.* 41(5): 700–706, 1995.

Bell PA, Chaturvedi S, Gelfand CA, Huang CY, Kochersperger M, Kopla R, Modica F, Pohl M, Varde S, Zhao R, Zhao X, and Boyce-Jacino MT. SNPstream® UHT: Ultra-high throughput SNP genotyping for pharmacogenomics and drug discovery. *BioTechniques* 32: S70–S77, 2002.

Bowtell DDL. Options available—From start to finish—For obtaining expression data by microarray. *Nat. Genet.* supplement 21: 25–32, 1999.

Buetow KH, Edmonson M, MacDonald R, et al. High-throughput development and characterization of a genomewide collection of gene-based single nucleotide polymorphism markers by chip-based matrix-assisted laser desorption/ionization time-of-flight mass spectrometry. *PNAS USA* 98(2): 581–584, 2001.

Chan A. Coupling different molecular techniques onto a porous microarray for diagnostic applications: The detection of deletions and duplications in Duchenne Muscular Dystrophy (DMD) using multiplexed amplifiable probe hybridization (MAPH) on PamChip. *First International Conference on Microarrays for Diagnostics*, March 21–22, 2002, San Diego.

Dill K, Montgomery DD, Wang W, Tsai JC. Antigen detection using microelectrode array microchips. *Analyt. Chim. Acta* 444: 66–78, 2001.

Garland PB, Serafinowski PJ. Effects of stray light on the fidelity of photo-directed oligonucleotide array synthesis. *Nucleic Acid Res.* 30(19)e99: 1–9, 2002.

Graves DJ. Powerful tools for genetic analysis come of age. *Trends Biotechnol.* 17(3): 127–134, 1999.

Gurevitch D, Dong XF, Pircher TJ, et al. A novel three-dimensional hydrogel-based microarray platform. *JALA* 6(4): 87–91, 2001.

Harvey MA. Nitrocellulose surfaces for quantitative protein microarrays. *Chips to Hits Conference* (International Business Communication), Boston, October, 2003.

Heller MJ, Forster AH, Tu E. Active microelectronic chip devices which utilize controlled electrophoretic fields for multiplexed DNA hybridization and other genomic applications. *Electrophoresis* 21: 157–164, 2000.

Hughes TR, Mao M, Jones AR, et al. Expression profiling using microarrays fabricated by an ink-jet oligonucleotide synthesizer. *Nat. Biotechnol.* 19: 342–347, 2001.

LeProust E, Zhang H, Yu P, Zhou X, Gao X. Characterization of oligo-deoxyribonucleotide synthesis on glass plates. *Nucleic Acid Res.* 29(10): 2171–2180, 2001.

Llorens F, Hummel M, Pastor X, et al. Multiple platform assessment of the EGF dependent transcriptome by microarray and deep tag sequencing analysis. *BMC Genomics* 12: 326, 2011.

Martel RR, Botros IW, Rounseville MP, et al. Multiplexed screening assay for mRNA combining nuclease protection with luminescent array detection. *ASSAY Drug Dev. Technol.* 1(1-1): 61–72, 2002.

Matson RS, Rampal JB. DNA arrays: Past, present, and future. *Am. Genomic/Proteomic Technol.*, April/May, pp. 37–44, 2003.

Matson RS, Rampal JB. Evaluation of a multiplexed immunoassay system for robustness. *42nd Annual Oak Ridge Conference*, April 22–23, 2010, San Jose, CA.

McCall MN, Murakami PN, Lukk M, Huber W, Irizarry RA. Assessing affymetrix GeneChip microarray quality. *BMC Bioinformatics* 12: 137, 2011.

Mendoza LG, McQary P, Mongan A, Gangadharan R, Brignac S, Eggers M. High-throughput microarray-based enzyme-linked immunosorbent assay (ELISA). *BioTechiques* 24(4): 778–788, 1999.

Nuwaysir EF, Huang W, Albert TJ, et al. Gene expression analysis using oligonucleotide arrays produced by maskless photolithography. *Genome Res* 12: 1749–1755, 2002.

Oliphant A, Barker DL, Stuelpnagel JR, Chee MS. BeadArray™ technology: Enabling an accurate cost-effective approach to high-throughput genotyping. *BioTechniques* 32: S56–S61, 2002.

Pease AC, Solas D, Sullivan EJ, Cronin MT, Holmes CP, Fodor SPA. Light-generated oligonucleotide arrays for rapid DNA-sequence analysis. *PNAS USA* 91: 5022–5026, 1994.

Schena M, Shalon D, Davis RW, Brown PO. Quantitative monitoring of gene expression patterns with a complementary DNA microarray. *Science* 270: 467–470, 1995.

Singh-Gasson S, Green RD, Yue Y, et al. Maskless fabrication of light-directed oligonucleotide microarrays using a digital micromirror array. *Nat. Biotechnol.* 17: 974–978, 1999.

Song Y, Boyer D, Leung I, et al. A² Microarray System: A novel multiplexed assay platform for cytokine profiling. *95th Annual Meeting American Association for Cancer Research*, Orlando, FL, March 27–31, 2004.

Tan PK, Downey TJ, Spitznagel EL Jr., et al. Evaluation of gene expression measurements from commercial microarray platforms. *Nucleic Acid Res.* 31(19): 5676–5684, 2003.

Tang K, Fu D-J, Julien D, Braun A, Cantor CR, Koster H. Chip-based genotyping mass spectroscopy. *PNAS USA* 96: 10016–10020, 1999.

Umek RM, Lin SW, Vielmetter J, et al. Electronic detection of nucleic acids—A versatile platform for molecular diagnostics. *J. Mol. Diagn.* 3(2): 74–84, 2001.

Wang DG, Fan JB, Siao CJ, et al. Large-scale identification, mapping, and genotyping of single-nucleotide polymorphisms in the human genome. *Science* 280: 1077–1082, 1998.

Wiese R, Belosludtsev Y, Powdrill T, Thompson P, Hogan M. Simultaneous multianalyte ELISA performed on a microarray platform. *Clin. Chem.* 47(8): 1451–1457, 2001.

Wheeler DL, Barrett T, Benson DA, et al. Database resources of the National Center for Biotechnology. *Nucleic Acid Res.* 31(1): 28–33, 2003.

Wu C, Zhao H, Baggerly K, Carta R, Zhang L. Short oligonucleotide probes containing G-stacks display abnormal binding affinity on Affymetrix microarrays. *Bioinformatics* 23(19): 2566–2572, 2007.

3 Supports and Surface Chemistries

INTRODUCTION

In this chapter we will survey the kinds of solid-supports (substrates) and surface chemistries that are currently being used in the creation of nucleic acid and protein microarrays. Which are the best supports and methods of attachment for nucleic acids or proteins? Does it make sense to use the same attachment chemistry or substrate format for these biomolecules? In order to begin to understand these kinds of questions it is important to briefly review how such biomolecules have been attached in the past to other solid-supports such as affinity chromatography media, membranes, or ELISA microtiter plates. However, there are a few unique properties and metrics, as well, that the microarray substrate does not share with its predecessors. Principal among these are printing, spot morphology, and image analysis. These are the subject of subsequent chapters.

SUBSTRATES

It is interesting to note that while column chromatography and centrifugation were developed for biomolecule purification and separation, many of the early diagnostic substrates for either nucleic acids or proteins were membranes. For the Southern transfer process, the membrane provided a convenient way to interrogate sequence in genomic DNA fragments (Southern, 1975). With the advent of polymerase chain reaction (PCR) it was possible to directly spot down cDNA amplicons onto membranes giving rise to the Southern dot blot format. In fact, the dot blot should be regarded as one of the earliest if not the first array format (albeit, a macroarray). Why did many abandon the membrane in favor of the glass substrate for DNA-microarrays? And, how is it that membranes cast upon glass substrates are now being used to prepare protein microarrays? We will address these questions in good time.

In the following sections the major kinds of substrates currently used for DNA and protein microarrays are discussed. Much of what is known regarding microarray surface chemistry and the immobilization of biomolecules comes from work with DNA microarrays. Therefore, many of the examples cited here will be from these studies. Zhu and Snyder (2003) in their review provide good insight into the manufacture and utility of protein microarrays. Here are some points to consider when choosing a substrate for protein microarrays:

1. The manufacture and processing of the protein microarray should be conducted in such a manner that the arrayed proteins remain in their native and active state. For most proteins this usually means the hydrated state in order to avoid surface denaturation. For antibodies arrays, which are perhaps more forgiving than other proteins, it has been our experience that while these could be stored cold and dry, it is most important to rehydrate the array prior to use. This process is in sharp contrast to the preparation of nucleic acid arrays in which strand melting or denaturation is necessary to achieve optimal binding to the solid support. And, while the hybridization process is well understood and can be controlled under thermodynamic principles, the folding and renaturation of proteins on planar (microarray) surfaces is under study.

2. Hydrogels and other porous 3D matrices that entrap water can offer an excellent milieu for maintaining proteins in the hydrated state. The higher surface area may also allow the immobilization of proteins at a higher density than planar or two-dimensional (2D) substrates. This in turn can lead to improved sensitivity and dynamic range. On the downside these polymeric materials can also exclude larger molecular weight biomolecules or slow diffusion and the exchange of buffers and other reagents.

3. Not all substrate materials will be applicable. While affinity chromatography or ion-exchange supports have been used quite effectively to purify native proteins, their conversion to a microarray substrate may not work. The printing down of proteins onto these surfaces may be hard to control in terms of spot uniformity and morphology. The exception is nitrocellulose that has been resurrected as a microarray support by casting it down on a glass slide (S&S). Poly-L-lysine (PLL) coated slides used so successfully for creating DNA microarrays were the first to be adopted for protein microarrays (MacBeath, 2000; Haab et al., 2001).

MEMBRANE SUBSTRATES

Use with nucleic acids: We will begin our discussion with examples from the late 1980s and early 1990s when membranes were first employed in the creation of nucleic acid arrays. At that time the terminology surrounding DNA analysis began to change. Dattagupta and co-workers first introduced the concept of reverse dot blot hybridization (Dattagupta et al., 1989, p. 85):

> We have developed a simple method of non-isotopically labeling sample nucleic acids, which are then hybridized simultaneously to an array of unlabeled, immobilized probes. This "reversed hybridization" procedure thus provides identification results after a single hybridization reaction.

So, instead of spotting down the target DNA (i.e., Southern dot blot), unlabeled DNA probes complementary to a target DNA were arrayed onto the membrane. The target DNA (e.g., cDNA) was then labeled and applied to the membrane for hybridization. Reverse hybridization allowed the simultaneous probing of the target against an allele specific oligonucleotide

(ASO) library. One no longer had to strip and re-probe the membrane. The "probe" came to mean the DNA attached to the solid-support, and the "target" as in the case of the Southern blot referred to the nucleic acid in the sample to be analyzed.

In this first study the arrays were constructed by hand spotting genomic DNA from various bacterial species onto nitrocellulose (NC) membrane. The DNA was denatured with sodium hydroxide and fixed to the NC membrane by baking under vacuum at 80°C. However, the attachment of short oligonucleotides onto NC was not practical. Later, nylon membranes were introduced to replace the fragile NC membrane. Single-stranded oligonucleotide probes could then be tethered to the charged nylon membrane by 3′ tailing with poly (dT) followed by ultraviolet (UV) cross-linking (Saiki et al., 1989). In 1991, Zhang and co-workers from Cetus introduced the use of amino linkers in which oligonucleotide probes were terminated at their 5′ end with amino spacers. The amino-modified probes were then covalently attached to a carboxylated nylon membrane (Biodyne C, Pall Corporation) via 1-ethyl-3(3-dimethylaminoethyl)carbodiimide (EDAC) activation (Zhang et al., 1991). The primary amine was found to be much more reactive than the aromatic secondary amines found in the bases, thus assuring oriented coupling of the majority of probes by the 5′ end. Probes tethered without the inclusion of spacers were fourfold less efficient in hybridization. Furthermore, the amino linker provided greater hybridization efficiency than the poly (dT) tailing method.

For longer double-stranded nucleic acids such as cDNA (~200 bp to 1500 bp), the positively charged nylon membrane easily sequestered the negatively charged strands by adsorption most probably involving both electrostatic and hydrophobic interactions with the support. The exact mechanisms of baking and UV cross-linking of nucleic acids to nylon or NC are not well understood but presumably involve some covalent interaction of the DNA and the support. Considerable study over the past several decades has been conducted on the use of the nucleic acid dot blot format. The fundamentals are aptly described by Anderson and Young (1985).

Much of the early work relied upon hand-spotting or manual application of probes using vacuum filtration devices such as the DotBlot apparatus (BioRad Laboratories) that allowed the formation of more uniform spotting of probes in the form of small dots or rectangular slots. The use of membranes for printed DNA arrays (often referred to as "grid" arrays) was subsequently developed. For example, Hans Lehrach and co-workers (Nizetic et al., 1991) developed a computer-controlled robotic system for the arraying of bacterial colonies (cosmid library) from 96 wells of a microtiter plate onto nylon membranes. Using this early robotic pin printer, a high-density grid of 9216 clones from 96 wells × 96 microtiter plates was constructed on a 22 cm × 22 cm filter. Drmanac et al. (1994) described the high-density arraying of PCR samples using a Biomek® 1000 robotic workstation adapted with a 96-pin tool (Bentley et al., 1992). Membrane-based nucleic acid grid arrays continue to be used for genomic analysis (Lane et al., 2001;

Hornberg et al., 2002). Note that grid arrays are also known as *macroarrays*, a term used to more formally differentiate them from microarrays in terms of spot size and density.

Use with proteins: Lueking et al. (1999) arrayed recombinant proteins on nitrocellulose membrane and screened these with different antibodies. Joos and co-workers (2000) printed down auto-antigens onto NC membrane and compared performance relative to silylated (aldehyde) and PLL glass slides. Protein arrays could be stored at room temperature for a month without significant loss in activity. Huang (2001) hand-spotted down IgG species and antibodies directed toward various cytokines onto membranes. The properties of various commercial membranes were assessed in terms of absorption as well as background and sensitivity levels based upon detection by enhanced chemiluminescence (ECL).

Advantages: Advantages include high binding capacity with multiple hybridization cycles possible.

Disadvantages: They include size limitations on immobilization of oligonucleotides, lower spot density due to spot diffusion, required large sample and rinse volumes, required blocking agents to reduce non-specific binding (NSB), and high intrinsic fluorescent background and light scattering issues.

Detection: Membrane substrates are detected by radioactivity (film) or phosphorimaging, chemiluminescence, and colorimetric reagents. See Figure 3.1 for a comparison of different detection results on membranes. In particular, membranes are well suited for colorimetric detection, while glass or plastic are less attractive alternatives.

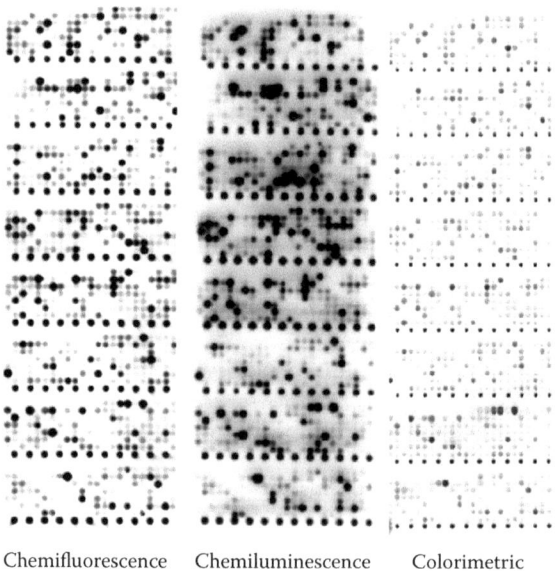

Chemifluorescence   Chemiluminescence   Colorimetric

FIGURE 3.1   Membrane-based signal detection formats. (Courtesy of Andrew Dubitsky, Pall Corporation, Port Washington, NY.)

> *Use with nucleic acids*: As a microarray substrate the use of glass begins with development of in situ generated oligonucleotide arrays (Fodor et al., 1991; Southern et al., 1992) and later with spotted arrays (Schena, 1995). Glass became the substrate of choice for the direct synthesis of oligonucleotides because of its relative inertness toward synthesis chemistries, the surface was uniform and impervious, and glasses with good optical properties (flatness, low intrinsic fluorescence) were available. These latter properties became very important in the development of the confocal scanner for reading the array in order to achieve a fluorescence signal with high sensitivity.

Covalent Attachment

Nucleic acids may be attached to glass by either noncovalent (electrostatic, hydrophobic interaction) or covalent means. The most common approach has been adsorption to poly-L-lysine (PLL) coated microscope slides or the second-generation aminopropyl-silane (APS) surface. PLL surfaces may be damaged under certain hybridization and stringency washing or stripping conditions (e.g., high salt and elevated temperatures). For this reason, Zammatteo et al. (2000) examined different covalent coupling conditions for nucleic acids on glass substrates. Silanization of glass with various silane coupling reagents resulted in the grafting of amine ($Si-O\sim NH_2$), carboxyl ($Si-O\sim COOH$), and aldehyde ($Si-O\sim CHO$) surface reactive groups (Figure 3.2). Likewise, cDNA amplicon (255 bp) probes were synthesized using chemically modified primers with 5′ ends terminating with either phosphate (PO_4-DNA), carboxyl (HOOC-DNA), or

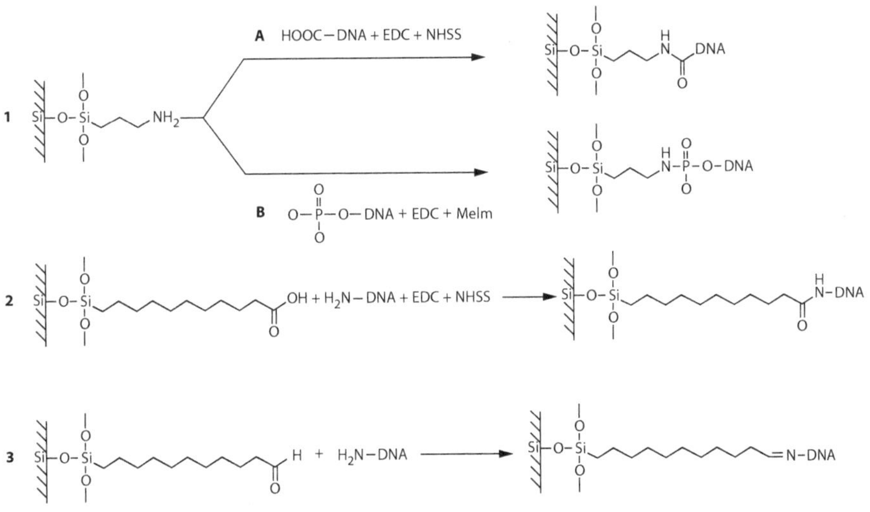

FIGURE 3.2   Grafting of various functional groups onto silanized glass. (From Zammatteo N, Jeanmart L, Hamels S, et al. Comparison between different strategies of covalent attachment of DNA to glass surfaces to build DNA microarrays. *Anal. Biochem.* 280: 143–150, 2000. With permission.)

amine groups (H_2N-DNA). Coupling of the modified amplicon probes to the modified glass surfaces was accomplished via carbodiimide-mediated reactions:

$$Si\text{-}O\sim NH_2 + HOOC\text{-}DNA + EDAC/NHS \rightarrow Si\text{-}O\sim NHCO\text{-}DNA + H_2O$$

or a Schiff's base reaction:

$$Si\text{-}O\sim CHO + H_2N\text{-}DNA \rightarrow Si\text{-}O\sim C = N\text{-}DNA + H_2O$$

The highest efficiency for probe attachment was found for the reaction of carboxylated probe with an amino-silane surface. However, the level obtained by simple adsorption (i.e., without the addition of EDAC) was equally as high (mean probe density, ~600 to 700 fmoles/cm²). Conversely, the attachment of an amine-modified probe to a carboxyl surface was much less efficient (mean probe density, ~300 fmoles/ cm²). The least efficient coupling was obtained using aldehyde surfaces (mean probe density, ~150 fmole/cm²). Based upon the loading concentrations of probe DNA (3 µM) the coupling reaction to the aldehyde supports was limiting and required a 60- to 300-fold excess (3000 nM versus 10 to 50 nM) relative to the other coupling chemistries. Yet, hybridization on the aldehyde surface was higher than on the amine or carboxylated surfaces. And, while the coupling of phosphorylated DNA was similar to that of the aldehyde surface, it too resulted in lower hybridization efficiency and higher nonspecific adsorption.

Do higher loadings of probe onto the support adversely affect performance or are other factors at play here as well? What about how the DNA probe is tethered to the surface? While there were differences in spacer arm length for the tethering via amine (3-carbon spacer) versus carboxyl or aldehyde (11-carbon spacer), this most likely had little direct impact because hybridization varied only about twofold for all surfaces regardless of the surface reactive group and probe combinations tested. However, probe density (or probe surface distribution) most surely plays an important role. Zammatteo et al. varied the loading onto the aldehyde surface from 10 nM to 3000 nM and determined that the highest coupling was at 500 nM input probe. The greatest hybridization efficiency was, however, obtained below the optimal loading at about 200 nM depending upon interpretation of the reported imprecision (signal, mean ± SD) (Figure 3.3).

If you follow the trend in DNA loading versus hybridization efficiency from these experiments, one would conclude that higher loadings lead to less efficient hybridization, while minimal loading leads to problems with nonspecific adsorption. One explanation is that the probe distribution may be too tightly packed, thereby preventing hybrid nucleation by steric hindrance of the incoming target. Reducing the loading avoids the issue of steric hindrance. However, the loading should also be titrated to minimize the contribution of nonspecific adsorption (Figure 3.4).

Therefore, while one should proceed to determine the optimal loading (input probe concentration versus bound probe), in general this does not correlate well with the optimal hybridization. It is prudent to carefully quantify that relationship (i.e., bound probe versus hybridization signal strength) as well as define the limits of nonspecific adsorption.

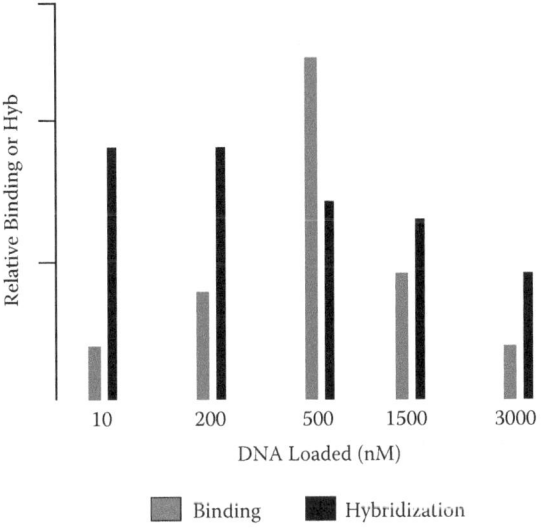

FIGURE 3.3 Oligonucleotide probe surface loading versus hybridization efficiency. (From Zammatteo N, Jeanmart L, Hamels S, et al. Comparison between different strategies of covalent attachment of DNA to glass surfaces to build DNA microarrays. *Anal. Biochem.* 280: 143–150, 2000. With permission.)

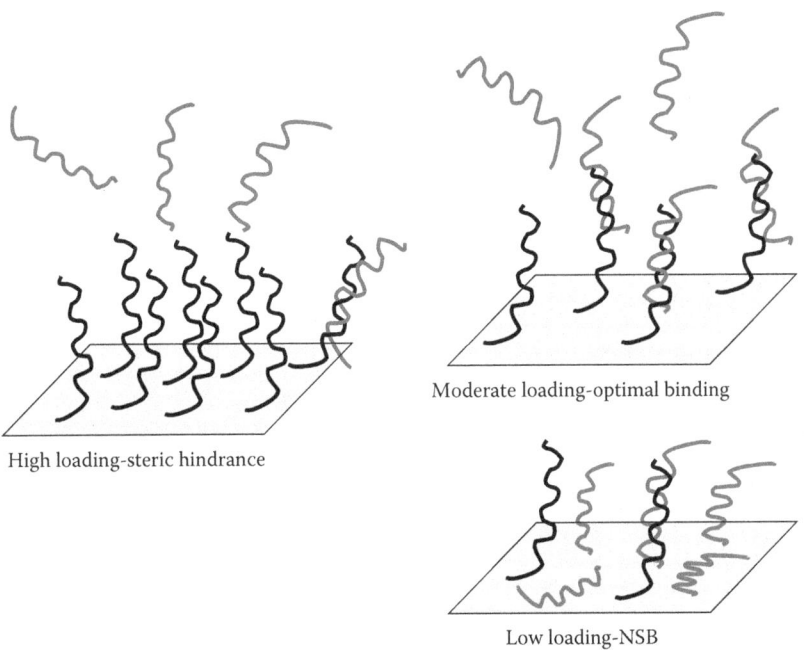

FIGURE 3.4 Substrate loading effects.

Adsorptive Attachment

PLL surfaces work reasonably well for creating cDNA microarrays, but others have questioned the suitability of this surface chemistry for immobilization of short oligonucleotides. However, as we have learned, covalent attachment chemistries can be problematic as well. In either case, if the oligonucleotide is constrained too close to the surface with multiple points of contact, it may not be able to fully participate in hybridization. Conversely, too few interactions with the surface may lead to loss during hybridization and washing steps. This is why the covalent attachment of oligonucleotides at their 3′ or 5′ terminus has the advantage. Yet, modified oligonucleotides and slide chemistries can be expensive.

Belosludtsev et al. (2001) have described an efficient process for preparing oligonucleotide microarrays based upon the adsorption of unmodified probes to aminosilanized glass slides. This is a two-step process: First, oligonucleotides are presented to the surface dissolved in water so that they remain in a fully denatured state and then dry down. The second step involves capping the residual surface amine groups. Recall from the work of Zammatteo et al. (2000) that the nonspecific adsorption of modified oligonucleotides to amino-silane surfaces was regarded as a significant problem. Capping with succinic anhydride is a common method for reducing backgrounds on amine surfaces (NH_3^+) by creating a negative ($NH-COCH_2CH_2COO^-$) surface charge, thereby repelling nonspecific nucleic acids. However, as these investigators discovered, too high a negative charge density will also repel incoming target DNA. To overcome this issue, a *double-capping* protocol was used in which partial capping with acetic anhydride (vapor phase, 50°C, 1 hour, vacuum oven) was followed by succinic anhydride (0.5 M in dimethylformamide (DMF), room temperature, 1 hour). This imparted a slight negative surface charge that was more favorable for hybridization while preventing significant levels of nonspecific adsorption.

The adsorptive process attaching unmodified oligonucleotide onto amino-silane glass was compared to the covalent attachment of amino-oligonucleotides to epoxysilane substrate. A set of 12mer capture probes for the codon 12 point mutations in K-ras were compared in the hybridization efficiency and specificity of the 152 bp amplicon. Comparable results between adsorption and covalent tethering were obtained (Figure 3.5). Interestingly, the adsorption-chemical capping method required only 20% of the probe loading used for covalent attachment (i.e., 5 μM unmodified oligo versus 25 μM amino-modified oligo).

Belosludtsev and co-workers proposed that the unmodified probes on the weakly cationic surface while prevented from diffusing off the surface because of electrostatic interaction, nevertheless are available for hybrid nucleation. They suggested that such probe behavior could be viewed similar to that observed in a liquid crystal matrix. This would be in sharp contrast to models describing the covalent attachment of short oligonucleotides as "oligo lawns" or monolayers of coiled probes (see Figure 3.6).

Call et al. (2001) from the Pacific Northwest National Laboratory (Richland, WA) have also studied the immobilization of unmodified oligonucleotides. Amine modified and unmodified oligonucleotides could be attached to epoxy-silane slides (covalent attachment) or acid-washed slides (noncovalent attachment) under the same conditions by printing in an alkaline-SDS buffer, pH 12 using a microarrayer. The

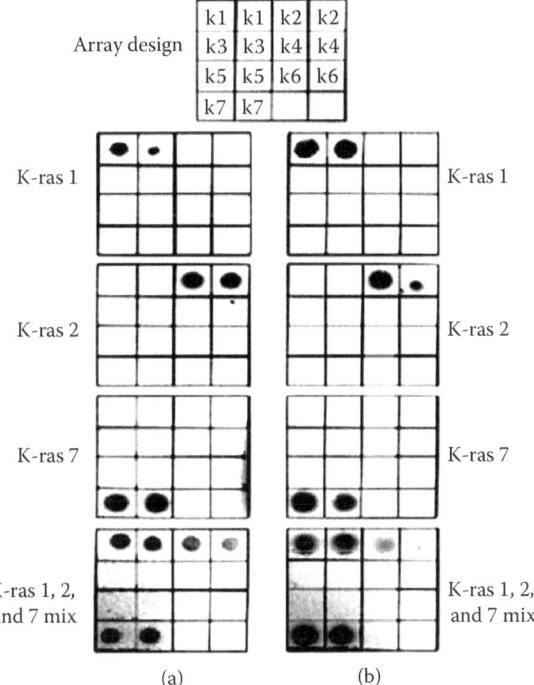

FIGURE 3.5 Comparison of adsorptive and covalent attachment of capture probes. (From Belosludtsev Y, Iverson B, Lemeshko S, et al. DNA microarrays based on noncovalent oligonucleotide attachment and hybridization in two dimensions. *Anal. Biochem.* 292: 250–256, 2001. With permission.)

slides were baked in a vacuum oven at 130°C for 30 to 60 minutes and then stored at 4°C. An optimal hybridization signal was achieved with unmodified oligos over that of amino-oligos on the epoxy activated surface using probe loadings of ~100 μM. When amine modified oligos attached to epoxy or hydroxyl slides were subjected to strong alkaline pH conditions, hybridization efficiency rapidly deteriorated on the acid-washed (hydroxyl) slides but not the epoxy slides (Figure 3.7). PNA (peptide nucleic acid) oligos that lack the backbone (PO_4^-) charge of standard oligonucleotides exhibited no loss in hybridization on either surface treated with alkaline. These results led the investigators to suggest that hydrogen bonding also plays an important role in nucleic acid attachment to glass surfaces. The effect of baking may further enhance binding by dehydration of the probes on the surface. Call et al. suggest that, "regardless of which attachment mechanisms are involved, the probes are probably oriented in flat, 'piled' conformation" (p. 378).

Use with proteins: Most protein microarray work has been done using the glass slide. Harvard's MacBeath and Schrieber (2000) nicely demonstrated the advantage in printing proteins onto aldehyde slides, and the use of bovine serum albumin (BSA) as a scaffold for presenting proteins on the substrate.

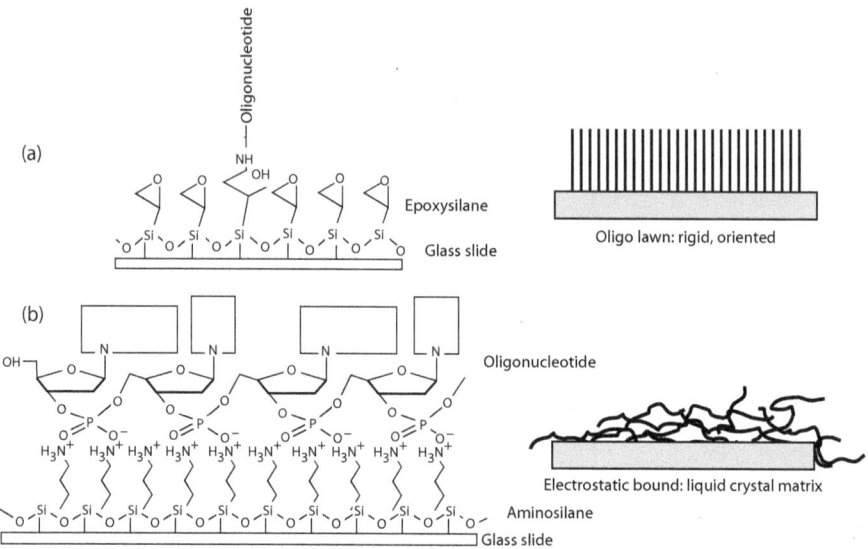

FIGURE 3.6 Models for oligonucleotide probe behavior on surfaces. (From Belosludtsev Y, Iverson B, Lemeshko S, et al. DNA microarrays based on noncovalent oligonucleotide attachment and hybridization in two dimensions. *Anal. Biochem.* 292: 250–256, 2001. With permission.)

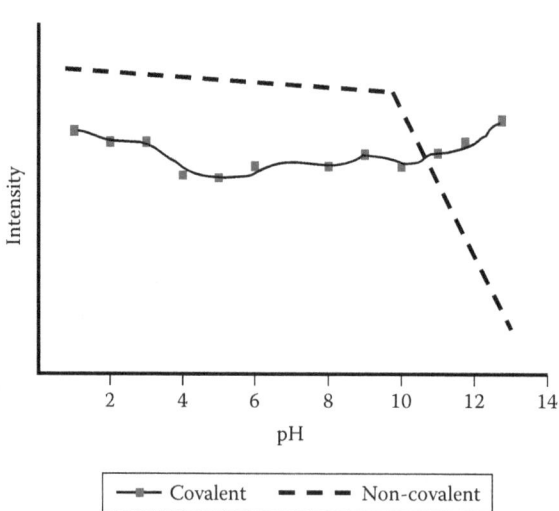

FIGURE 3.7 Removal of noncovalent bound probes under alkaline pH conditions. (From Call DR, Chandler DP, Brockman F. Fabrication of DNA microarrays using unmodified oligonucleotide probes. *BioTechniques* 30(2): 368–379, 2001. With permission.)

This was followed by an impressive study from Pat Brown's laboratory (Stanford University) on the characteristics of 115 antibody and antigen pairs on PLL microarrays (Haab et al., 2001). Wiese et al. (2001) printed down monoclonal antibodies on a 96-well Teflon masked silanized (amino-silane) glass plate and performed micro-ELISAs for prostate-specific antigen (PSA) and IL-6.

Delehanty and Ligler (2002) sought to preserve biotinylated antibody immobilized onto an avidin coated slide. For printing of the capture antibodies, a deposition buffer including 10 mM phosphate, 10 mM NaCl, 10 mM sucrose, and 0.1% BSA (w/v) was employed. Antibody was dispensed (piezo-type; BioChip Arrayer I, Packard Bioscience) at 10 µg/mL. BSA was added as a carrier protein (1 mg/mL) in an effort to reduce the loss of antibody activity by surface denaturation or mass losses due to nonspecific adsorption onto the printing parts and source plate. Sucrose was added to maintain antibody hydration. Glycerol has been successfully used to protect proteins from dehydration.

Seong (2002) compared silylated (aldehyde) and silanated (amine, epoxy) slides from several commercial sources on the performance of an antigen (IgG) microarray. In addition, the efficiencies of PBS, pH 7.4 and carbonate, pH 9.6 printing buffers were compared. While the various slides and surface chemistries showed differences in their binding isotherms, they ultimately reached similar levels of saturation. Silylated (aldehyde) slides showed comparable loading in either buffer system. Apparently, tethering of antibody to the surface by Schiff's base formation of the surface aldehyde and lysine residues on the protein is applicable over a broad pH. However, carbonate buffer increased binding of proteins on silanated surfaces. The interaction on silanated slides is thought to be the result of a combination of electrostatic and hydrogen bonding. Amino-silane surfaces are positively charged at neutral pH. Printing under alkaline conditions shifts the net charge on the protein making it more negative, thereby increasing the likelihood of electrostatic interaction with the surface.

Angenendt et al. (2003) reasoned essentially that because surface charge on proteins is variable, it would be unlikely that a single surface chemistry would prove to be universal for the arraying of all proteins while preserving their native, active state. The group examined eight different coatings and compared their performances for suitability as an antibody microarray or as a protein microarray. According to the authors, "one surface for both antibody and protein microarray applications could not be found" (p. 251).

Advantages: Advantages include optical transparency and flatness; low intrinsic fluorescent backgrounds and light scattering; ultra-flatness and impervious surface which allows high-density array construction.

Disadvantages: They include lower probe binding capacity, blocking agents required to reduce NSB, and the need for an expensive scanner for increased sensitivity.

Detection: Primarily a fluorescence-based confocal scanner is necessary.

PLASTIC SUBSTRATES

Use with nucleic acids: For higher-throughput applications, injection molded plastic microtiter plates have been the format of choice for automated assay development. Thermoplastics such as polystyrene, polycarbonate, and polypropylene are used for a variety of purposes including storage and assay plates, lids, pipette tips, and Eppendorf PCR tubes. The polystyrene plate is used for cell culture and the ELISA. Polycarbonate reagent bottles are popular, while polypropylene storage plates and PCR tubes are standard. The covalent immobilization of DNA into micro-wells has been described. Rasmussen et al. (1991) used aminated polystyrene plates (CovaLink NH, Nunc) in which a secondary amino group tethered via a spacer arm was grafted to polystyrene. Phosphorylated oligonucleotides or plasmid DNA were attached to the polystyrene by carbodiimide-mediated coupling of the nucleic acid's 5′-terminal phosphate group to the secondary amine. Hamaguchi and co-workers (1998) from the Hitachi Chemical Research Center (Irvine, CA) explored the use of plastic micro-plates for mRNA capture using immobilized oligo(dT) that allowed them to perform RT-PCR directly in the wells. Initially, rigid and transparent polystyrene plates were used (GenePlate™, Hitachi Chemical) but were found not to be heat stable under PCR thermo cycling conditions, especially the required 94°C denaturation step. For that reason, polypropylene micro-plates were substituted (GenePlate-PP, initially based upon Nunc's GeNunc PP plate) and became the basis for the GenePlate product (RNAture, Irvine, CA). Surface oligonucleotide in each well was approximately 20 pmole and could capture close to 50% of the sample RNA (e.g., well-bound mRNA ~2 ± 0.4 ng from 5 ng globin mRNA [40%]; 27.3 ± 4 ng per 50 ng [54.7%], and 216 ± 26.2 ng per 500 ng [43.2%]). From 0.5 μg liver total RNA (5.83 ng mRNA) well capture was 2.8 ± 0.3 ng (48.0%), while at higher sample inputs (5 to 50 μg) the recoveries were substantially reduced.

Such thermoplastics have also been used as DNA-microarray substrates (Matson, 1995; Shchepinov et al., 1997; Beier, 1999) as well as in the construction of protein microarrays in micro-wells (Matson et al., 2001; Moody et al., 2001). Pierce introduced the SearchLight™ series of microarray-based ELISA assays immobilizing capture antibodies in a low-density array format into polystyrene micro-wells. Beckman Coulter (www.beckmancoulter.com) originally offered the A²™ MicroArray System (acquired by QuantiScientifics, Irvine, CA) based upon a 6 × 7 array of capture oligonucleotides printed in the bottom of a 96-well polypropylene plate (A² Plate, QuantiScientifics). Complementary oligonucleotides conjugated to the user's antibody provide a convenient method for the creation of custom microarrays. The plate is Society for Biomolecular Screening (SBS) compliant for automation using liquid-handling robotics. Greiner bio-one (Germany) introduced the HTA™ Plate (High-Throughput microArraying Plate) and HTA™ Slides for the printing of microarrays onto the polystyrene surface (see www.gbo.com/bioscience).

Liu and Rauch (2003) from Motorola investigated oligonucleotide probe attachment onto four plastic surfaces: polystyrene (PS), polycarbonate (PC), poly (methylmethacrylate) (PMMA), and polypropylene (PP). They utilized three different immobilization processes: SurModic's surface modification solution (allows attachment of adsorbed reactive groups to the surface by photoactivation of polymers at 254 nm); Pierce Reactive-Bind coating solution; and CTAB (cetyltrimethylammonium bromide, a cationic detergent). Not surprisingly, the microarray performance on these plastics varied. A number of different properties of plastics were studied relative to the utility of these surfaces in the preparation of DNA-microarrays and subsequent detection. Intrinsic fluorescence backgrounds were measured at 532 nm (Cy3 excitation) and 635 nm (Cy5 excitation) relative to glass. In terms of the Cy3/Cy5 ratio, PC was 20-fold higher in background fluorescence, as were PS (5.5-fold), PMMA (7.7-fold), and PP (9.3-fold) over that of glass's intrinsic background. PMMA had the lowest relative backgrounds at Cy3 = 6.9- fold and Cy5 = 0.9- fold, while PC was highest overall with backgrounds for Cy3 = 107.2-fold and Cy5 = 5.3-fold. Plastics also vary in their relative hydrophobicity (measured as the wetting contact angle), and this property can have a profound effect on the ability to spot down oligonucleotides of uniform spot diameter and morphology. The suitability of the various surface modifying agents is also dependent upon good wetting in order to uniformly cover the surface with the reactive groups. In that regard, Reacti-Bind failed to provide a useful surface for printing oligonucleotide microarrays. Reacti-Bind actually increased the contact angle on these plastics, and this may be the reason for its poorer performance. The spotting behavior (spot size, morphology) of 5′ amino-oligonucleotide probe on the various modified surfaces reflected differences in the surface wetting properties among these plastics. Increased wetting (lower contact angle) resulted in a spreading of the droplet and an increase in spot diameter (Figure 3.8). The SurModics surface that provided the greatest degree of wetting and spot uniformity also produced microarrays with the highest hybridization efficiency in terms of the limit of detection (LOD) in sensitivity. Of the plastics studied PMMA proved to be the best substrate for the immobilization of the 21mer 5′ amino-oligonucleotide probe used in these studies based upon CTAB and the Surmodics processes. However, all substrates performed very well with the Surmodics process with LODs ranging from 12 to 100 pM. CTAB worked best below its CMC (critical micelle concentration), but this was more difficult to control under spotting conditions due to evaporative water loss that could shift the concentration at or near the CMC. This is an important point to consider in utilizing detergents for printing.

Use with proteins: In 1991, Roger Ekins convinced Boehringer-Mannheim to pursue commercial development of his Microspot™ technology (Ekins, 1998). Antibody microarrays were constructed on single-well polystyrene carriers by ink-jet printing. Unfortunately, no product was commercialized.

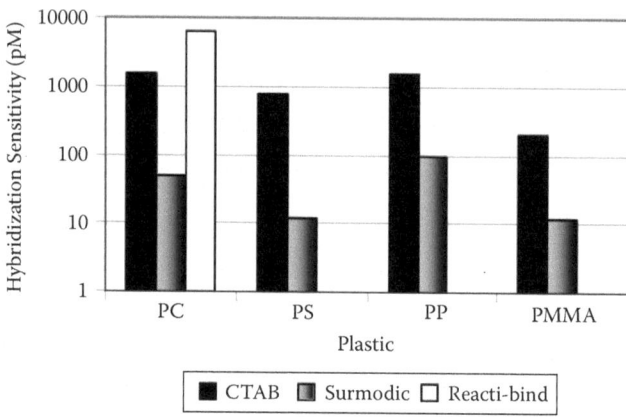

FIGURE 3.8　Hybridization efficiency on various plastics with reactive surfaces. (From Liu Y, Rauch CB. DNA probe attachment on plastic surfaces and microfluidic hybridization array channel devices with sample oscillation. *Anal. Biochem.* 317: 76–84, 2003. With permission.)

At about the same time, Beckman Instruments (now Beckman Coulter) had begun an array-based product development program focused on the use of modified plastics. Silzel and co-workers (1998) and Matson et al. (2001) from Beckman Coulter were among the first to pursue printing of antibodies onto a plastic surface in a microarray format. Silzel et al. immobilized biotinylated monoclonal antibodies onto an avidin coated polystyrene surface and performed micro-ELISA-based isotyping of IgG species. Matson et al. printed down monoclonal antibody microarrays in the bottom of an acyl fluoride surface activated micro-well plate and demonstrated a multiplexed micro-ELISA for cytokines (Figure 3.9).

Other early work includes that of Moody et al. (2001) who spotted anti-cytokine monoclonals onto the bottom of a polystyrene microtiter plate (Maxisorp, Nalge Nunc) and measured cytokine levels in stimulated peripheral blood mononuclear cells. Finally, although not strictly a microarray, the micro-well array system developed by Michael Snyder's group at Yale University to measure kinase activity is a simple and elegant approach (Zhu et al., 2000). The "protein chip" is composed of micro-wells fabricated in a flexible elastomer of poly(dimethylsiloxane) (PDMS) substrate by a molding process. The PDMS

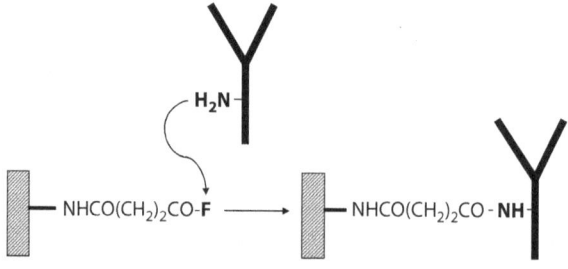

FIGURE 3.9　Acyl fluoride coupling chemistry.

micro-well array is mounted onto a glass slide and activated using an epoxy-silane to which the protein may be attached. Well features were about 1.4 mm diameter and 300 μm deep allowing a volume of approximately 300 nL. The micro-well protein chip should be widely applicable.

Advantages: They include inexpensive material; moldability into a variety of shapes; and resistance to chemicals, especially to salts, acids, and bases.

Disadvantages: They include flatness and optical clarity issues making them largely unsuitable for confocal scanner.

Detection: Detection includes primarily fluorescence-based CCD camera detectors.

PHYSICAL FEATURES

HYDROGELS

Arguments have been made that planar, 2D surfaces are not the optimal physical structure for immobilizing nucleic acids or proteins. The proposition is that because of the lower surface capacity the 2D surface has a limitation in providing both high sensitivity and dynamic range. On the other hand, glass slide microarrays appear to work fine for gene expression analysis, while the confocal scanner seems to offer enough sensitivity. Nevertheless, it is very likely that the arraying of proteins in hydrogel matrix would offer some advantage in preserving the protein's native conformation in the hydrated state. Much of the pioneering work in this area has come out the laboratory of the late Andrei Mirzabekov at the Engelhardt Institute of Molecular Biology, Moscow (Khrapko et al., 1989, 1991). Later in collaboration with Argonne National Laboratory (ANL) his groups perfected the fabrication processes for hydrogel-based microarrays or MAGIChips™ (microarrays of gel-immobilized compounds on chips) for both nucleic acids and proteins (Guschin et al., 1997; Vasiliskov et al., 1999). Essentially, various polyacrylamide gels were produced based upon the common polyacrylamide gel electrophoresis (PAGE) gel casting systems (e.g., acrylamide: bis-acrylamide: sodium persulfate: tetramethylethylene-diamine (TEMED)). Following polymerization the cast gel was exposed to aqueous hydrazine hydrazide to create surface hydrazide reactive groups for attachment of aldehyde-oligonucleotide probes. The aldo-oligonucleotide was prepared from oligo containing a 3′ terminal 3-methyluridine. The terminal ribose was oxidized to di-aldehyde in the presence of sodium periodate. Reaction of the aldo-oligonucleotide with gel hydrazide resulted in a covalent bond immobilizing the oligonucleotide to the gel (Figure 3.10). The process was later refined for photo polymerization of both nucleic acids and proteins. The issue regarding slowed diffusion or exclusion of targets because of limited porosity was addressed by substituting diallyltartardi-amide as the cross-linker in place of bis-acrylamide (Guschin et al., 1997). Finally, photo-induced copolymerization of biomolecule (oligonucleotide; protein) with gel components was introduced that allowed cross-linking of different probes in different gel pads. This was accomplished by the use of a physical mask and movable

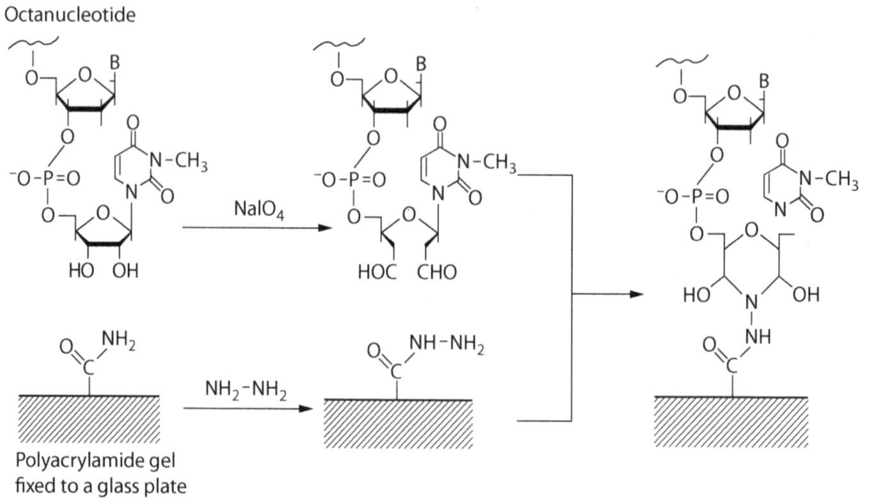

Octanucleotide

FIGURE 3.10 Immobilization of aldehyde oligonucleotides in hydrazide gel. (From Khrapko KR, Lysov YP, Khorlyn AA, Shick VV, Florentiev VL, Mirzabekov AD. An oligonucleotide hybridization approach to DNA sequencing. *FEBS Lett.* 256: 118–122, 1989. With permission.)

diaphragm in front of the light source. Photo polymerization was conducted at 254 nm under a modified fluorescent microscope. Allyl-oligonucleotides were employed with methylene blue as the free radical initiator for cross-linking. In the case of proteins, acryloyl-streptavidin was first immobilized so that biotinylated proteins could be applied to the gel pad (Vasiliskov et al., 1999). One of the major drawbacks with the gel pad approach was that separate pads needed to be manufactured rather than coating the entire slide with the gel and then printing down the microarray. Note that Motorola Life Sciences and Packard Biosciences (now Perkin Elmer Life Sciences) established a development partnership with ANL to commercialize the technology in 1998 but later abandoned the technology in favor of the SurModics hydrogel introduced in 1999 (3D-Link™). Motorola introduced the CodeLink™ microarray product based upon the SurModics PhotoLink chemistry. Perkin Elmer sold a similar product under the trade name of HydroGel™ 3-D.

Biocept introduced the 3D HydroArray™. The array was produced by mixing the oligonucleotide (or protein, etc.) into a pre-polymer solution (polyethylene glycol based). The solution is then arrayed onto a glass slide (amino-silane, GAPS2™, Corning) and the droplets cured to form a hydrogel. They estimate 10^{10} to 10^{11} probes per 300 micron diameter spot.

SURFACE CHEMISTRIES

LINKERS

While linkers (spacers) were added to oligonucleotides to increase the number of available interactions possible with the membrane surface in order to assure

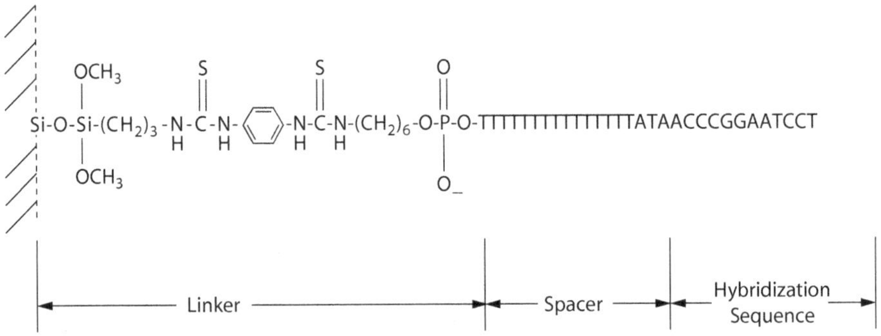

FIGURE 3.11 Tethering oligonucleotides with a poly(dT) spacer arm. (From Guo Z, Guilfoyle RA, Thiel AJ, Wang R, Smith LM. Direct fluorescence analysis of genetic polymorphisms by hybridization with oligonucleotide arrays on glass supports. *Nucleic Acid Res.* 22(24): 5456–5465, 1994. With permission.)

attachment, the use of linkers with glass substrates was largely for the opposite reason. That is, the tethering of oligonucleotides by spacers was done to reduce interaction and provide access of the incoming target. This is because of the excessive number of surface silanol groups that are available to form hydrogen bonds with the oligonucleotide, thereby tightly sequestering the biomolecule to the surface resulting in steric interference with probe-target hybrid formation. Lloyd Smith and coworkers (Guo et al., 1994) derivatized amino-silane glass slides with 1,4-phenylene diisothiocyanate in order to covalently attach 5′ amino modified probe oligonucleotides. Direct attachment of the probe to the support through this linkage failed to support hybridization of the PCR targets, while the inclusion of poly (dT) spacer elements between the coupling agent and the oligonucleotide probe resulted in efficient capture of the incoming target (Figure 3.11). The minimal spacer length appeared to be 6 nt, and the hybridization signal was found to increase linearly up to 15 nts (i.e., dT15) (Figure 3.12).

Joos et al. (1997) tethered a 35mer oligonucleotide to an amino-silane glass coverslip with or without a 15 nt spacer (15mer hetero-oligonucleotide sequence) and compared hybridization efficiency for the labeled complementary 35mer target nucleic acid (Figure 3.13). Under conditions of excess solid-phase capture, oligonucleotide hybridization efficiencies ranged from 23 to 64% (no spacer) and 26 to 74% (with spacer) (Table 3.1). While these results were rather inconclusive, the authors believed that the spacer offered some advantage. In an earlier study from Ken Beattie's group (then at the Houston Advanced Research Center) a similar conclusion was reached using a 9mer oligonucleotide and triethylene glycol phosphoryl repeat unit as the spacer (Beattie et al., 1995). An optimal probe density leading to the greatest hybridization efficiency was achieved at a probe loading of about 5 μM. Higher loadings were found to reduce hybridization efficiency to substantially background nonspecific adsorption levels (Figure 3.14). Additionally, poly (dT) units were examined for their effect when distributed between the glycol spacers. In this study, PCR products (rendered ssDNA by removal of the biotinylated 5′ primer strand by capture to a streptavidin spin-column) ranging between 142 nt to 1300 nt were hybridized to the

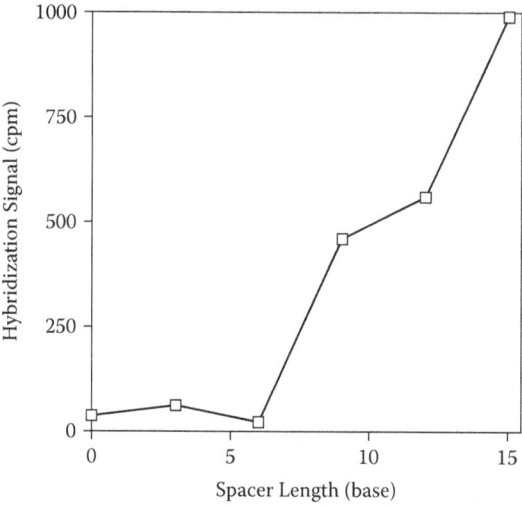

FIGURE 3.12 Poly (dT) spacer arm effect on hybridization efficiency. (From Guo Z, Guilfoyle RA, Thiel AJ, Wang R, Smith LM. Direct fluorescence analysis of genetic polymorphisms by hybridization with oligonucleotide arrays on glass supports. *Nucleic Acid Res.* 22(24): 5456–5465, 1994. With permission.)

9mer probe. Although the image quality was rather poor, there appeared to be some benefit to including spacer arms for the efficient hybridization of the longer target sequences (624 nt and 1300 nt). Nevertheless, even the "directly" attached 9mer that was tethered to the support via about 7 to 10 atom rotatable bonds showed some hybridization to these targets. The difficulty in assessing the influence of spacers on hybridization efficiency is that factors other than length may be equally as important. Relative hydrophobicity, electrostatic charge effects (spacer arm as well as surface charge), and wetting of the particular solid-support used in the studies may contribute to the overall hybridization event. Keep in mind that the nature of the target molecule (e.g., oligonucleotide, cDNA, RNA) as well as the concentrations of both the tethered probe and incoming target are also important considerations.

Ed Southern's group at Oxford University undertook a systematic study on the influence of spacer molecules (Shchepinov et al., 1997). In this case, however, rather than amino-silane glass they chose to use aminated polypropylene, a largely hydrophobic support material (Matson et al., 1995). In these studies hydrophilic, zwitterionic, and uncharged spacer arms were used to tether the probes (Figure 3.15). When the spacer units included phospho-propanediol, di- or tri-ethyleneglycols (i.e., $-(OCH_2CH_2)n\text{-}OPO_2^-$), there was a 50- to 150-fold enhancement in hybridization relative to the oligonucleotide directly coupled to the support. Optimal spacer unit length for the mono-, di-, tri-glycols was reported to be in the range of 8 to 10 unit repeats corresponding to 45 to 90 atom lengths depending upon the repeat unit utilized. Beyond $n = 10$ units hybridization decreased to that of a directly tethered probe at $n = 30$ spacer units (Figure 3.16). Longer spacers may in fact fold back onto the surface. Certainly, charge variation among these spacers also had some effect. The

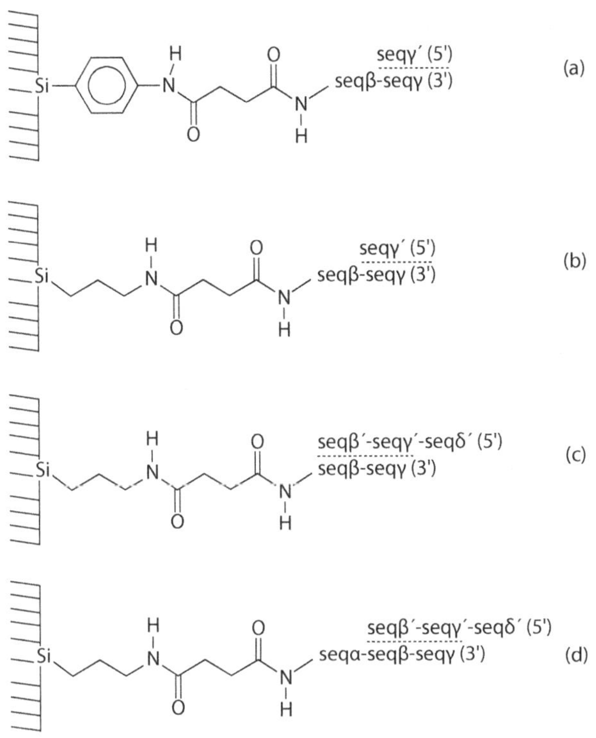

FIGURE 3.13 Coupling of a 35-mer probe with and without a 15-mer hetero-oligonucleotide spacer. (From Joos B, Kuster H, Cone R. Covalent attachment of hybridizable oligonucleotides to glass supports. *Anal. Biochem.* 247: 96–101, 1997. With permission.)

introduction of amphiphilic spacers of both negative ($\sim$OPO$_3^-$) and positive ($\sim$NH$_3^+$) into the spacer-arm backbone resulted in less dramatic increases in hybridization efficiency. An optimal spacer unit of $n = 3$ to 4 was observed, while additional units $n = 5$ to 7 led to a decrease.

Another approach largely borrowed from the extensive work done on chromatographic supports is the use of polymeric coatings. These offer some advantage in masking out the support surface properties while increasing the functional group capacity, and most importantly effectively removing the probe far away from the surface and freely exposing it to the bulk solution.

Beier and Hoheisel (1999) created a dendrimeric linker based upon stepwise condensation of di-amines with acryloyl (chloride) groups generated from reaction with the surface of either amino-silane glass or aminated polypropylene. This stepwise process resulted in a glass (or polypropylene) surface modified with a mixture of dendrimers with a multiplicity of terminal amine groups (Figure 3.17). The application of homobifunctional cross-linking agents (e.g., disuccinimidylcarbonate [DSC]) transformed these end groups into various isothiocyanates, NHS-esters, or imidoesters that were in turn reactive toward amino-oligonucleotides. One particular dendrimer system was reported to have increased the loading capacity 10-fold over

TABLE 3.1

Efficiency of Hybridization to Attached Oligonucleotides

Number of Target Molecules[a]		Number of Probe Molecules Applied in Hybridization		
Applied	Attached[b]	10^{10}	10^{11}	10^{12}
		5′-Succinyl-seqβ-seqγ		
10^{13}	2.7×10^{12}	3.8×10^9 (38%)	3.7×10^{10} (37%)	6.4×10^{11} (64%)
10^{12}	3.9×10^{11}	3.2×10^9 (32%)	3.9×10^{10} (39%)	2.0×10^{11} (51%)
10^{11}	4.1×10^{10}	2.3×10^9 (23%)	1.3×10^{10} (32%)*	3.7×10^{10} (90%)*
10^{10}	4.1×10^9	2.2×10^9 (5%)	8.9×10^8 (22%)*	2.2×10^9 (54%)*
		5′-Succinyl-seqα-seqβ-seqγ		
10^{13}	n.d.	2.6×10^9 (26%)	5.8×10^{10} (58%)	7.4×10^{11} (74%)
10^{12}	n.d.	4.6×10^9 (46%)	5.8×10^{10} (58%)	3.5×10^{11}
10^{11}	n.d.	4.0×10^9 (40%)	3.4×10^{10}	7.8×10^{10}
10^{10}	n.d.	1.1×10^9	4.0×10^9	7.4×10^9

Source: Joos B, Kuster H, Cone R. Covalent attachment of hybridizable oligonucleotides to glass supports. *Anal. Biochem.* 247: 96–101, 1997. With permission.

[a] Succinylated target oligonucleotides attached to 12-mm diameter aminopropyl-modified glass coverslips.

[b] Based on data from Joos et al. Table 1, last column; n.d., not determined.

[c] Hybridization with 32P-labeled probe seqβ-seqγ-seqδ according to Joos et al. Figures 2C and 2S, respectively. Results average number of hybrids found in duplicate determinations. Percentages are expressed relative to amounts of probe applied in hybridization when immobilized target was in excess, or relative to immobilized target (marked by *) when the probe was in excess.

direct coupling of the oligonucleotide to the surface. Unfortunately, the presumed improvement in hybridization efficiency was not discussed in quantitative terms. Reusability of the array was a primary objective of the work. Here again quantitative information was not provided, but as the authors point out the array withstood more than seven actual hybridization-stripping cycles as well as 30 simulated cycles (i.e., in the absence of a target).

Benters et al. (2002) at the University of Bermen, Germany, utilized starburst dendrimers to create polymeric surface coatings for glass substrates in an effort to increase sensitivity for DNA microarrays (Figure 3.18). Amino-silane or epoxy-silane chemistries were applied to glass substrates. The amino-silanes were further derivatized with terminal NHS or isothiocyanate groups for coupling of the starburst dendrimer. The starburst dendrimer, PAMAM (polyamidoamine; Aldrich) containing 64 primary amines could be covalently immobilized to the glass surface by reaction with the three different chemistries: epoxide, NHS, or isothiocyanate linkage. Subsequently, the remaining PAMAM amines could be converted to terminal NHS reactive groups for coupling of 5′–amino oligonucleotides (18 to 24mers). All PAMAM surfaces were significantly more efficient in hybridization compared to poly-L-lysine, epoxy-silane or NHS, and isothiocyanate modified amino-silane

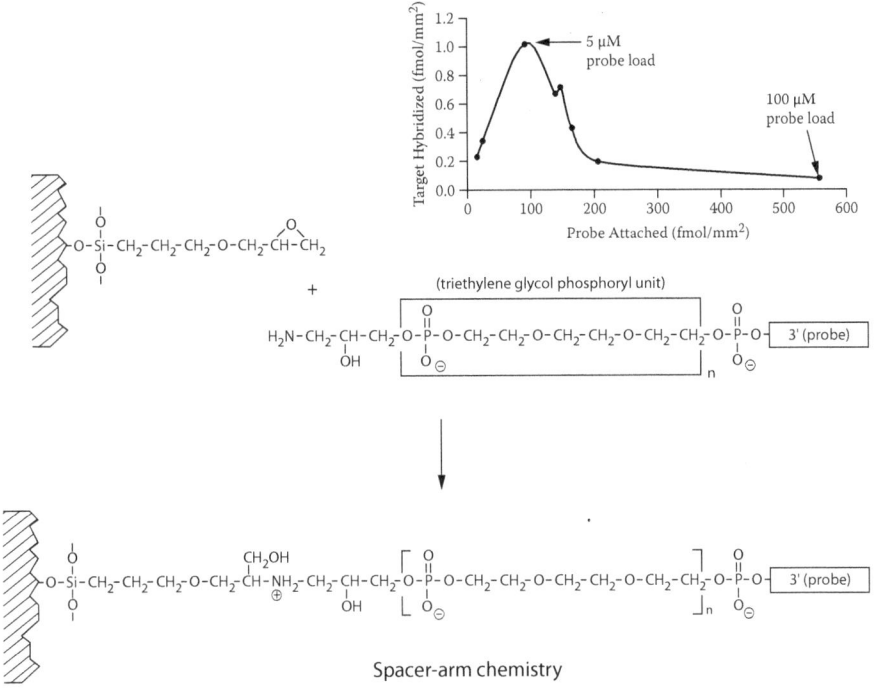

FIGURE 3.14 Probe density versus hybridization efficiency. (From Beattie WG, Meng L, Turner SL, Varma RS, Dao DD, Beattie KL. Hybridization of DNA targets to glass-tethered oligonucleotide probes. *Mol. Biotechnol.* 4: 213–225, 1995. With permission.)

slides. Up to 10 hybridization-regeneration cycles were accomplished using the optimal starburst surface.

Researchers at Incyte Genomics in collaboration with Samuel Sawan at the University of Massachusetts examined the utility of cross-linked polymer systems for improving oligonucleotide loading capacity in an effort to improve hybridization performance (Lee et al., 2002). Starting with aminopropyl silanated glass (APS-slides), the amino groups were activated using cyanuric chloride (referred to as CC-slides). The APS and CC slides were considered to be 2D surface chemistries. Reaction with polyethylenimine (branched polymer, average molecular weight ~25,000) converted these slides into PEI-slides. A final round with cyanuric chloride (P-slides) provided multiple sites for attachment of amino-oligonucleotide probes (Figure 3.19). The PEI and P slides were denoted as three-dimensional (3D) surface chemistries. All surfaces were characterized in terms of dynamic contact angles, Fourier transform infrared (FTIR) spectra, surface binding uniformity (based upon signal histogram for amino-reactive fluorescent dye), probe oligonucleotide binding capacity, and hybridization. The amount of amino-oligonucleotide probe covalently attached to each surface was directly proportional to the loading concentrations applied. In contrast, hybridization efficiency decreased at the higher loadings. Optimal loading for hybridization appeared to be at 10 μM oligonucleotide. The 3D surfaces performed quantitatively better than the 2D surfaces in terms of hybridization, while both types

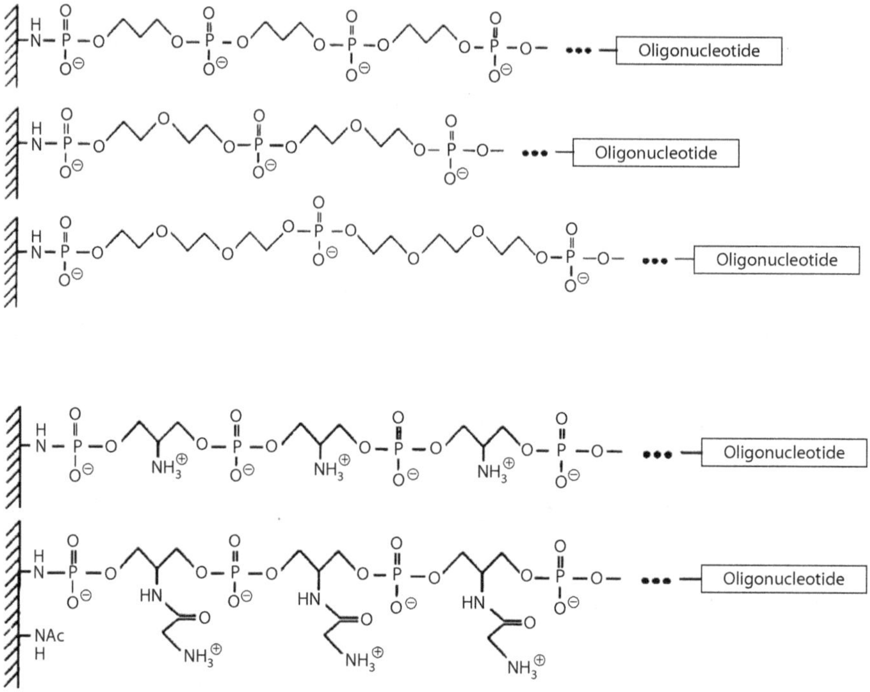

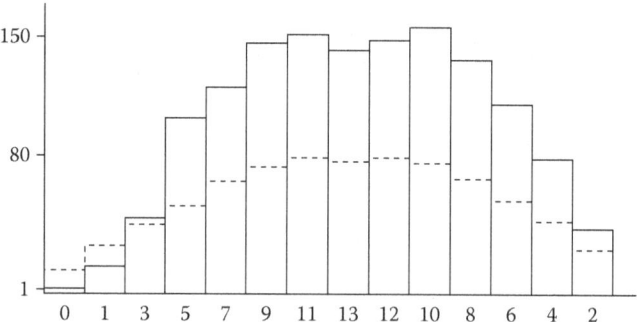

FIGURE 3.15 Oligonucleotide probe attachment with charged and uncharged spacer arms. (From Shchepinov MS, Case-Green SC, Southern EM. Steric factors influencing hybridization of nucleic acids to oligonucleotide arrays. *Nucleic Acid Res.* 25(6): 1155–1161, 1997. With permission.)

FIGURE 3.16 Impact of spacer arm length on hybridization efficiency. (From Shchepinov MS, Case-Green SC, Southern EM. Steric factors influencing hybridization of nucleic acids to oligonucleotide arrays. *Nucleic Acid Res.* 25(6): 1155–1161, 1997. With permission.)

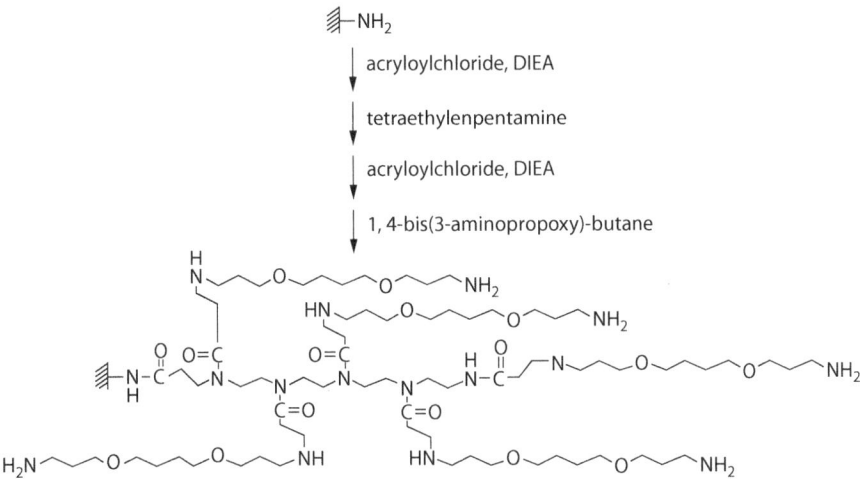

FIGURE 3.17 Creation of dendrimer spacers. (From Beier M, Hoheisel JD. Versatile derivatization of solid support media for covalent bonding on DNA-microchips. *Nucleic Acid Res.* 27(9): 1970–1977, 1999. With permission.)

of activated surface features (CC versus P) had similar binding capacity. All surfaces (APS, CC, PEI, and P) bound more oligonucleotide and produced higher hybridization signals than poly-L-lysine slides (Figure 3.20). Taking into consideration the mean signal and variance (SD) the relative rankings would be for probe immobilization, P > PEI > CC >> APS >> poly-L-lysine, and hybridization, P > PEI > CC > APS > poly-L-lysine.

Taylor et al. (2003) at Virginia Commonwealth University examined three commonly used glass slide surface chemistries (poly-L-lysine, epoxy-silane, aminopropyl-silane) as well as a dendrimer structure (DAB) similar to that described by Benters et al. (2002) (Figure 3.21). Slides derivatized with separate surface chemistries were arrayed with the same cDNA (~600 bp, GAPDH amplicon) and 30mer GAPDH oligonucleotide probes organized into subgrids for cDNA and probe, respectively (Figure 3.22). Each subgrid contained spots varying in DNA loading concentration in the range of 0.1 ng/µL to 500 ng/µL (cDNA) and 0.1 ng/µL to 1000 ng/µL (probe) along with random and nonhomologous control sequences. A rather extensive series of studies were conducted on these surfaces to measure hybridization performance (Alexa 555-labeled GAPDH amplicon) under different blocking conditions (unblocked versus BSA versus succinic anhydride). Spot signal intensity, spot quality, and background fluorescence were evaluated. The following observations can be made:

1. cDNA probes were more efficient at hybridization (i.e., produced higher spot signal intensity than oligonucleotide probes) but with greater signal variability. Neither the signal intensity nor its variability could be linked to a particular surface chemistry at a given loading.

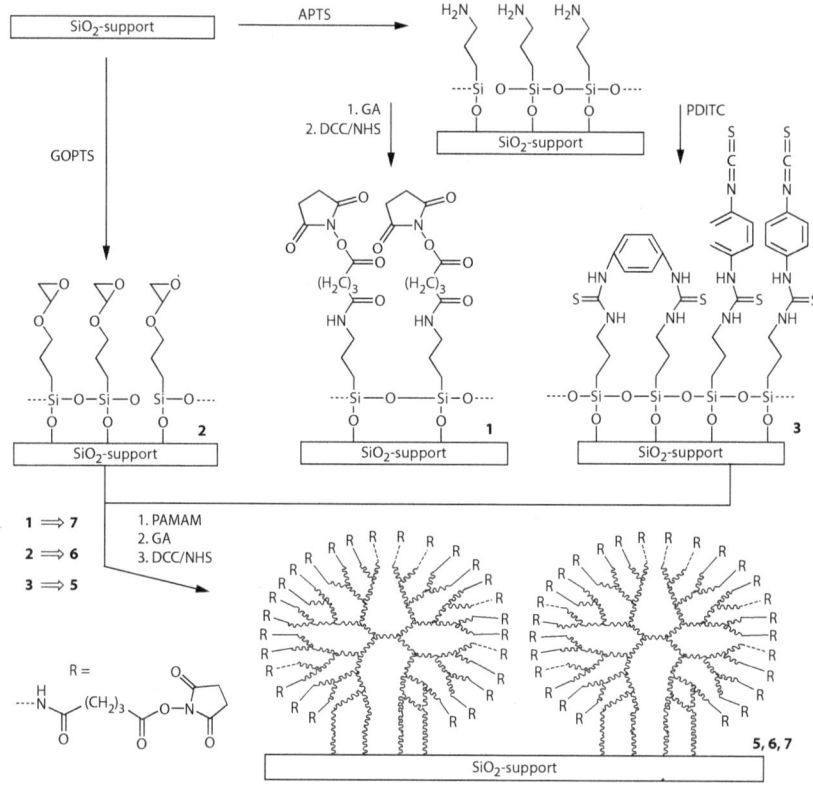

FIGURE 3.18 Polymeric dendrimer coatings on glass. (From Benters R, Niemeyer CM, Drutschmann D, Blohm D, Wohrle D. DNA microarrays with PAMAM dendritic linker systems. *Nucleic Acid Res.* 30(2) e10: 1–7, 2002. With permission.)

2. Oligonucleotide probe signal CVs were in general lower than those of the cDNA probes; however, this was somewhat dependent upon the surface chemistry and blocking condition.

3. Epoxy-silane surfaces provided the most efficient hybridization (S/B, signal/background ratio) for either cDNA or oligonucleotide (Figure 3.23):
   a. Optimal loading, cDNA ~ 0.1 to 0.5 μg/μL
   b. Optimal loading, 30mer ~ 0.01 to 0.1 μg/μL

4. DAB dendrimer surfaces showed no improvement in hybridization efficiency or spot variability relative to the other surface chemistries evaluated. This is in contrast to the findings by Benters et al. (2002) and Beier and Hoheisel (1999) on dendrimers where such scaffolds worked well.

5. BSA blocked slides exhibited reduced backgrounds over those blocked with succinic anhydride. However, succinic anhydride blocked epoxy slides appeared to be less variable. Microarrays created on unblocked epoxy-silane glass exhibited the greatest hybridization (S/B) efficiency.

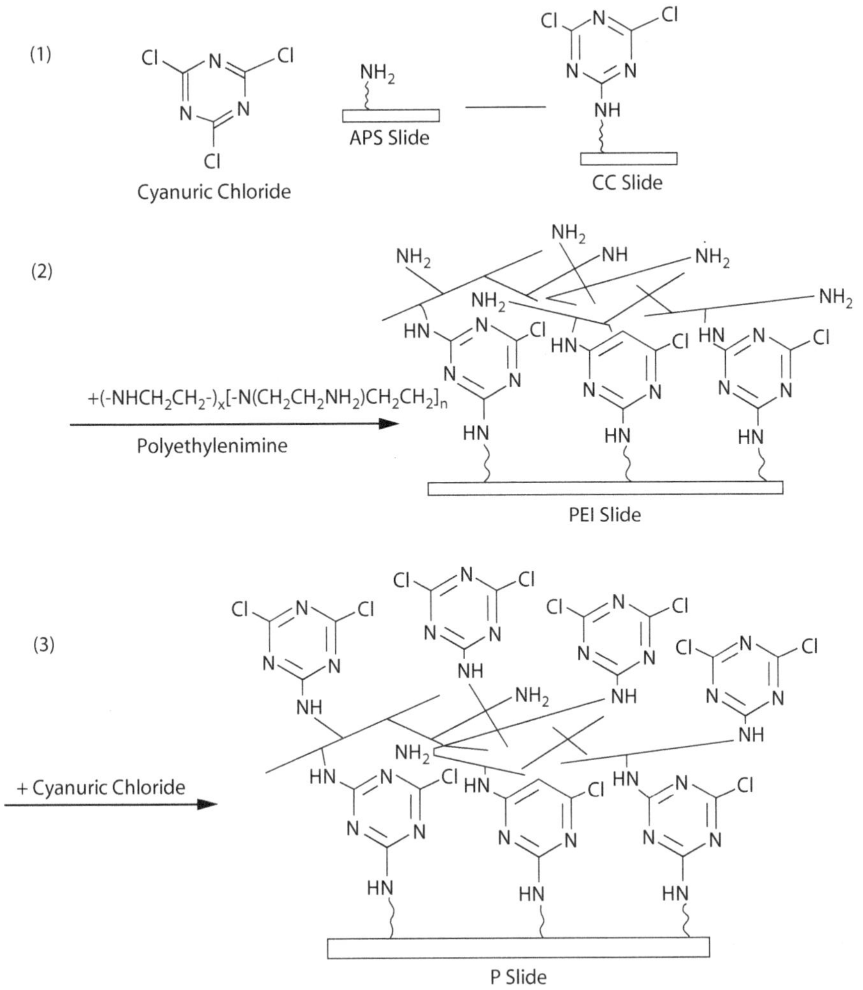

FIGURE 3.19 Process for preparation of activated polymeric surfaces. (From Lee PH, Sawan SP, Modrusan Z, Arnold LJ Jr., Reynolds MA. An efficient binding chemistry for glass polynucleotide microarrays. *Bioconjugate Chem.* 13: 97–103, 2002. With permission.)

So, in general the inclusion of a spacer arm between the solid phase and the oligo-nucleotide probe can result in an improvement in the hybridization efficiency. The problem arises in selecting the spacer unit that obviously must be carefully matched up with the surface properties and those of the probe. Linker (spacer-arm) selection is more likely than not to be an empirical process.

REACTIVE GROUPS

We can denote and diagram covalent coupling reactions between an activated substrate (S*) and a reactant (R), where R represents in this case the capture

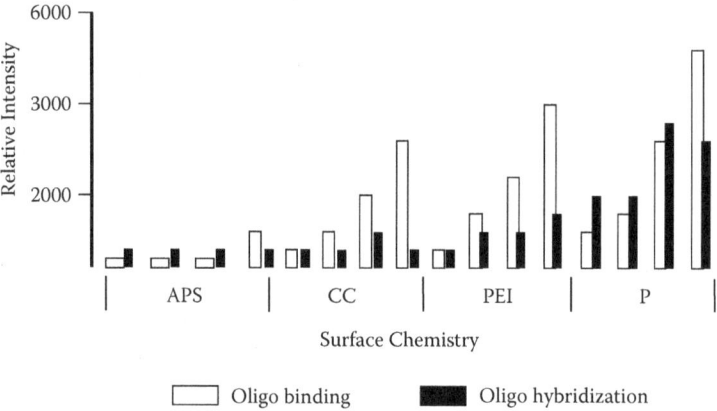

FIGURE 3.20   Oligonucleotide loading versus hybridization on activated surfaces. (From Lee PH, Sawan SP, Modrusan Z, Arnold LJ Jr., Reynolds MA. An efficient binding chemistry for glass polynucleotide microarrays. *Bioconjugate Chem.* 13: 97–103, 2002. With permission.)

probe (e.g., oligonucleotide or antibody) by functional groups. For example, Figure 3.24 shows the steps in preparing a glass slide for the attachment of an antibody. The substrate (S) is reacted with amino-silane to form an activated substrate (S*). Oxidized antibody (R) containing reactive aldehyde groups (R-CHO) can be covalently immobilized to the amino-silane slide by a Schiff's base reaction:

$$S^*\text{-}NH_2 + O = CH\text{-}R \rightarrow S\text{-}NH = CHR + H_2O$$

However, most coupling chemistries do not go to completion so that the substrate will contain a mixture of functional groups capped with attached probe (SR) while others remain free (S*). These residual reactive functional groups need to be capped or blocked in some manner to reduce nonspecific binding to the microarray. Residual surface amines may be capped by reaction with succinic anhydride. This renders the support neutral (SR). Using this abbreviated nomenclature we can describe common surface modifications for microarray substrates.

Most immobilization chemistries for microarrays currently rely upon derivatization of the substrate with amine reactive functional groups (e.g., aldehyde, epoxide, *N*-hydroxysuccinimide, or NHS-ester). While there are many available surface reactive chemistries to choose from, it is important to keep in mind that these must be compatible with a printing process. Ideally, the biomolecule should react completely and rapidly with the substrate in order to achieve good spot formation. It is also critical that the probe remain or be recoverable in its active state following printing. If too reactive a chemistry is employed, there is the possibility for excessive cross-linking that can hinder performance by reducing the number of rotatable bonds in the probe. The best substrate is one that will present the probe to the solution phase with as much rotational freedom as possible so that it can undergo favorable binding with

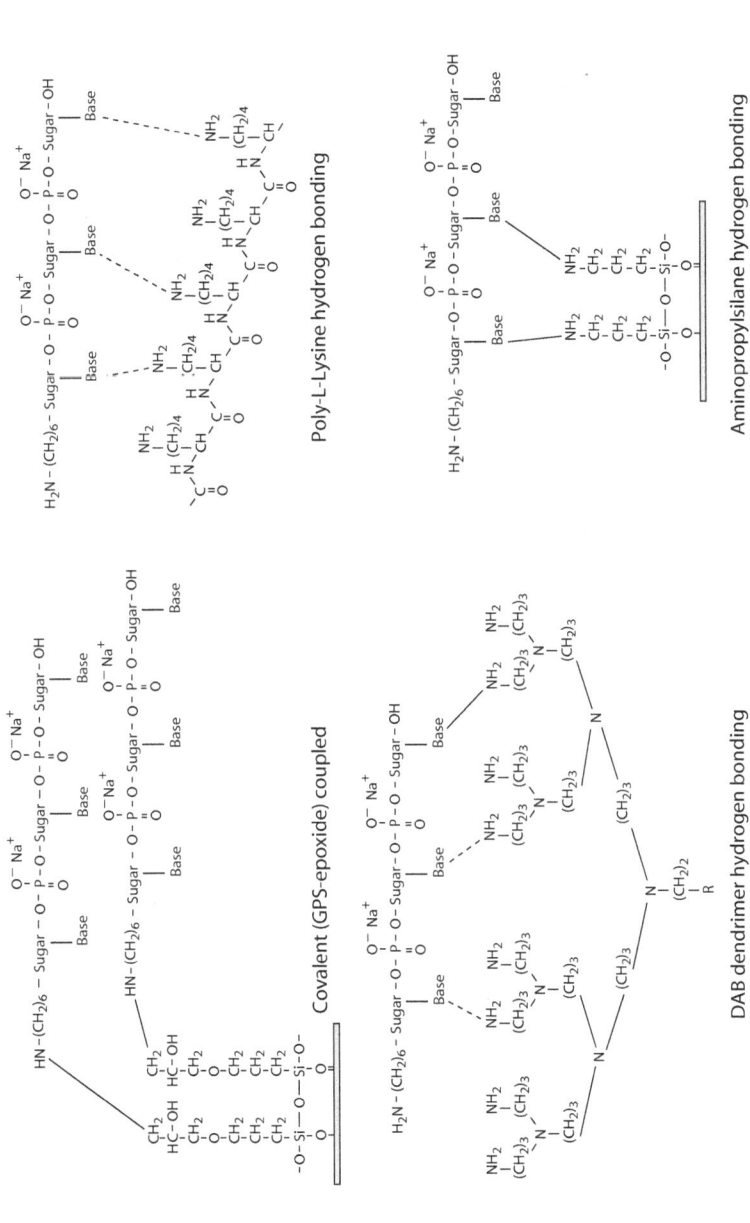

FIGURE 3.21 Abbreviated surface chemistry structures: PLL, epoxysilane, APS, and DAB. (From Taylor S, Smith S, Windle B, Guiseppi-Elie A. Impact of surface chemistry and blocking strategies on DNA microarrays. *Nucleic Acid Res.* 31(16) e87: 1–19, 2003. With permission.)

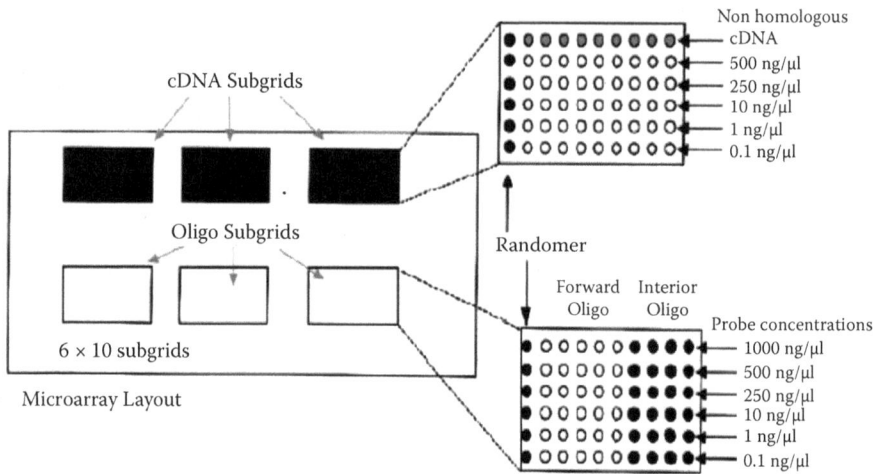

FIGURE 3.22 Array map for glyceraldehyde 3-phosphate dehydrogenase (GAPDH) DNA and 30mer probe. (From Taylor S, Smith S, Windle B, Guiseppi-Elie A. Impact of surface chemistry and blocking strategies on DNA microarrays. *Nucleic Acid Res.* 31(16) e87: 1–19, 2003. With permission.)

the incoming target molecule. The binding should approximate free solution association. Table 3.2 lists common coupling chemistries employed for probe (nucleic acid, protein) attachment useful for microarrays. Many of these have been discussed in previous sections. Additional information about specific coupling chemistries for proteins can be found in the protocols and discussion regarding solid-phase reagents (Matson, 2000).

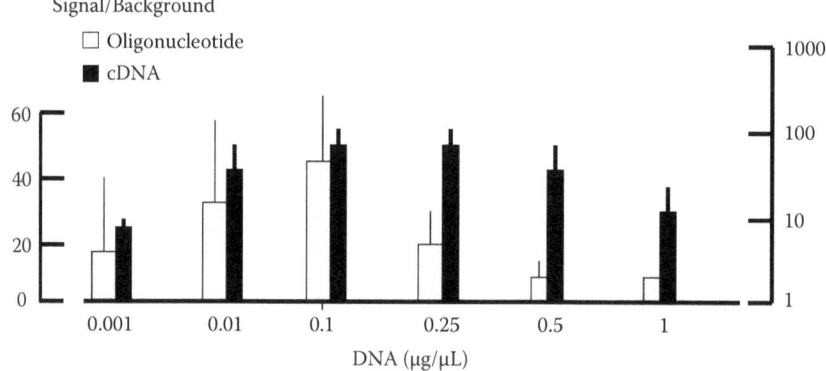

FIGURE 3.23 Hybridization signal to background (S:B) levels for cDNA versus oligonucleotide probe. (From Taylor S, Smith S, Windle B, Guiseppi-Elie A. Impact of surface chemistry and blocking strategies on DNA microarrays. *Nucleic Acid Res.* 31(16) e87: 1–19, 2003. With permission.)

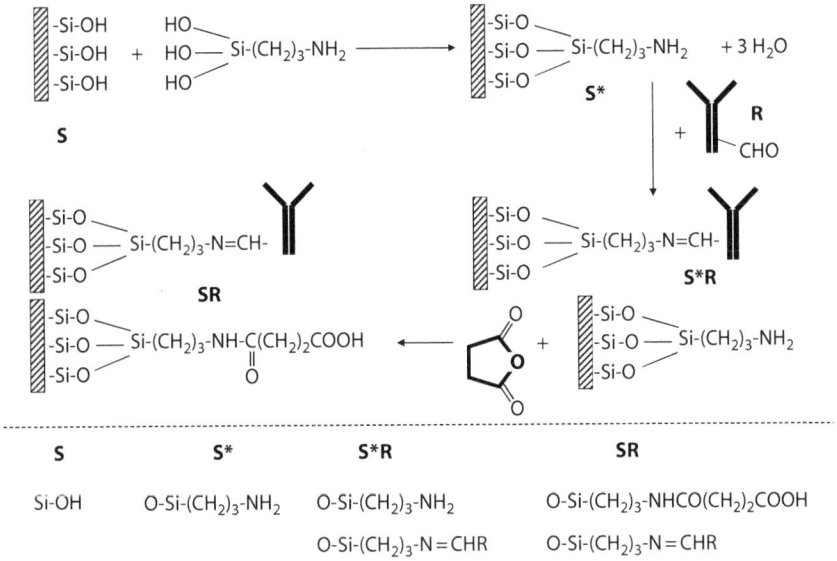

FIGURE 3.24 Steps in the preparation of aminopropyl silane (APS) surface, coupling of oxidized antibody, and blocking of residual reactive surface groups.

PREPARATION OF GLASS SUBSTRATES FOR DERIVATIZATION

Glass must be thoroughly cleaned to be useful as a microarray support. The cleaning process must be able to remove surface contaminants such as oils and grease as well as particulates. The other important reason to clean the slide is to reactivate the surface so that there are plenty of surface hydroxyls (Si-OH) for subsequent chemical derivatization. The following protocols have been used to clean glass slides. Also included are the functionalization processes to create amine, carboxyl, epoxide, aldehyde, NHS, and polyethylenimide (PEI) surface chemistries.

See Beattie et al. (1995) for attachment of oligonucleotides by epoxide:

1. Soak the slide for 30 to 60 minutes in 1 N nitric acid.
2. Rinse in water and then dry.
3. Sonicate for 10 minutes in each of the following solvents:
   a. Hexane
   b. Acetone
   c. Absolute ethanol
4. Dry the slides.
5. Soak the slides at 80°C for 5 hours in epoxy-silane solution made up of
   a. Xylene (anhydrous): glycidoxypropyltrimethoxysilane: *N,N*-diisopropyl ethylamine (24:8:1 v/v).
   b. Wash slides three times in THF.
6. Dry and store slides desiccated under vacuum.

TABLE 3.2

Common Coupling Chemistries Employed for Nucleic Acid and Protein Probe Attachments

S	S*	R	S-R	Conditions
]-COOH + NHS	]-COO-NHS	RNH_2	]-COO-NHR	pH 6.5-8.5
]-NH$_2$	Poly-L-lysine	RCHO	]-N=CHR	pH 8-9
]Si-OH+APS	]-NH$_2$		]-NHCH$_2$R	Hydrogen Reduction
]Si-OH + GPTS Zhu (2000)	]-CH—CH \O/	RNH_2	]-CH(OH)CH-NHR	pH 9-10
		RSH	]-CH(OH)CH-SR	pH 7.5-8.5
		ROH	]-CH(OH)CH-OR	pH 11-12
]-CONH$_2$ + NH$_2$NH$_2$ Khrapko (1991)	]-CONHNH$_2$	RCHO	]-CONH=CHR	pH 5-6
	]-CHO	RNH_2	]-CH=NR	pH 8-9
]Si-OH Silane-SH/GMBS Delehanty (2002)	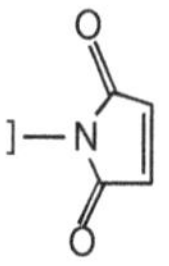	RSH	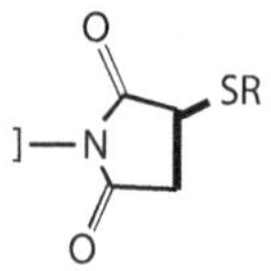	pH 6.5-7.5

] = solid-phase

See Beier and Hoheisel (1999) for attachment of dendrimer linkers from amine:

1. Soak the slides overnight in 10% sodium hydroxide.
2. Rinse them sequentially:
   a. Water
   b. 1% hydrochloric acid
   c. Water
   d. Methanol
3. Sonicate the slides 15 minutes in 3% aminopropyltrimethoxysilane-95% methanol.
4. Rinse them in 100% methanol.
5. Rinse them with water.
6. Dry the slides under nitrogen stream.
7. Bake them for 15 minutes at 110°C.
8. An acylation/amination reaction starts with the amine slide prepared above to prepare the dendrimer.

See Zammatteo et al. (2000) for preparation of carboxylic acid and aldehyde slides:

1. Soak slides for 30 minutes in hydrogen peroxide (30%):18 M sulfuric acid, (33:66 v/v).
2. Rinse slides with distilled water.
3. Soak them in boiling water for 10 minutes.
4. Dry them under argon stream.
5. Immerse slides in 1 mM TETU in toluene for 1 hour (TETU = 2′,2′,2′-tri-fluoroethyl-11-(trichlorosilyl) undecanoate).
6. Sonicate them three times for 10 minutes in fresh toluene soaks.
7. Dry them under argon stream.
8. Prepare functional groups:
   a. Carboxylic acid:
      i. Soak TETU slides in 8 M hydrochloric acid at 95°C for 2 hours.
      ii. Sonicate them three times with distilled water.
      iii. Dry them under argon stream.
   b. Aldehyde:
      i. Soak TETU slides in 20 mM LiAlH$_4$ in anhydrous ether for 2 hours.
      ii. Soak them in 10% hydrochloric acid for 1 hour, rinse them two times with water, rinse acetone, then dry at 120°C for 10 minutes.
      iii. Soak "hydroxyl" slides in 100 mM pyridinium chlorochromate in dichloromethane (anhydrous) for 2 hours, rinse three times with dichloromethane for 10 minutes each, soak them in water for 5 minutes, dry them under argon, and then store aldehyde slides under vacuum.

See Belosludtsev et al. (2001) for vacuum amine and epoxy silanization protocols:

1. Clean slides in ultrasonic bath with detergent for 2 minutes.
2. Rinse them three times with water.
3. Rinse them two times with methanol.
4. Dry them at 40°C for 30 minutes.
5. Place slides on a rack into a vacuum oven.
   a. Place a Petri dish containing 3 mL silane + 3 mL xylene:
      i. 3-Aminopropyltrimethoxysilane (amine) or
      ii. 3-Glycidoxypropyltrimethoxysilane (epoxide)
   b. Incubate overnight under vacuum (25 inches, Hg) at 70 to 80°C.

See Benters et al. (2002) for preparation of carboxyl and NHS activated surfaces:

1. Clean slides by ultrasonic treatment in chloroform.
2. Soak in fresh "piranha" solution for 30 minutes: sulfuric acid (conc.):hydrogen peroxide (2:1 v/v).
3. Soak slides with stirring in 3-aminopropyltrimethoxysilane (APS) solution for 2 hours: APS:water:ethanol (2:3:95 v/v).
4. Soak "amine" slides in a saturated solution of glutaric acid in DMF overnight.

5. Rinse thoroughly with DMF.
6. Incubate "carboxyl" slides with 1 M each NHS-DCC (*N,N'*-dicyclohexylcarbodiimide) in DMF solution for 1 hour.
7. Rinse with DMF.

This protocol was described for immediate use. Otherwise, I recommend that the NHS slides be dried under an argon stream, sealed, and stored at 4°C until needed for printing.

Always allow slides to warm to room temperature before opening to avoid moisture condensation that can inactivate the surface chemistry.

See Lee et al. (2002) for preparation of PEI-coated slides:

1. Clean the slides in an ultrasonic bath with
   a. 1% SDS
   b. 4% HF
2. Dry the slides in an oven at 110°C.
3. Immerse the slides in APS, 0.05% w/v in 95% ethanol.
   a. Rinse several times with 95% ethanol.
   b. Oven dry at 110°C.
4. Immerse "amine" slides in a stirred slurry of cyanuric chloride (12.7 g/L)-sodium carbonate (25g/L) in hexane at 4°C for 1 hour with sonication.
   a. Rinse with hexane.
   b. Air dry the slides.
5. Immerse the slides in 0.1% aqueous polyethylenimine (25,000 average M.wt.) for 1 hour at 4°C with sonication.
6. Sonicate the slides in deionized water and air dry.

VARIATION IN THE PERFORMANCE OF GLASS SLIDE-BASED ANTIBODY MICROARRAYS

Seurynck-Servoss et al. (2007) undertook a rather extensive examination of the performance of commercially available glass slides used for the immobilization of antibodies. While some of these slides may no longer be available, the study provides insight into the various surface chemistries used for protein tethering.

The group from the Pacific Northwest Laboratory (Richland, WA) took 16 different slides from 4 different manufacturers: Erie Scientific (now part of Thermo Scientific), Full Moon BioSystems (Sunnyvale, CA), Schott (Germany), and NSB Postech (Korea). All slides were arrayed with 16 subarray replicates using a noncontact dispenser (GeSiM NanoPlotter 2.1). The same set of antibodies was applied to each slide and 23 ELISAs performed. Standard curves were prepared by threefold serial dilution. Biotinylated detection antibodies were applied to complete the sandwich immunoassay that was then detected from tyramide (TSA) amplification using streptavidin-Cy3 as the fluorescent reporter.

The kinds of slide surfaces assessed were aminosilane, epoxysilane, poly-L-lysine, aldehyde silane, cast cellulosic polymer, hydrogels, and dendritic polymer coatings. The following summarizes the general observations:

1. *Spot Characteristics*—No significant differences were found between spot diameters and spot morphology across the various slides. The same print buffer, PBS, was used in all cases. All slides were also blocked in PBS-casein and rinsed in PBST (PBS + 0.05% Tween 20). Therefore, these wetting/blocking conditions may have masked out any differences in surface properties among the slides, except for a few noted exceptions (below).

2. *Background Issues*—With the exception of two slides, the slide noise surrounding spots as well as the level of nonspecific binding on spots were found to be low (Figure 3.25).

3. *LOD* (lower limit of detection)—This is usually taken as the lowest dose response detectable above the mean blank signal + 2 SD associated with spot background. In this case, a relative LOD was assigned by comparison to the median LOD from all slides for a given assay. Thus, with the exception of three slides that rated as poor, the majority of slides exhibited superior performance. The median LOD ~3 pg/mL.

4. *Reproducibility*—Examining replicate assays on different days demonstrated similar levels of reproducibility among the slides.

5. *Covalent Attachment or Physical Adsorption*—It was found that "noncovalent binding chemistries performed comparable to the best of the covalent chemistries in all categories" (Seurynck-Servoss et al., 2007, p. 7).

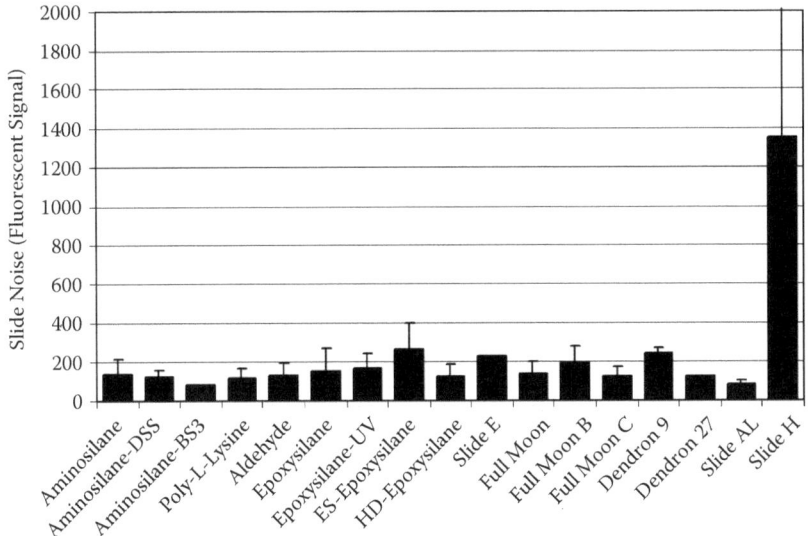

FIGURE 3.25 Evaluation of slide noise for the various slide types. (From Seurynck-Servoss SL, White AM, Baird CL, Rodland KD, Zangar RC. Evaluation of surface chemistries for antibody microarrays. *Anal. Biochem.* 371(1): 105–115 2007. With permission.)

6. *3D or 2D Surfaces*—Slides with hydrogel or dendron coatings (3D surface) exhibited the worst-case scenario with high backgrounds and poor spot morphology leading to reduced sensitivity and poor reproducibility. Not surprising, the conclusion reached was that these high surface area matrices are more difficult to clear of nonspecific binders, resulting in relatively higher backgrounds.

Finally, the slides were ranked based upon the above features as good, average, or poor. In terms of overall performance there were three slide chemistries suggested by the authors as most appropriate for the development of slide-based antibody microarrays for ELISA:

1. Aminosilane coated slides—Physical adsorption of proteins with or without cross-linking
2. Poly-L-lysine coated slides—Adsorption by electrostatic interaction
3. Aldehyde silane coated slides—Covalent attachment of proteins by Schiff's base formation

COMPARISON OF DIFFERENT SURFACE CHEMISTRIES FOR THE IMMOBILIZATION OF AUTO-ANTIGENS

Balboni et al. (2008) examined 24 slide surfaces for the immobilization of auto-antigens. These were printed in a microarray format using contact array printing with quill-style pins (ChipMaker 2, Telechem; Silicon Spotting Pins, Parallel Synthesis). The auto-antigens were prepared in PBS for printing at 200 µg/mL. Microarrays contained 10 auto-antigens and 10 fluorescent dye labeled protein "doped" controls with replicates varying in concentration, at mixtures and positions to create 726 features. The resulting microarray slides were then blocked in PBST containing 3% fetal calf serum.

Following an initial screen, 10 slides were selected for further study. Intra-slide and inter-slide spot CVs were determined with a cut-off of 30%. In particular, three slides stood out as most suitable for auto-antigen assay development: FAST (cast nitrocellulose, Whatman-now Maine Manufacturing, Inc.); PATH (ultra-thin nitrocellulose coating, GenTel BioSciences, now manufactured by Grace Bio); and SuperEpoxide2 (ArrayIt Inc., Sunnyvale, CA). These slides were evaluated in a final round using human serum positive control sera on a 690 feature microarray format.

Intraslide CVs, LOD, NSB, and S/N levels were determined. The FAST slides ranged from 3.7 to 20.5% CV; PATH (5.0 to 44.3%); and SuperEpoxy2 slides, 3.5 to 53.5%. While FAST had the highest signal-to-noise ratio, the other slides were deemed acceptable for use. All slides exhibited acceptable backgrounds and detection limits.

CLICK CHEMISTRY AS AN IMMOBILIZATION STRATEGY

Krishnamurthy et al. (2010) provide an interesting approach to studies involving the strategy for peptide immobilization on native cell surfaces in order to affect cellular function. This is often referred to as *remodeling*. For example, the pentapeptide, IKVAV, is an extracellular matrix cell adhesive peptide involved in the regulation of cellular function in pancreatic islets such as increased insulin secretion. This is mediated by interaction (sequestering) of IKVAV with an integrin (cell surface receptor protein). The problem encountered is that introducing such probes (e.g., covalent labeling surfaces for cell imaging or immobilization to receptors) can at times unwarily disrupt normal cellular physiology.

In this study, a derivative of the pentapeptide was prepared that would allow rapid cell surface attachment by covalent coupling without a significant disruption of cellular function. The coupling chemistry is based upon a variation of "click" chemistry referred to as *strain promoted azide-alkyne cycloaddition* (Agard et al., 2004). Click chemistry generally requires a catalyst, such as Cu (I), to achieve cycloaddition. The use of a strained cyclic ring structure removes the need for the catalyst. More importantly, it presents a more tolerable condition for living cells.

Here, cyclooctyne (CyO) maleimide was linked to a cysteine terminated pentapeptide via a thiol-maleimide conjugation (Figure 3.26) for synthesis of CyO-IKVAV *alkyne*. The peptide also included glycine spacers and a biotin for labeling purposes, as CK(biotin)-GG-IKVAV. In order to couple to the cell surface a biocompatible PLL-g-PEG copolymer with reactive azido groups was prepared and referred to as PP-N$_3$ (*azide*).

The local concentration of peptide on the cell surface is known to affect intergin receptor activity. A microarray "surface" model was therefore developed to better

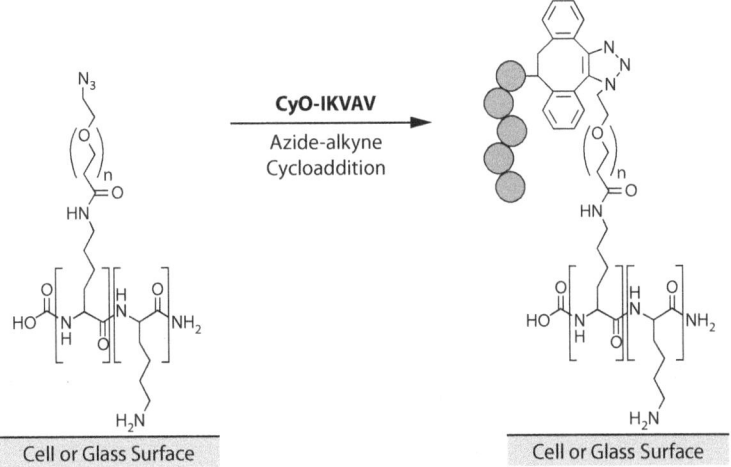

FIGURE 3.26 Peptide immobilization on abiotic and living surfaces using strain promoted azide-alkyne cycloaddition. (From Krishnamurthy VR, Wilson JT, Cui W, et al. Chemoselective immobilization of peptides on abiotic and cell surfaces at controlled densities. *Langmuir* 26(11): 7675–7678, 2010. With permission.)

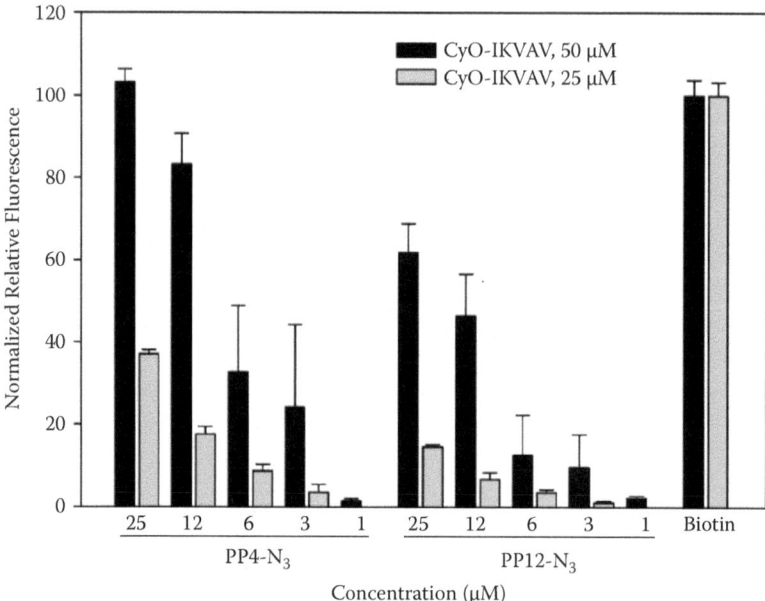

FIGURE 3.27 Normalized relative fluorescence intensities of CyO-IKVAV peptides immobilized by cycloaddition. (From Krishnamurthy VR, Wilson JT, Cui W, et al. Chemoselective immobilization of peptides on abiotic and cell surfaces at controlled densities. *Langmuir* 26(11): 7675–7678, 2010. With permission.)

understand issues of peptide density and steric hindrance. In particular, PEG spacers in PP-N_3 were studied for their relative effect on immobilization of the peptide. In this case, a PEG_4 spacer was found to promote greater immobilization of the peptide over that of a PEG_{12} spacer unit (Figure 3.27).

Canalle et al. (2011) compared two "click" chemistries for immobilization of proteins for ELISA-based assays. Copper catalyzed cycloaddition (azide-alkyne 1,3 dipolar cycloaddition or CuAAC) and strain promoted cycloaddition (SPAAC) that do not involve catalyst were evaluated for the tethering of a citrullinated peptide known to be associated with rheumatoid arthritis (RA). The arginine analog of the peptide served as a negative control. RA is characterized by severe joint inflammation. As a consequence, a diminase enzyme released from dead cells converts arginine residues of proteins into citrulline. Thus, citrullinated (cit) proteins may serve as a useful biomarker of the disease state.

In order to validate the use of click chemistries for ELISA it was necessary to prepare solid-phase alkynes in microplates. This was accomplished by coating wells with methacrylate copolymers having pendant alkyne groups. Thus, wells were coated with either methacrylate polymerized with propargyl (prop-2-yn-1-yl) groups for CuAAC couplings or with BCN (bicyclo [6.1.0] non-4-yne) methacrylate for SPAAC (Figure 3.28).

Next, the peptides required the addition of a terminal azido group that would be available to participate in the cycloaddition reactions. Fmoc azido norleucine

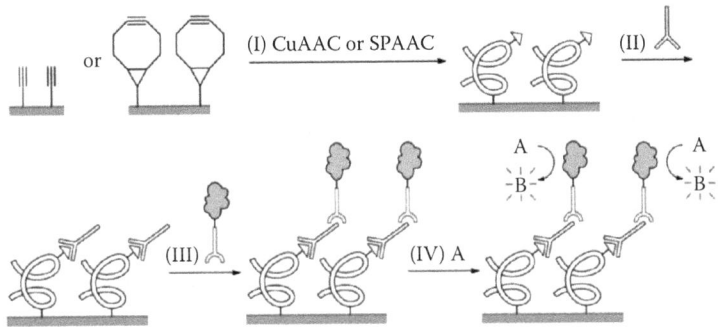

FIGURE 3.28 Surface immobilization of a diagnostic peptide using click reaction and subsequent ELISA. (From Canalle LA, Vong TH, Adams PHHM, et al. Clickable enzyme-linked immunosorbent assay. *Biomacromolecules* 12: 3692–3697, 2011. With permission.)

was synthesized from Fmoc-lysine (BOC) and then incorporated into the arg- and cit-peptides during standard solid-phase peptide synthesis to create peptides with azide on the C-terminus.

The click-based coupling of azido-peptides to solid-phase alkynes using the CuAAC or SPAAC approach was evaluated using an ELISA model. Following peptide tethering, the monoclonal human anticitrulline antibody (MQR2.101, ModiQuest Research) was applied to wells containing either cit- or arg-peptides. Anti-human IgG (rabbit monoclonal) conjugated with HRP (P0212, Dako) was used as the reporter antibody with signal generated by TMB reagent.

The results were very clear. Detection of cit-peptide tethered by CuAAC immobilization was achieved. However, the detection of the negative control arg-peptide also occurred at the same level of signal intensity. SPACC coupling resulted in a positive signal for cit-protein and a negligible measure of arg-protein (Figure 3.29). The reason for the high false positive with CuAAC is believed to be due to Cu interference. It is possible for Cu^{+2} to form coordination complexes with protein amide bonds by the well-known Biuret reaction. Copper-containing proteins are known to catalyze peroxidase-mediated development of TMB (Pugia et al., 2000).

OXYGEN PLASMA-MEDIATED MODIFICATION OF DVD-R DISKS FOR TETHERING OF OLIGONUCLEOTIDES

Tamarit-Lopez et al. (2011) describe the surface activation of compact disks by oxygen plasma for the immobilization of oligonucleotides. Compact disks are made from polycarbonate plastic, a material that is difficult to chemically modify without damage or alteration of the optical properties required for reading of the disk. To solve this problem oxygen plasma was used to introduce functional groups.

DVD-R disks were placed in a microwave plasma reactor (PVA Tepla 200 Plasma System; 2.45 GHz, 100 W). Oxygen was introduced at 120 Pa and power applied for

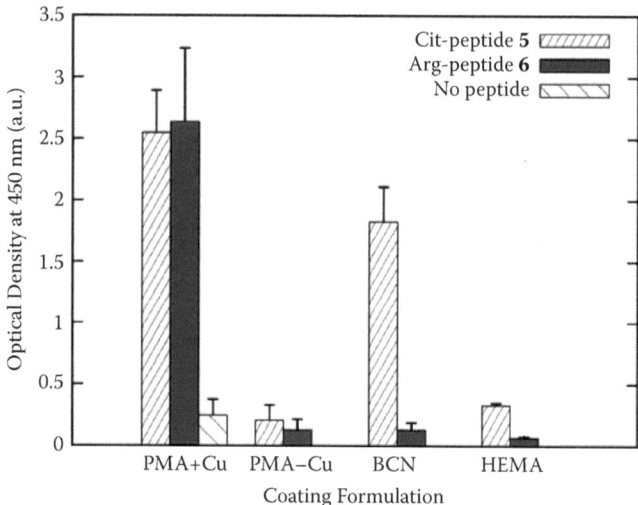

FIGURE 3.29 Comparison of ELISA using Cit-peptide or Arg-peptide coatings. (From Canalle LA, Vong TH, Adams PHHM, et al. Clickable enzyme-linked immunosorbent assay. *Biomacromolecules* 12: 3692–3697, 2011. With permission.)

30 seconds to achieve surface modification (Figure 3.30). The modified disks were stored dry under vacuum prior to use. The surface modification was assessed by standard techniques for surface evaluation:

1. Contact angle decreased from <79° to 16° within 30 seconds of treatment indicating the presence of hydrophilic groups on the surface. Stored disks showed a minimal increase to <21° after 1 month.
2. X-ray photoelectron spectra (XPS) confirmed the presence of carbonyl species on the surface.
3. Crystal Violet dye analysis determined the presence of 1.2×10^{-9} moles/cm^2 of ionizable surface group, presumably attributable to the formation of carboxylic acids based upon XPS analysis.

Amino-terminated oligonucleotide probes were immobilized to the treated disks by carbodiimide active ester coupling. The oligo probe (5 μM) was prepared in 50 mM HEPES, pH 7, containing 40% glycerol to reduce spot spreading. For printing purposes, 20 mM EDC and 20 mM NHS were added to the oligo probe and the print solution immediately used for arraying. 50 nL drops were dispensed to the disk surface using a noncontact dispenser (AD1500, BioDot Inc., Irvine, CA), and the relative humidity was maintained at 90%. The resulting microarray included 500 micron diameter spots that were arranged in a 6 × 5 array. The coupling was allowed to proceed for 16 hours at 4°C with controlled humidity.

Hybridization of digoxigenin labeled complementary targets (21 to 26 bp) as well as a similarly labeled PCR amplicon (151 bp) were conducted at 37°C for 1 hour in 3X SSC hybridization buffer. A colloidal gold anti-digoxigenin conjugate (Aurion,

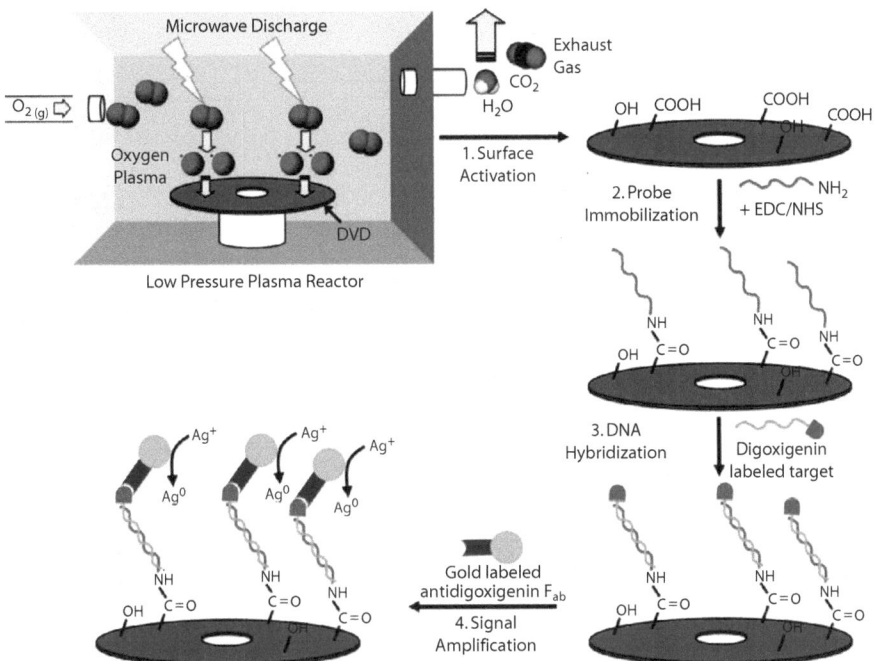

FIGURE 3.30 Oxygen plasma activation of DVD polycarbonate surfaces. (From Tamarit-López J, Morais S, Puchades R, Ángel Maquieira A. Oxygen plasma treated interactive polycarbonate DNA microarraying platform. *Bioconjugate Chem.* 22: 2573–2580, 2011. With permission.)

The Netherlands) was applied for 1 hour. Signal amplification was achieved in the presence of a silver enhancer (Sigma-Aldrich) for 18 minutes, followed by a water rinse to stop the reaction and achieve optimal signal-to-noise ratio.

The density of immobilized probes on the activated DVD-R disk was estimated to be approximately 2 pmoles per cm^2. Target hybridization was estimated to plateau at 200 nM, achieving a maximum DNA hybrid (duplex) density of 0.93 pmoles per cm^2 at a hybridization efficiency of 46%. The probe density values were obtained by extrapolation of spot intensity (measured using a CCD camera) standard curves for a serially diluted Cy5 labeled tracer oligo probe that was included during printing. Hybrid density was measured from extrapolation of Cy5 labeled oligo target complementary to a probe immobilized at varying density on the disk array. While these represent indirect measures of surface probe-target density, they are within the "ballpark" from other reported studies. However, the calculated occupancy appears to be much higher than would have been expected.

This study clearly demonstrates the ability to surface modify polycarbonate by oxygen plasma in a manner useful for the immobilization of oligonucleotides. The performance of the oligonucleotide array in the capture and discrimination of SNPs reaching S/N~13 as well as in the detection of a 151 bp PCR amplicon at a LOD of 2 nM (1 CFU/mL sensitivity) offers credence to the development of an interactive DVD disk-based molecular analysis platform.

CONSTRUCTION OF LIPID BILAYER MICROARRAYS

Joubert et al. (2009) created lipid bilayers on quartz slides in an array format using a continuous-flow microspotter (CFM) system. The CFM is a PDMS block having an array of vertical channels that make contact with a planar substrate at discrete locations to produce 400 micron square spots arrayed as 48 features at a pitch of 875 microns. Each spot is individually addressable allowing for the creation of different lipid bilayer compositions at each location. In order to produce the lipid bilayer it was necessary to use bis-SorbPC, a photopolymerizable glycerol-3-phosphocholine capable of forming a stable lipid bilayer base on planar substrates. The lipid bilayer array is stabilized by UV polymerization using a mercury lamp (4500 µW/cm²; 254 nm; 15 minutes). Blocking for nonspecific binding is accomplished using bovine serum albumin (2 mg/mL in PBS).

Other lipophilic materials may be doped into the bis-SorbPC bilayer. To demonstrate the binding characteristics within the lipid bilayer array, brain ganglioside (GM_1) was doped into the base lipid at a mole percentage of 0 to 10% to study binding of cholera toxin subunit b (CTb). GM_1 is associated with lipid rafts on the host cell surface. The B subunit of cholera toxin binds to GM_1 and is transported into the cell by endocytosis. However, the A subunit of cholera toxin is responsible for toxicity. It forms a complex with five B subunits, AB_5, and is transported as discussed.

Alexa 488 CTb conjugate (Invitrogen) was applied to the GM_1 array and the epifluorescence signal response detected by CCD camera (Photometrics-Roper Scientific). The relative intensity from CTb binding was directly proportional to the level of GM_1 within the bilayer (Figure 3.31).

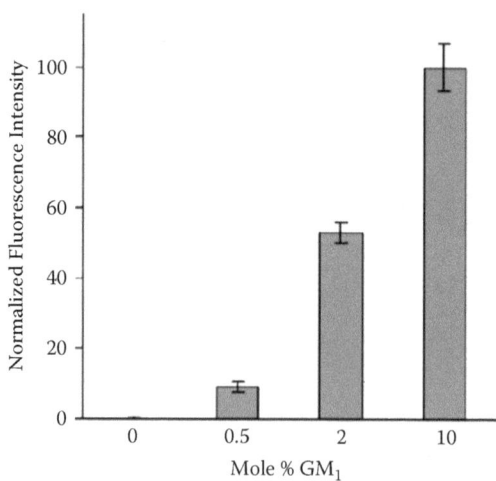

FIGURE 3.31   CTb binding in GM_1 lipid bilayer. (From Joubert JR, Smith KA, Johnson E, et al. Stable, ligand-doped, poly(bis-SorbPC) lipid bilayer arrays for protein binding and detection. *ACS Appl. Mater. Interfaces* 1(6): 1310–1315, 2009. With permission.)

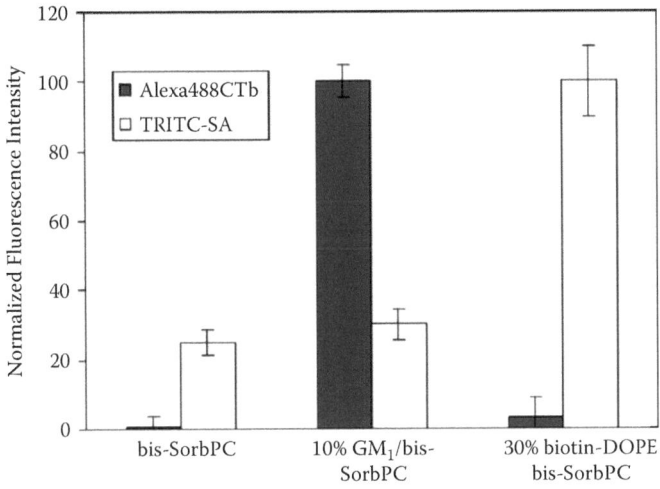

FIGURE 3.32 Detection of multiple analytes in GM_1-poly(bis-SorbPC).

Detection of multiple analytes was shown by the doping of GM_1 into lipid bilayer spots with the doping of a biotinylated dioleoyl glycerol-3-phosphoethanolamine (biotin-DOPE) lipid into other spots. Biotin-DOPE was detected with the binding of streptavidin-TRITC fluorescent dye, while binding of GM1-CTb Alexa 488 was observed as before (Figure 3.32). Reasonable levels of specificity were found although nonspecific binding of streptavidin dye was relatively high. Apparently, UV irradiation leads to a significant (90%) destruction of the biotin specific binding activity so that the normalized signal for nonspecific binding appears higher. Repeating this experiment with photo-stable binding pairs would have resolved the issue.

Finally, perhaps of greater importance, was the demonstration of the ability to reuse these arrays following 6 M urea denaturation and removal of the bound CTb protein. Regeneration was possible through at least three cycles with no loss in GM_1 ligand binding capacity.

SUMMARY

Both proteins and nucleic acids may be immobilized to a variety of solid supports. For high-density microarrays the glass slide is the preferred substrate because of its flatness and optical properties. A higher spot resolution is also possible on nonporous glass as opposed to the porous membrane primarily due to a reduction in diffusion at the surface-liquid interface. However, keep in mind that spot (droplet) diffusion can occur on most substrates by the action of surfactants and other wetting agents including proteins. Control of spot size and morphology is required in order to achieve reproducible and reliable results with microarrays.

Membranes such as nitrocellulose (NC) supported on glass may be more applicable for protein microarrays than glass substrates. The essential ingredient for protein is water. Protein hydration reduces the likelihood for surface denaturation.

Hydrophilic membranes allow protein to be adsorbed and maintained in its hydrated state. Hydrogel is also an important milieu for consideration, especially for proteins because of the gel's ability to retain a high water content. Such 3D surfaces (hydrogels, membranes, other porous substrates) also may provide greater probe density allowing for increases in sensitivity and dynamic range over 2D surfaces such as planar glass substrates.

Plastic substrates are being utilized for both nucleic acid and protein microarrays. In particular, plastic microarrays have been introduced for high-throughput applications based upon the microtiter plate that is a standard automations platform.

The nucleic acid microarray is now a well-established tool. The attachment of oligonucleotides and cDNA to surfaces has been studied in some detail. Linkers and extended spacer-arms allow nucleic acid probes additional degrees of freedom to interact with target molecules in the bulk solution, thereby improving hybridization efficiency and target detection. Linkers can also reduce unwanted interactions of the probe with the surface. However, complex linking systems such as dendrimers that can potentially increase the probe density as well as block nonspecific adsorption are not always the optimal choice. Probe density is an important element in the design of a microarray, whether for proteins or nucleic acids. For nucleic acids we learned that higher probe densities do not necessarily lead to increased hybridization efficiency. Rather there is an optimal probe surface density (distribution) that needs to be determined.

Substrates for the creation of protein microarrays were initially selected from those used for DNA arrays (e.g., the poly-L-lysine glass slide). As first demonstrations these substrates proved to be sufficient for antibody microarray studies. However, we have discovered that not all proteins will behave well or similarly on a particular substrate material. New solid-phases applicable for the immobilization of biomolecules (DNA, protein, peptide, glycan, etc.) need to be considered.

REFERENCES

Agard NJ, Prescher JA, Bertozzi CR. A strain-promoted [3 + 2] azide-alkyne cycloaddition for covalent modification of biomolecules in living systems. *J. Am. Chem. Soc.* 126: 15046–15047, 2004.

Anderson MLM, Young BD. Chapter 4: Quantitative filter hybridization in *Nucleic Acid Hybridization—A Practical Approach*, Hames BS, Higgis SJ, eds., IRL Press, Oxford, 1985.

Angenendt P, Glokler J, Murphy D, Lehrach H, Cahill DJ. Toward optimized antibody microarrays: A comparison of current microarray support materials. *Analyt. Biochem.* 309: 253–260, 2002.

Angenendt P, Glokler J, Sobek J, Lehrach H, Cahill DJ. Next generation of protein microarray support materials: Evaluation for protein and antibody microarray applications. *J. Chromatography A* 1009: 97–104, 2003.

Balboni I, Limb C, Tenenbaum JD, Utz PJ. Evaluation of microarray surfaces and arraying parameters for autoantibody profiling. *Proteomics* 8(17): 3443–3449, 2008.

Beattie WG, Meng L, Turner SL, Varma RS, Dao DD, Beattie KL. Hybridization of DNA targets to glass-tethered oligonucleotide probes. *Mol. Biotechnol.* 4: 213–225, 1995.

Beier M, Hoheisel JD. Versatile derivatization of solid support media for covalent bonding on DNA-microchips. *Nucleic Acid Res.* 27(9): 1970–1977, 1999.

Belosludtsev Y, Iverson B, Lemeshko S, et al. DNA microarrays based on noncovalent oligonucleotide attachment and hybridization in two dimensions. *Anal. Biochem.* 292: 250–256, 2001.

Benters R, Niemeyer CM, Drutschmann D, Blohm D, Wohrle D. DNA microarrays with PAMAM dendritic linker systems. *Nucleic Acid Res.* 30(2) e10: 1–7, 2002.

Bentley DR, Todd C, Collins J, et al. The development and application of automated gridding for efficient screening of yeast and bacterial ordered libraries. *Genomics* 12: 534–541, 1992.

Call DR, Chandler DP, Brockman F. Fabrication of DNA microarrays using unmodified oligonucleotide probes. *BioTechniques* 30(2): 368–379, 2001.

Canalle LA, Vong TH, Adams PHHM, et al. Clickable enzyme-linked immunosorbent assay. *Biomacromolecules* 12: 3692–3697, 2011.

Dattagupta N, Rae PMM, Huguenel ED, et al. Rapid identification of microorganisms by nucleic acid hybridization after labeling the test sample. *Anal. Biochem.* 177: 85–89, 1989.

Delehanty JB, Ligler FS. A microarray immunoassay for simultaneous detection of proteins and bacteria. *Anal. Chem.* 74: 5681–5687, 2002.

Drmanac S, Drmanac R. Processing of cDNA and genomic kilobase-size clones for massive screening, mapping and sequencing by hybridization. *BioTechniques* 17(2): 328–336, 1994.

Ekins RP. Ligand assays: From electrophoresis to miniaturized microarrays. *Clin. Chem.* 44(9): 2015–2030, 1998.

Fodor SPA, Read JL, Pirrung MC, Stryer L, Lu AT, Solas D. Light-directed, spatially addressable parallel chemical synthesis. *Science* 251: 767–773, 1991.

Guo Z, Guilfoyle RA, Thiel AJ, Wang R, Smith LM. Direct fluorescence analysis of genetic polymorphisms by hybridization with oligonucleotide arrays on glass supports. *Nucleic Acid Res.* 22(24): 5456–5465, 1994.

Guschin D, Yershov G, Zaslavsky A, et al. Manual manufacturing of oligonucleotide, DNA and protein microchips. *Anal. Biochem.* 250: 203–211, 1997.

Haab BB, Dunham MJ, Brown PO. Protein microarrays for highly parallel detection and quantitation of specific proteins and antibodies in complex solutions. *Genome Biol.* 2(2): research0004.1–0004.13, 2001.

Hamaguchi Y, Aso Y, Shimada H, Mitsuhashi M. Direct reverse transcription-PCR on oligo (dT)-immobilized polypropylene microplates after capturing total mRNA from crude cell lysates. *Clin. Chem.* 44(11): 2256–2263, 1998.

Hornberg JJ, de Hass RR, Dekker H, Lankelma J. Analysis of multiple gene expression array experiments after repetitive hybridizations on nylon membranes. *BioTechniques* 33(1): 108–117, 2002.

Huang R-P. Detection of multiple proteins in an antibody-based protein microarray system. *J. Immunol. Methods* 255: 1–13, 2001.

Joos B, Kuster H, Cone R. Covalent attachment of hybridizable oligonucleotides to glass supports. *Anal. Biochem.* 247: 96–101, 1997.

Joos TO, Schrenk M, Hopfl P, et al. A microarray enzyme-linked immunosorbent assay for autoimmune diagnostics. *Electrophoresis* 21: 2641–2650, 2000.

Joubert JR, Smith KA, Johnson E, et al. Stable, ligand-doped, poly(bis-SorbPC) lipid bilayer arrays for protein binding and detection. *ACS Appl. Mater. Interfaces* 1(6): 1310–1315, 2009.

Khrapko KR, Lysov YP, Khorlin AA, et al. A method for DNA sequencing by hybridization with oligonucleotide matrix. *DNA Seq.* 1: 375–388, 1991.

Khrapko KR, Lysov YP, Khorlyn AA, Shick VV, Florentiev VL, Mirzabekov AD. An oligonucleotide hybridization approach to DNA sequencing. *FEBS Lett.* 256: 118–122, 1989.

Krishnamurthy VR, Wilson JT, Cui W, et al. Chemoselective immobilization of peptides on abiotic and cell surfaces at controlled densities. *Langmuir* 26(11): 7675–7678, 2010.

Lane S, Birse C, Zhou S, Matson R, Liu H. DNA array studies demonstrate convergent regulation of virulence factors cph1, cph2 and efg1 in *Candida albicans*. *J. Biol. Chem.* 276(52): 48988–48996, 2001.

Lee PH, Sawan SP, Modrusan Z, Arnold LJ Jr., Reynolds MA. An efficient binding chemistry for glass polynucleotide microarrays. *Bioconjugate Chem.* 13: 97–103, 2002.

Lueking A, Horn M, Eickhoff H, Bussow K, Lehrach H, Walter G. Protein microarrays for gene expression and antibody screening. *Anal. Biochem.* 270: 103–111, 1999.

Liu Y, Rauch CB. DNA probe attachment on plastic surfaces and microfluidic hybridization array channel devices with sample oscillation. *Anal. Biochem.* 317: 76–84, 2003.

MacBeath G, Schreiber SL. Printing proteins as microarrays for high-throughput function determination. *Science* 289: 1760–1763, 2000.

Matson RS, Rampal JB, Pentoney SL, Anderson PD, Coassin P. Biopolymer synthesis on polypropylene supports: Oligonucleotide arrays. *Anal. Biochem.* 224: 110–116, 1995.

Matson RS. Chapter 5: Solid-phase reagents in *Immunoassays: A Practical Approach*, Gosling JP, ed., Oxford University Press, 2000.

Matson RS, Milton RC, Cress, MC, Rampal JB. Microarray-based cytokine immunosorbent assay. *Oak Ridge Conference*, Poster No. 20, 2001.

Moody MD, Van Arsdell SW, Murphy KP, Orencole SF, Burns C. Array-based ELISAs for high-throughput analysis of human cytokines. *BioTechniques* 31(1): 1–7, 2001.

Nizetic D, Zehetner G, Monaco AP, Gellen L, Young BD, Lehrach H. Construction, arraying, and high-density screening of large insert libraries of human chromosomes X and 21: Their potential use as reference libraries. *PNAS USA* 88: 3233–3237, 1991.

Pugia MJ, Lott JA, Wallace JF, Cast TK, Bierbaum LD. Assay of creatinine using the peroxidase activity of copper-creatinine complexes. *Clin. Biochem.* 33(1): 63–70, 2000.

Rasmussen SR, Larsen MR, Rasmussen SE. Covalent immobilization of DNA onto polystyrene microwells: The molecules are only bound at the 5′ end. *Anal. Biochem.* 198: 138–142, 1991.

Saiki RK, Walsh PS, Levenson CH, Erlich HA. Genetic analysis of amplified DNA immobilized sequence-specific oligonucleotide probes. *PNAS USA* 86: 6230–6234, 1989.

Schena M, Shalon D, Davis RW, Brown PO. Quantitative monitoring of gene expression patterns with a complementary DNA microarray. *Science* 270: 467–470, 1995.

Seong S-Y. Microimmunoassay using a protein chip: Optimizing conditions for protein immobilization. *Clin. Diagn. Lab. Immunol.* 9(4): 927–930, 2002.

Seurynck-Servoss SL, White AM, Baird CL, Rodland KD, Zangar RC. Evaluation of surface chemistries for antibody microarrays. *Anal Biochem.* 371(1): 105–115, 2007.

Shchepinov MS, Case-Green SC, Southern EM. Steric factors influencing hybridization of nucleic acids to oligonucleotide arrays. *Nucleic Acid Res.* 25(6): 1155–1161, 1997.

Silzel JW, Cerecek B, Dodson C, Tsong T, Obremski RJ. Mass-sensing, multianalyte microarray immunoassay with imaging detection. *Clin. Chem.* 44(9): 2036–2043, 1998.

Southern EM. Detection of specific sequences among DNA fragments separated by gel electrophoresis. *J. Mol Biol.* 98: 503–517, 1975.

Southern EM, Maskos U, Elder JK. Analyzing and comparing nucleic acid sequences by hybridization to arrays of oligonucleotides: Evaluation using experimental models. *Genomics* 13: 1008–1017, 1992.

Tamarit-López J, Morais S, Puchades R, Ángel Maquieira A. Oxygen plasma treated interactive polycarbonate DNA microarraying platform. *Bioconjugate Chem.* 22: 2573–2580, 2011.

Taylor S, Smith S, Windle B, Guiseppi-Elie A. Impact of surface chemistry and blocking strategies on DNA microarrays. *Nucleic Acid Res.* 31(16) e87: 1–19, 2003.

Vasiliskov AV, Timofeev EN, Surzhikov SA, Drobyshev AL, Shick VV, Mirzabekov AD. Fabrication of microarray of gel-immobilized compounds on a chip by copolymerization. *BioTechniques* 27(3): 592–605, 1999.

Wiese R, Belosludtsev Y, Powdrill T, Thompson P, Hogan M. Simultaneous multianalyte ELISA performed on a microarray platform. *Clin. Chem.* 47(8): 1451–1457, 2001.

Zammatteo N, Jeanmart L, Hamels S, et al. Comparison between different strategies of covalent attachment of DNA to glass surfaces to build DNA microarrays. *Anal. Biochem.* 280: 143–150, 2000.

Zhang Y, Coyne MY, Will SG, Levenson CH, Kawasaki ES. Single-base mutational analysis of cancer and genetic diseases using membrane bound modified oligonucleotides. *Nucleic Acids Res.* 19(14): 3929–3933, 1991.

Zhu H, Snyder M. Protein chip technology. *Curr. Opin. Chem. Biol.* 7: 55–63, 2003.

Zhu H, Klemic JF, Chang S, et al. Analysis of yeast protein kinases using protein chips. *Nat. Genet.* 26: 283–289, 2000.

4 Arraying Processes

Out, damned spot! out, I say!—One: two: why, then, 'tis time to do't.

The Tragedy of Macbeth **by William Shakespeare,**
Lady Macbeth speaks

INTRODUCTION

The array is very simply a collection of small spots on a surface organized in a particular geometric pattern. All microarrays share this feature. What is placed on the surface differentiates the microarray's utility. We can create arrays of DNA, proteins, carbohydrates, small molecules, or even cells if we choose and then use these arrays to look at some aspect of biology on a global scale (e.g., examining the expression of the entire yeast genome at once). However, to do this in a quantitative manner we need to rely upon high-sensitivity labeling schemes and sophisticated detection systems in order to see the spots, and complicated algorithms to arrange the data in a meaningful way in order for us to make a conclusion about our experiment. All of this ultimately depends upon the equality of these little spots.

What do I mean by spot equality? Let's start here. First, ask yourself: What do I wish to accomplish with this microarray? Most likely you want to use the array to measure an event that will be informative of a specific biological response that is of interest. There are basically two levels of data that can be obtained. The first is a rather qualitative yes/no or +/– examination of the array image (i.e., did the spot light up or not?). The second level provides more quantitative information by the assignment of numerical values to the spot intensity. This allows us to answer the question: How bright is the spot? Normally, you would really like to know how bright is the spot relative to other spots in the array so that you can ascribe some notion concerning biological relationships.

This is where spot equality becomes important to you. If I am to compare signal intensities among spots, then I would like these spots to behave similarly to one another. The reality is that spot equality is difficult to achieve. At first glance of the microarray, it may be difficult to understand this kind of imprecision. After all, they all look to be of the same uniform size and shape. And, if we were able to rely upon the use of spot diameter as our principal quantitative measure, then we would find great comfort. Unfortunately, while spotted arrays can be produced with control of the spot diameter within 5 to 10% CV, this parameter alone has little bearing on the performance of the array. What lies within the spot must be controlled. Thus, spot morphology and the spatial arrangement of the biomolecule capture agents within the spot become important concerns. Ideally, we would like each spot to contain the same number of molecules properly oriented and in their native conformation to assure optimal and equivalent levels of binding to their cognate targets. That would be spot equality.

In the sections to follow we explore different approaches to producing microarrays that achieve spot equality.

CREATING SPOTTED MICROARRAYS

It is important to understand the context of the term *spotted* microarrays. Here we refer to the creation of arrays in which the spots are produced by contact printing methods or by the dispensing of droplets onto the surface. In this particular case, the biomolecules within these spots are presynthesized (or of a native form) prior to spotting. This excludes arrays produced by in situ means such as by the photolithographic or other related processes used to produce tethered biomolecules from monomers or precursors. For example, Affymetrix's arrays are produced by flooding a photo-deprotected region with active monomers, while Agilent arrays are produced by spotting active monomer on acid deprotected sites.

We will use the spotting of oligonucleotides here as a general example of the processes involved. The oligonucleotide array was originally introduced to accomplish sequencing by hybridization by means of a number of formats including both in situ and ex situ based tethering to a variety of substrates (see Chapter 3, "Supports and Surface Chemistries"). Before we attempt to discuss specific protocols for the spotting of oligonucleotides, it is worthwhile to understand something about the fundamental processes involved. There are four important factors to consider in the spotting of ex situ synthesized probes onto a surface:

1. Substrate
2. Probe (ink) composition
3. Environment
4. Printing mechanics

SUBSTRATES

The primary substrate for spotted arrays is the glass slide. The salient physical-chemical features of the microarray slide are optical clarity (including low fluorescence background), flatness, and coating uniformity. Although the standard microscope slide was originally used, most commercial manufacturers have replaced these with higher-grade glasses in order to reduce the number of irregularities in the surface features and improve optical quality. You should use slides specifically selected for microarray applications. These are available as ultra-cleaned (an important consideration) but untreated for those who wish to prepare their own surfaces or with a variety of precoated surface chemistries (e.g., lysine, aldehyde, or epoxide). The density of reactive groups and surface coating uniformity are difficult to control. So, if the slide lot-to-lot consistency is most important, consider using commercially available slides that are quality controlled.

Print Buffer

Perhaps the least understood factor in the process of microarraying is the print buffer composition (i.e., probe ink). This should not be too much of a surprise. After all, manufacturers of computer printers offer the consumer a multitude of different inks (closely guarded trade secrets) to be used with a particular printer for use on a specific kind of paper. In fact, it can be argued that the ink is perhaps the most important piece of the consumable product stream for this manufacturing sector. There are a number of factors that influence the ability to reliably print down oligonucleotide probes onto a surface such as glass. The most important ones are represented in Figure 4.1.

First, consider oligonucleotide structure. Unmodified nucleic acid probes may be viewed as negatively charged polymers (i.e., they are polyanions) under normal physiological conditions. In that state they readily adsorb to positively charged surfaces such as a glass slide coated with poly-L-lysine because this polycation would carry a net positive charge. Oligonucleotides are also able to bind other counter-ions such as metal ions present on, for example, printing pins. The purine and pyrimidine bases participate in hydrogen bonding and van der Waals interactions allowing for adsorption of nucleic acids onto a variety of materials. For example, oligonucleotides bind extremely well to polystyrene surfaces. Microtiter plates of molded polystyrene are often used for the storage of oligonucleotides or as source plates during the printing process.

The ink is composed of the probe, buffer, and most often a wetting agent that allows uniform deposition of the oligonucleotide to the substrate surface to control spot size and spot morphology. Other additives to the ink may be present to prevent or slow evaporation in an effort to control spot size. Fluorescent or other dyestuffs are sometimes included to monitor the printing efficiency. The ink can therefore represent a complex matrix for the probe. Selection of the ingredients must be undertaken with extreme care, keeping in mind that these components may interact with the various surfaces they come in contact with as well as the

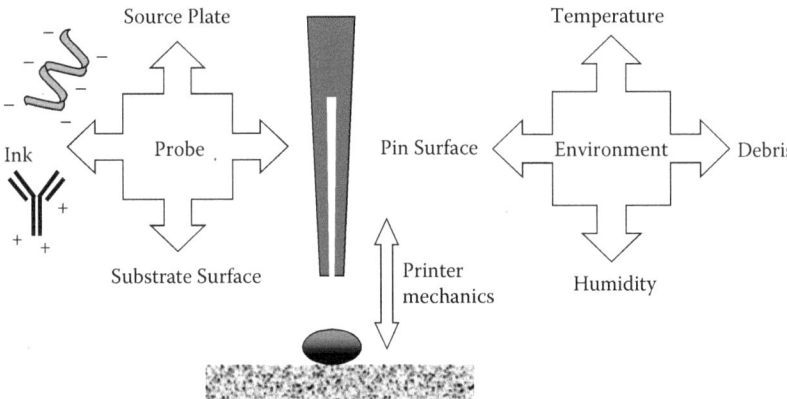

FIGURE 4.1 Factors influencing spotting.

oligonucleotide probe. One of the most common effects associated with improper selection of the printing ink is contribution to the background signal following hybridization, or in other cases ingredients may cause precipitation of the probe during the printing process. A detailed discussion of print buffer selection is provided in the protocols section.

THE PRINTING ENVIRONMENT

Environmental conditions are also very important. For both quill pin printing and noncontact dispensers (e.g., piezo or solenoid-based) it is highly recommended that all buffers and inks be filtered prior to use to reduce the level of dirt or other debris that could clog the system. During printing the temperature and humidity (i.e., relative humidity) should be held within a specific range (e.g., 20°C; relative humidity, 50 to 65%) to avoid print variation. Avoid wide swings in temperature and humidity during the print run. For instance, do not locate the printer next to an air-conditioning vent or where the room airflow is subject to frequent change. Most microarray printers are now housed in boxes that are set up to control humidity, and many are provided with high-efficiency particulate air (HEPA) filtration to remove particulates. However, a word of caution: Attempting to run the HEPA filter along with the humidifier may be counterproductive if the humid air is exhausted by the HEPA system. Check on the design of the environmental chamber to verify whether or not you should be running both during the print run. Substrates should be kept clean, preferably sealed and opened just prior to use. Gloves should be worn to avoid unwanted fingerprints or other contamination from the hands. If possible, the printing operation should take place in a clean room environment in which the printer is compartmentalized (e.g., class 10,000 clean room with the printer housed within a class 100 to 1000 HEPA filtered box). Finally, pins should be cleaned and inspected to be free of any obstruction to the tip prior to use. Manufacturers have specific recommendations on how best to clean their pins. Some vendors such as TeleChem International (www.arrayit.com) offer cleaning solutions and equipment (sonic bath, pin holders, etc.) for their pins and print-heads. In our laboratory we routinely inspect pins under a microscope before and after cleaning and keep a photographic image as a record.

The manner in which the spotting pin makes contact with the substrate surface is dependent not only upon the design of the pin, but also upon the design of the print-head and printer mechanism. There is now a variety of microarray printers commercially available as well as a number of different pin designs and print-head mechanisms. While printers may accept a variety of print-heads with appropriate adapter hardware, there is a very close match between pins and print-heads. Such physical tolerances make it very difficult to mismatch pins with print-heads and achieve good spot quality.

MICROARRAY PRINTING MECHANISMS

The printing mechanics are obviously different between dispense and pin printers. Moreover, there are also contact and noncontact pin modes (Figure 4.2). We can refer

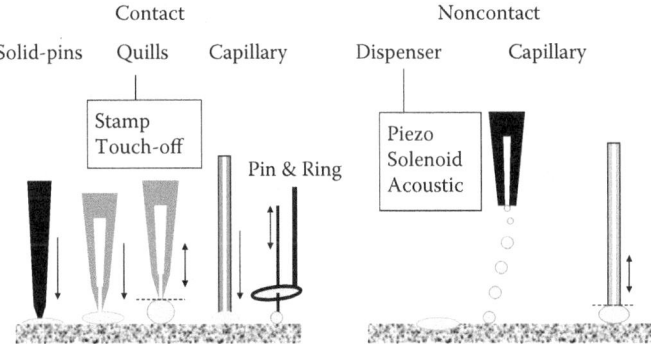

FIGURE 4.2 Microarray printing mechanisms.

to these as stamp and touch-off printing depending upon whether or not the pin actually touches the surface. The most common microarray printing technique employs a split pin (quill) in order to deposit the probe. The fundamental process is simple enough. Just fill the quill's capillary with the printing solution and deliver a very small droplet to the surface upon taping the tip. This is best accomplished using a computer-controlled x,y plotter with z-axis motion control. However, the "devil is in the details," and many have learned that the repetitive and precise printing of small droplets onto a glass slide can be difficult. Commercialization of the microarrayer has been very important for the expansion of microarray-based applications. Today, most commercial arrayers meet the resolution and repeatability criteria necessary for producing medium- to high-density microarrays. Where arrayers begin to differentiate are in such areas as cost, ease of use, deck capacity, and high-throughput robustness issues. Table 4.1 presents a summary of different types of arrayers and their throughput capabilities. Listed is a selection of instruments from bench-top models for low capacity users to higher-throughput printers enabling custom microarray manufacturing. Unfortunately, there has been little standardization in design of the pin and print-heads, making it difficult to directly compare the performance between the products (Figure 4.3). Also, as we discussed in Chapter 2 there has been some consolidation in the industry, and few of the companies described are no longer in business such as Genometrix who was an early entry into the high-throughput custom microarray marketplace. Others including well-established companies like GeneMachines and Cartesian have survived as divisions under larger companies. Their products are still supported, and newer models are being introduced. More importantly, the table shows the many different kinds of technologies that can be used to produce microarrays. We discover later in this chapter a continued emergence of new printing technologies to further the adoption and advancement of the microarray field.

The original quill pin developed in Brown's lab at Stanford (see the "Print Tip Gallery" anthology of quill pins at the following Web site at http://cmgm.stanford.edu/pbrown/mguide/tips.html) was designed to strike the surface of the glass slide with enough force to eject a droplet from the capillary (see U.S. Patents 5,807,522 and 6,110,426). To do so required overcoming the surface tension forces at the

TABLE 4.1
Arrayer Types

Manufacturer	Trade Name	Print-Head	Pin/ Dispenser Types	Number of Pins/ Tips	Number of Slides on Deck	Approximate Print Rates[a] (spots/sec)
Affymetrix	Model 417	Ring and pin	Solid pin	4	20	1.7
Cartesian	ProSys	SynQuad	Solenoid tip noncontact	8	9	2.4
Cartesian	PixSys 7500	TeleChem	Quill	4	6	3
GeneXP	BioGridArrayer	TeleChem	Quill	8	15	5.2
Genetix	QArray2	HPLF	Quill	12	15	8.4
GeneMachines	OmniGrid 300	TeleChem	Quill	48	300	83.6
Genospectra	N/A	Custom	Fiber optic capillary bundle noncontact	10,000	3000	116
IMTEK	TopSpot/P	Nozzel	Noncontact	96	300	120
Genometrix	N/A	Custom	Capillary tube bundle contact	256	15	427

[a] Print rates based upon start to finish runs estimated from vendor specifications or supplied data with a certain number of slides.

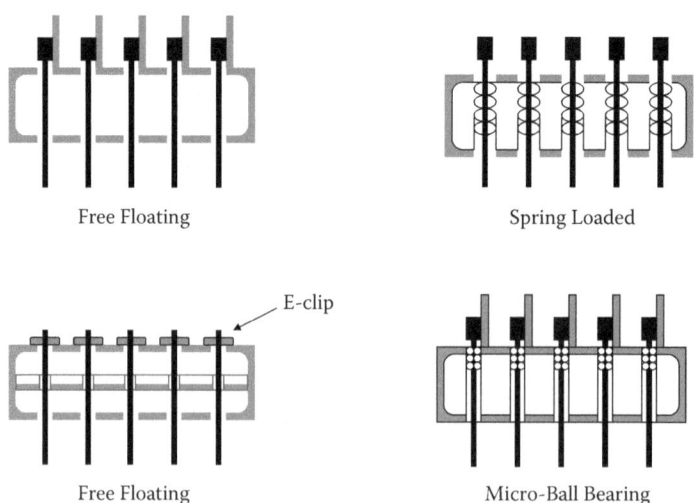

FIGURE 4.3   Print-head designs.

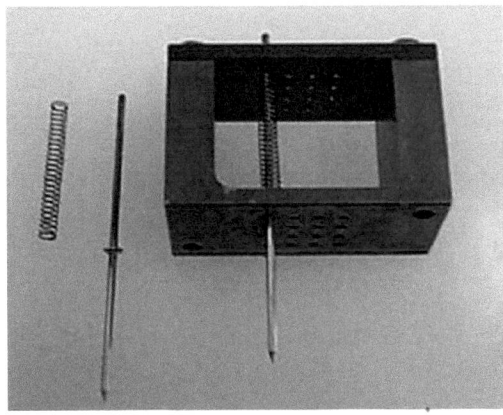

FIGURE 4.4   Majer Precision print-head and MicroQuill® pins. (From Majer Precision Engineering, Inc., Tempe, AZ. With permission.)

meniscus. The Stanford quill was commercialized by Majer Precision (Tempe, AZ) who was responsible for the fabrication of some of the more successful pin designs. Majer Precision also offered a specifically designed holder in which the quills are spring loaded, assuring an even striking of the substrate surface and reducing the wear on the pin tips. The holder is produced from a stress-resistant aluminum alloy that maintains the necessary structural tolerances. It is nickel coated to guard against corrosion and permits a low coefficient of friction (embedded polytetrafluoroethylene [PTFE]) with the quill (Figure 4.4). Genetix, Inc. (United Kingdom) also designed a special holder to reduce friction. The quill shafts actually rode on opposing micro-ball bearings, thereby further reducing contact with the holder body (Figures 4.5a, 4.5b).

In contrast to this stamping action is that of the touch-off strategy employed with the TeleChem quills (see U.S. Patent 6,101,946 for design features). Unlike the pointed pin tips used for stamping, these pins are flat at the tip (Figure 4.6). The force associated with the acceleration of the pin toward the surface is used to break the meniscus, allowing a droplet to just make contact with the substrate. Relying upon the adhesive forces on the substrate to be greater than those on the pin allows the substrate to capture the droplet before the meniscus can reform. In this case, the pin does not strike the surface in order to propel the droplet but rather touches the droplet down onto the surface. However, this is a subtle difference between the two processes, and it is not uncommon for the novice to overdrive the TeleChem pin into the slide. The most likely outcome is a reduced lifetime for the pin and a loss in spot quality during printing.

Solid pins are also used for microarray printing. These types of pins have been adopted from gridding applications where they were primarily used to transfer colony plaques from agar or microtiter plates onto nylon membranes for cDNA-based gene expression analysis (Figure 4.7). Such blots were the forerunners of the slide-based microarray. While automated grid blotting may be used to create microarrays, the linear accuracy requirements for these robots are not as strict as those found on

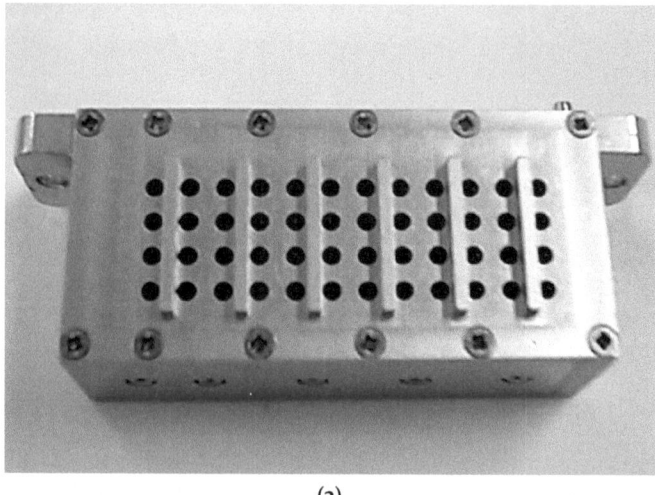

(a)

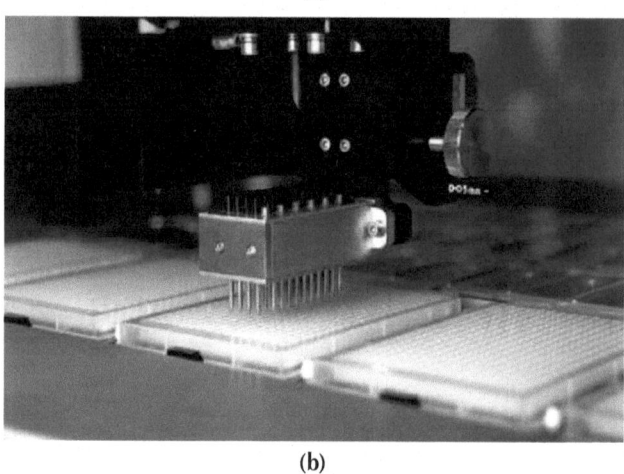

(b)

FIGURE 4.5   (a) Genetix high-precision low friction (HPLF) Print head. (Photo courtesy of Genetix Limited, United Kingdom.) (b) HPLF with quills. (Photo courtesy of Genetix Limited, United Kingdom.)

microarray printers. This limits their use to low-density arraying where the spot center-to-center spacing is on the order of 500 microns. However, advances in solid pin technology have progressed with new surface features and pin shapes that provide precision printing comparable to quill pins. The main advantages of these new generations of solid pins are uniformity in spot formation and morphology primarily due to a more controlled droplet pick up. Because they lack the capillary slit, they are much easier to clean and maintain. The main disadvantage is that each print requires a re-inking of the pin from the source plate. This uses up a lot more time than that of the quill pin which can perform multiple prints before refill. However, unlike the

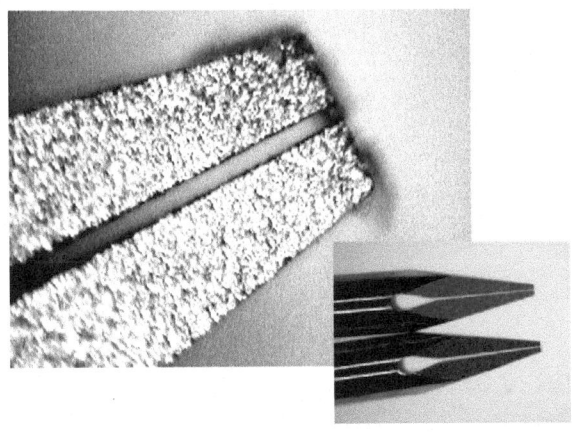

FIGURE 4.6 TeleChem's ChipMaker™ quill: close-up view. (From TeleChem International, Inc., Sunnyvale, CA.)

FIGURE 4.7 Biomek 2000 Laboratory Automation Workstation equipped with high-density replicating tool (HDRT) and transfer nails.

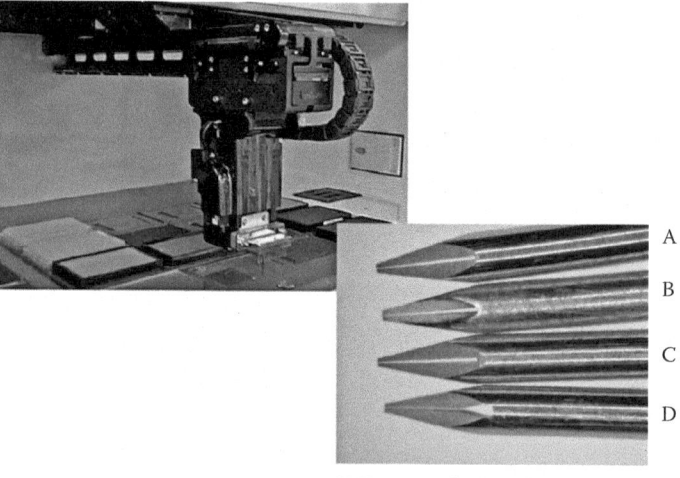

CMP-10 standard quill pins (A, B, C)
CMP-10B quill pin (D)

FIGURE 4.8   CMP™ quill pin types. (From TeleChem International, Inc., Sunnyvale, CA.)

quill pin, the solid pins do to not require pre-printing. Additional time can be saved by using an array of pins in the print-head.

A unique format that employed solid pins was that originally offered by Genetic MicroSystems (renamed the Affymetrix 417 arrayer following the acquisition of GMS by Affymetrix in February 2000). Commonly referred to as the *pin and ring* system, the microarrayer used a ring to capture a film of print buffer (much the same way as a bubble wand). Once loaded with the probe solution the print-head was positioned over the substrate and a small amount of liquid was delivered from the tip of a solid pin which pierced through the film ring. The film ring reformed after drawing the pin through it, and the process was repeated. Arrays produced with the pin and ring were of high quality. The main disadvantages of such a system were the need to use considerable print buffer in the ring and the inability to use a large number of such pins in the print-head, thereby reducing throughput. The pin and ring arrayer is no longer available from Affymetrix.

In addition to spotting using the split or quill pin is capillary printing. The most notable process was that described Genometrix, Inc. (no longer doing business [*nldb*]). In this case, open capillary tubes were bundled to create a single head that would strike the surface to simultaneously deposit up to 256 spots. Each capillary terminated into a separate well of a 384-well source plate. Pressure was applied across the plate forcing liquid into the capillary. Droplets were formed at the open end and released from the capillary following contact with the substrate. The printing process was very rapid allowing for the placement of tens of thousands of spots within a few minutes. A similar approach was introduced by GenoSpectra (*nldb*, Hayward, CA). Here, however, the capillary did not make direct contact with the surface but allowed the droplet to first touch the substrate and break away as the capillary retracts from the surface.

As for spotting precision, state-of-the-art microarrays such as the MicroGrid II (BioRobotics) can achieve 10 micron spot (*x,y*) resolution at a repeatability of about 1 micron. This allows for relatively high-density printing onto glass slides. However, control of the Z height travel and acceleration is also very important in producing good spot quality, especially in printing high-density arrays. The BioRobotics system employed a "Soft Touch" in which pins are slowed (Z deceleration) upon approach to the substrate to ensure a controlled spot size. Most contact microarrayers utilize pin heads that permit the pin to float. As the pin strikes the surface it can quickly rebound (float), thereby compensating for any surface irregularities encountered during printing. For example, the TeleChem pin head is composed of a precision-bored brass block and holds stainless steel pins at a close tolerance. The brass permits low-friction travel of the steel pins. The pin motion is controlled by the Z-travel of the print-head and gravity. On the other hand, the Majer Precision print-head employs spring action to control the pin movement. Each pin actually is seated in a spring coil, and this allows a similar "soft touch" landing and gentle but controlled rebound of each pin upon striking the surface. The Genetix pin head seats each pin between micro-ball bearings to achieve low-friction travel and precise rebound.

MICROARRAY PINS

A limited variety of quill spotting pins are available in the marketplace. These are essentially improvements on the original designs provided by Brown's group at Stanford (see above Web site). What has changed are the materials used to construct these pins. Surface finish and alloy mix appear to have some effect on printing performance. Thus, quills used to print DNA may not be optimal for the printing of proteins. Currently, TeleChem (Arrayit Corporation) remains the major manufacturer of quill pins under the trade names Chipmaker™ and Stealth. (For updated pin styles see www.arrayit.com.) These quills are constructed of stainless steel (Figure 4.9). Majer Precision offered quills under the trade name, MicroQuill™. These are produced from a very hard 17-4 stainless steel that increases the lifetime of the tip and is corrosion resistant (see Figure 4.4). Point Technologies offers tungsten-coated quill pins under the trademark, PT Accelerator™ Pins (Figure 4.9). Tungsten was selected because of its superior strength and hardness relative to stainless steel and titanium. The tungsten pins are electro-polished for optimal fluid transfer. Point sharpness can be controlled down to about 25 microns. The electrochemical pointing (ECP) process leaves the surface extremely smooth and clean by chemically removing any rough features and surface impurities from the metal (Figure 4.10). There is no grinding or other mechanical polishing involved. These processes can leave abrasions and metal fragments on the pin. Such highly reflective surfaces reduce fluid surface tension with the metal, making the pin essentially hydrophobic. For custom applications PT can apply a proprietary "Zonal" texturing process that creates different microsurfaces or zones on the pin (Figure 4.11). For example, the tip can be rendered hydrophilic while the pin shaft is made hydrophobic to create a surface tension barrier. These features are accomplished by a

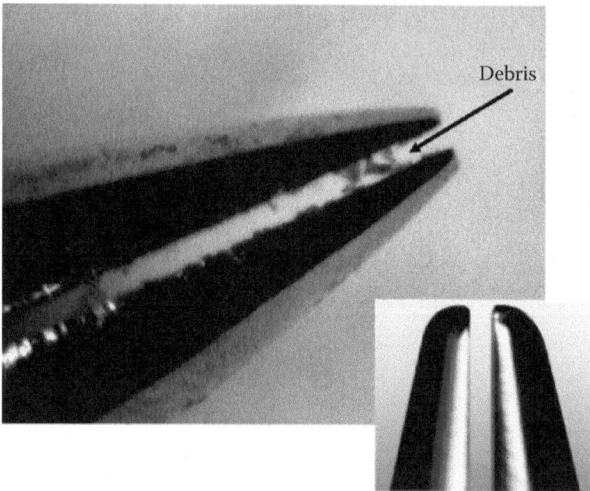

FIGURE 4.9   PT Accelerator® split pins. (From Point Technologies, Inc., Boulder, CO.)

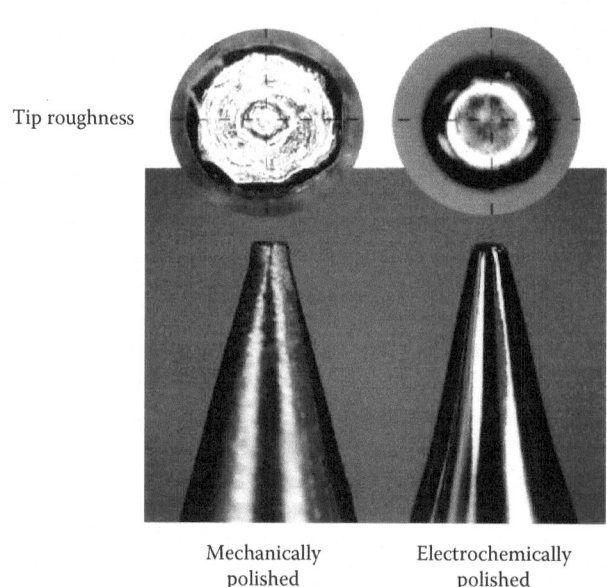

FIGURE 4.10   PTS Accelerator® solid pins. (From Point Technologies, Inc., Boulder, CO.)

combination of electrochemical etching and applied polymer coatings. The end result provides a means to control spot diameter and delivery without increasing the pin size. Unfortunately, Point Technologies was acquired in 2006 by American Medical Instruments Holdings, Inc., and the microarray pin products were discontinued shortly thereafter.

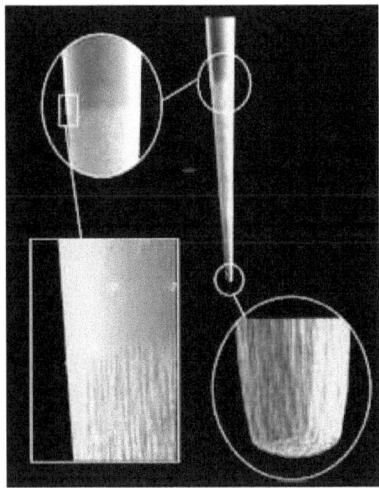

FIGURE 4.11   Zonal pin texturing. (Courtesy of Point Technologies, Inc., Boulder, CO.)

Parallel Synthesis Technologies (Santa Clara, CA) offers silicon micro-contact split pins produced by plasma etching of silicon wafers. This process produces essentially identical sets of pins. There is no need to match pins. Silicon pins are extremely flat allowing for a very close fit (5 to 10 microns) of the pin's shaft and a micromachined collimator that is housed in the print-head block. This takes any wobble out of the pin strike, producing a highly uniform, precise spotting. Because of the flatness of the pins requiring only a 5 micron clearance within the collimator, it is possible to employ print-heads capable of holding 96 to 1536 pins for high-density arraying. The silicon pins provide quantitative volume delivery such that pre-printing is unnecessary. The pins can be cleaned using a butane torch. While the hardness of silicon and reduced mass of the pins mean less tip wear, these pins are extremely fragile and must be handled with great care to prevent breakage.

OTHER APPROACHES

For the creation of lower-density microarrays, several groups have quite successfully employed robotic gridding devices to print onto membrane substrates. The arraying of cDNA (and proteins) onto membranes in this manner is well documented and still in practice. For example, Lane et al. (2001) used a standard Biomek®2000 equipped with a 384-solid pin HDRT (high-density replicating tool) used for gridding to create an array of 3456 cDNAs on a nylon membrane representing about 1000 *Candida albicans* genes in triplicate. A single run could produce up to 10 such blots using the available work surface. Others such as Macas et al. (1998) have successfully adapted the Biomek 2000 (Beckman Coulter, Inc.), a commonly used liquid handling robot, to prepare microarray slides using a specially constructed print-head and quill pins. Up to 28 microscope slides could be placed on the work surface for printing. The Biomek's HDRT tool head was adapted to accept microarray quill pins held between

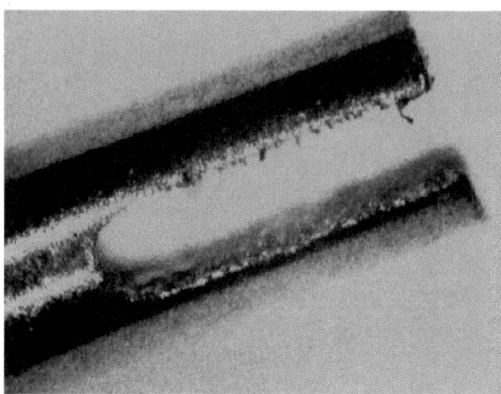

FIGURE 4.12   V&P Scientific slot pin, close-up view. (From V&P Scientific, Inc., San Diego, CA.)

two parallel plates with holes drilled on 9 mm centers to dip into a 96-well source plate. The quill pins were spring loaded similar in design to that of Majer Precision. Using Tool Command Language (Tcl) the Biomek was reprogrammed for use as a microarrayer. Up to about 3000 elements could be spotted onto each microscope slide (125 micron spot diameter at a 500 micron spacing) using this system. The positional accuracy for $n = 768$ repeats on a 500 micron center was measured to be at a standard deviation of $x \pm 52.6$ microns, $y \pm 63.7$ microns. While the Biomek 2000 was found to be adequate for low-density printing applications, the Biomek FX series robot offered higher precision and has been used to create higher-quality microarrays (R. Matson, unpublished). V&P Scientific (http://www.vp-scientific. com/pin_tools.htm) offers a range of adapters, print-heads, and slotted pins that can be used with the major commercially available liquid handling robots (Figure 4.12). These print-heads can also be transformed into manual gridding devices if only a few microarrays are needed or where the cost of a robotic system is not warranted.

George and co-workers (2001) from the Lawrence Berkeley National Laboratory introduced the use of ceramic capillaries for printing with claimed improvements in the consistency of spot morphology. The researchers compared the distribution of spot diameters created by the ceramic tips (K & S MicroSwiss) relative to that of a stainless steel quill pin (MicroQuill 2000, Majer Precision, Inc.). The ceramic capillaries (Figure 4.13) exhibited a more uniform distribution of spot diameters than those of the quill (Figure 4.14). The mean spot diameters closely matched the size of the ceramic tip (e.g., a 50 micron tip produced a 56 micron spot, while a 132 micron tip deposited a 130 micron spot). The 50 micron tip was used to print a 12 × 12 array of labeled cDNA on 100 micron center-to-center spacing. While the stainless steel quills performed well, there was apparent tip wear following 100,000 stampings. The ceramic tips are reported to perform with a lifetime beyond 225,000 deposits. However, ceramic tips require greater care in leveling the print-head to the substrate surface in order to prevent damage. Also, because they are ceramic, the drying times may need to be extended.

FIGURE 4.13 Ceramic capillary pin in holder. (From George RA, Woolley JP, Spellman PT. Ceramic capillaries for use in microarray fabrication. *Genome Res.* 11: 1780–1783, 2001. With permission.)

The "trench pen" was created by Stephen Quake (now at Stanford University) for high-density microarray printing (Reese et al., 2003). Using optical lithography to etch away photolithographic resists on stainless steel foils, trench pens were produced with a rectangular geometry of 6 microns in depth, 30 microns wide, with 30 micron sidewalls at the tip. To add structural support the features anterior to the tip were expanded out such that the width of the sidewalls increased to 120 microns with a trench width set at 90 microns. Because the trench pen was designed with considerable flexure, it serves as a miniature shock absorber and does not require any external fixturing (such as springs or ball bearings) in order to reduce shock from striking the substrate. In order to achieve capillary action, the trench was coated with hydrophilic polyurethane. Rectangular spots are produced that range 10 to 30 microns in width by 20 to 140 microns in length depending upon the design. In one example, an array of more than 2500 dye spots was printed onto a 3.2 mm × 3.2 mm square corresponding to ~25,000 spots per cubic centimeter. Such densities are about twofold

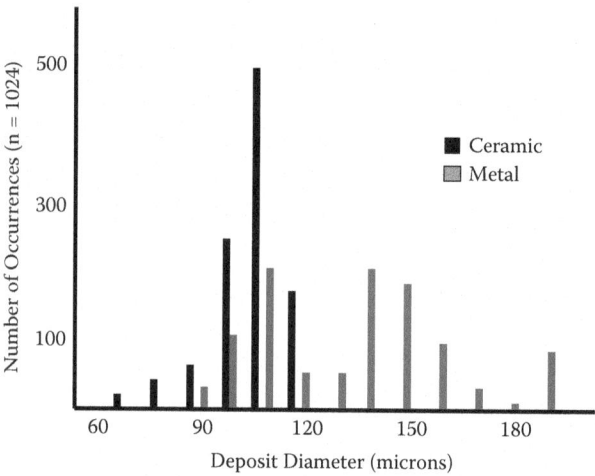

FIGURE 4.14 Comparison between spot diameter distributions for metal and ceramic capillary pins. (From George RA, Woolley JP, Spellman PT. Ceramic capillaries for use in microarray fabrication. *Genome Res.* 11: 1780–1783, 2001. With permission.)

higher than other state-of-the-art microarray fabrication processes. According to the authors, an individual pen prints 5 to 20 spots per inking. Undoubtedly, higher throughputs will be required for consideration as a viable printing device. In terms of performance relative to a conventional quill from Majer Precision, the trench pen showed spot size variance in the range of 10 to 20% CV, while the quill pin exhibited a spot CV ~14%. Hybridization signal variation from these trench pen printings ($n = 3$ arrays, 72 spots per array) ranged from 1.6 to 4.3% CV, while the quill produced ~3% CV in signal. Thus, the trench pen performed at similar levels of print quality to those of a more conventional quill pin.

There are, of course, noncontact printing devices useful for the construction of microarrays (see Figure 4.2). These are micro-dispensers that eject droplets by several different mechanisms (e.g., solenoid, piezoelectric, heated jet, acoustical wave). Perhaps the best-known commercial dispensers are the syringe driven solenoid pump (e.g., Cartesian; BioDot) and piezo systems (e.g., Scenion AG, Germany).

IMTEK (Institute of Microsystem Technology, University of Freiburg, Germany) developed a microfluidic dispenser device for the simultaneous printing from 24, 96, or 384 channels (Daub, 2002). The devices were composed of a reformat plate of microfluidic channels that terminate into nozzles (Figure 4.15). The plate reservoir is first filled with printing ink and then loaded automatically onto a print-head. The print-head piezo actuator forces the fluid from the nozzles. Droplets are formed and jetted down onto the surface (Figure 4.16). For example, 209 micron diameter spots can be printed down onto a glass slide reproducibly at CV <2% spot diameter or CV <5% signal intensity. The TopSpot E printer can produce 200 slides per hour at a 96 element density (5.3 spots/sec), while the TopSpot P unit can produce 300 slides per hour at a 1440 element density (120 spots/sec) (Figure 4.17). BioFluidix GmbH (www.biofluidix.com), a commercial spin-off of IMTEK, continues to offer these products.

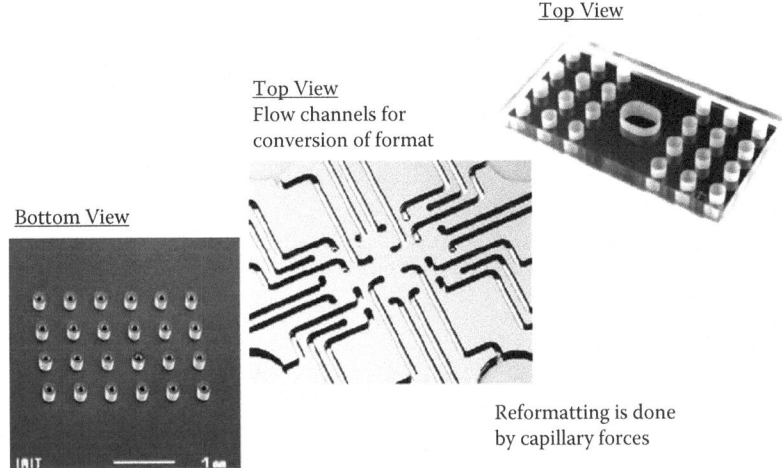

FIGURE 4.15   TopSpot reformatting. (Courtesy of Martina Daub, IMTEK, Germany.)

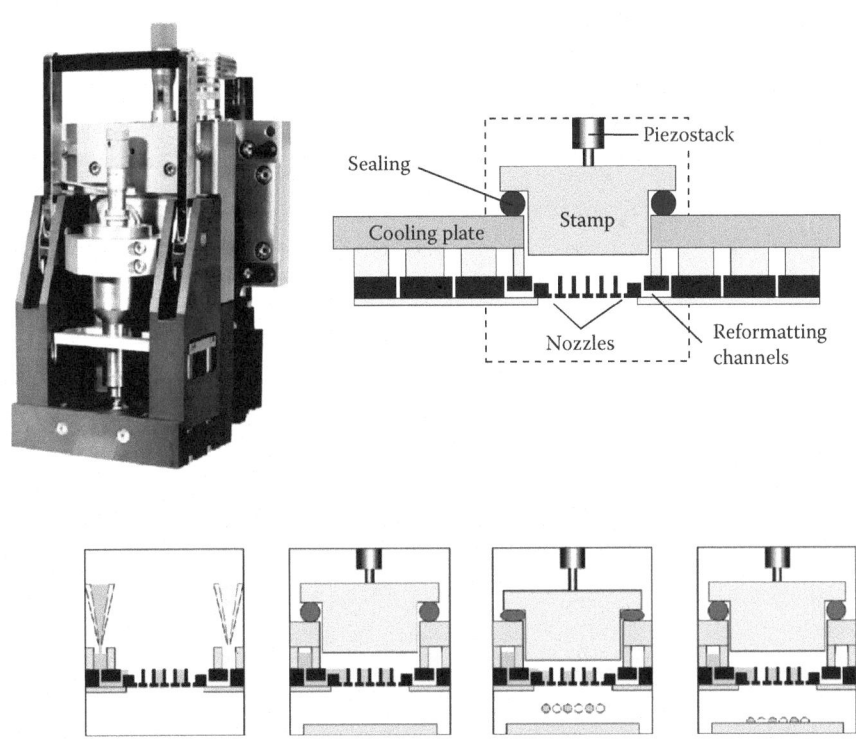

FIGURE 4.16   TopSpot actuator mechanism. (Courtesy of Martina Daub, IMTEK, Germany.)

TopSpot/E   TopSpot/M   TopSpot/P

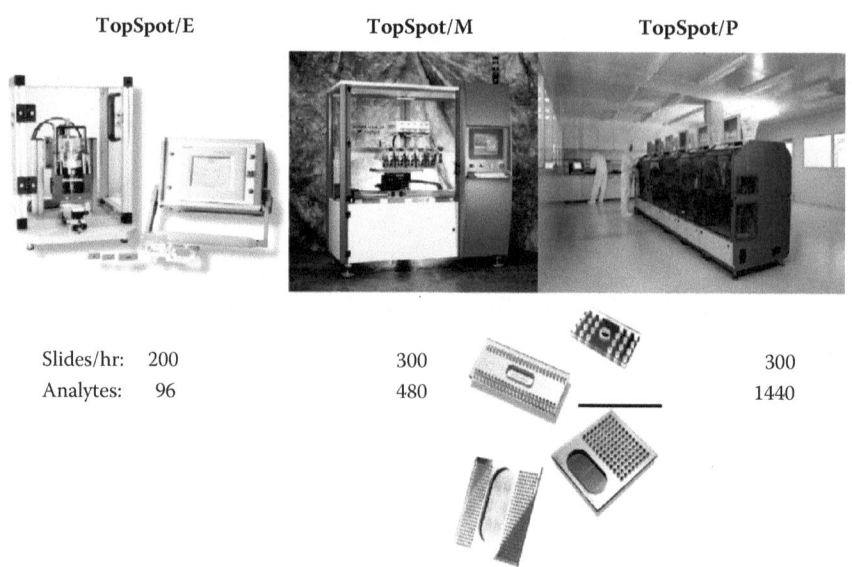

| Slides/hr: | 200 | 300 | 300 |
| Analytes: | 96 | 480 | 1440 |

FIGURE 4.17   TopSpot platform scalability. (Courtesy of Martina Daub, IMTEK, Germany.)

Another approach is using a modified commercial ink-jet printer to dispense bio-logical reagents. Over a decade ago we took an HP DeskJet printer, emptied the ink out of the print cartridge, and replaced it with either DNA or protein solutions (R. Matson, unpublished; Silzel et al., 1998). We were able to print arrays down onto either membranes or activated films without much trouble. The problem we encountered was primarily a high rate of missing spots (drop-out). Moreover, in order to create larger array formats a more versatile multi-head printer would have been required along with a low-volume ink reservoir. Stimpson et al. (1998) transformed an Apple StyleWriter II (Apple Computer) for the thermal ink-jet printing of oligonucleotides. They too disassembled an ink cartridge and replaced the ink with oligonucleotides dissolved in phosphate buffered saline (PBS) containing 7% isopropanol for printing onto membranes. Okamato et al. (2000) at the Canon Research Center, Kanagawa, Japan, reported on the fabrication of glass slide microarrays using a modified Canon Bubble Jet printer. As they point out, the key to printing biologics is to avoid dena-turation by heat or shearing stress forces. At least for DNA an ink formulation was identified that allowed the printing of 10 nt to 300 bp nucleic acids at suitable concen-tration for optimal hybridization performance on the microarray. The ink was made up of a combination of glycerin, urea, thiodiglycerol for wetting, and acetylenol in order to control viscosity. In one instance, a 5′-thiolated 18mer oligonucleotide (the substrate was activated with maleimide groups) was prepared in this ink at 8 μM and transferred onto the print-head (BC-62, Canon). The printer (BJC-700J, Canon) permitted the simultaneous dispensing of six inks (24 pL/droplet) onto the glass sub-strate. Approximately 70 micron diameter spots were printed in an 8 × 8 pattern at a resolution of 100 × 100 dpi. That corresponds to roughly four spots per millimeter or 70 micron spots spaced at ~180 microns.

In order to print down 64 different probes on the substrate, two BC-62 print-heads were aligned. This allowed the simultaneous firing of up to 12 jets and permitted the construction of a p53 gene mutation array. Both point mutations and SNP analysis were demonstrated using this array. However, one reported drawback to Bubble Jet printing was the need to change out print solutions after each dispense to create the 64 element microarray. Nevertheless, with disposable cartridges this is not a difficult task provided the operation is relatively fast for switching out cartridges and there are no alignment issues.

PRINTER PERFORMANCE

The printing of microarrays requires careful attention and vigilance. It is highly recommended that the performance features of the arrayer be well understood. This is best accomplished by designing experiments that will measure the robustness of the arrayer (i.e., at what point does the print quality begin to deteriorate). These experiments should be able to differentiate between mechanical or software issues and those related to pin performance. Arrayer performance is commonly measured in terms of its positional accuracy or resolution and repeatability. For example, a typical manufacturer's specification may state a positional resolution (x,y) at 10 microns with a repeatability of 1 micron (x,y). What this means is that the arrayer can print down a spot within 10 microns of the desired position and return to within 1 micron of that x,y coordinate most often. For printing down of, for example, 100 micron spots, such levels of resolution are easily met. However, how does this arrayer perform after printing 10,000 spots on 100 slides? To answer such a question one must obtain data across the print run and measure variance, typically in the form of intra- and inter-spot CV (coefficient of variation). Is the last slide printed as well as the first slide printed? How many missing spots are there? Do spots begin to merge or do the diameters increase or decrease across the print run? Are there systematic variations? These issues are not easily resolved, and one must be very careful not to confuse the arrayer's robustness with that of the pin's performance. For instance, a loss in spot quality or the disappearance of spots may be attributed to a clogged quill pin, while randomly missing spots could be due to a failure in either the pin or the arrayer mechanism. For example, if the pin failed to float, it might not touch down on the slide surface or pick up enough print buffer from the source plate. Or if the z travel of the arrayer became erratic the pin may not completely touch down, failing to deposit the droplet. However, it would be most difficult to imagine adjacent spots suddenly appearing within 10 microns of each other from pins set in a print-head on 4.5 mm centers. Most likely this would be a mechanical or software failure. In any respect, it is a good idea to understand the limitations of both the arrayer and the pin performance prior to proceeding with the production of the microarrays to be used in your studies.

PIN PERFORMANCE

While the arrayer's controller mechanism and the geometry of the print-head determine the spacing between spots (often expressed as the center-to-center distance or pitch), it is primarily the physical characteristics of the pin that will determine the spot size. That is, we can rate pins based upon the delivery of a particular volume (e.g., 1 nL) corresponding to a spot size range (e.g., 120 to 130 micron diameter). There are certain limitations to printing with quill pins. Aside from the common problems with clogging of the capillary or damage to the tip, one must appreciate that these instruments are individually machined to a specific tolerance (i.e., no two pins are a perfect match in performance). For example, consider the printing profiles for two pins from the same vendor as shown in Figure 4.18. In this experiment, the two pins A and B were loaded into the print-head, dipped once into print buffer (containing a fluorescent dye), and the number of printed spots each could deliver from the single inking counted. Spot intensity per unit diameter was used as a measure of pin delivery. The first thing that you will notice is that the initial spotting pattern (Zone 1) is rather erratic. Zone 1 is usually reserved for pre-printing in which a certain number of spottings are performed on a substrate prior to initiating the construction of the actual array. This process provides a means to remove any excess ink from pins (see Pin A) or to permit clearing of the capillary or tip (see Pin B) of any obstruction such as salt buildup or air bubbles or simply to fully wet out the tip for optimal delivery. Pre-printing is generally performed elsewhere such as on a sacrificial slide, and then printing is resumed onto a new slide at a different location on the arrayer's deck. Each pin should be assessed for the number of pre-prints required. In this particular example, the pins required about 150 to 200 pre-prints before achieving uniform delivery (Zone 2). Notice that both pins within Zone 2 are now performing at about the same level that would be ideal for producing several microarrays. Depending upon the number of replicates on each microarray, Zone 2 spotting (150 dispenses) could be used for construction of about 50 slides ($n = 3$ replicate spots per slide) per inking. Beyond Zone 2 is the post-print stage (Zone 3). This is essentially where the pin runs out of ink. This is evident more so for Pin B, while Pin A appears to deliver far beyond that limit. However, because most likely more than a single pin would be used in printing, the number of spots per inking would be necessarily dictated by the lowest performing pin.

Obviously, in the above example we were using higher loading capacity pins. In addition, we were printing on a hydrophobic plastic substrate. So while the profile shown in Figure 4.18 would still apply for the more hydrophilic glass slide, we would also expect that the number of pre-prints could be greatly reduced as explained below.

The number of pre-prints that a particular pin requires is going to depend upon several parameters. First, if the substrate is hydrophilic it may allow for increased wetting of the surface, thereby drawing out more liquid from the capillary but more importantly from the outside of the pin. The net effect would be a reduction in the number of pre-prints. Depending upon the buffer this might also reduce the effective capacity of the quill. In contrast, if the substrate were rather hydrophobic, there would be a tendency to deposit less fluid by wicking action. This may also reduce the number of pre-prints because the deposited spots would be smaller and more uniform at an earlier stage in the printing process. Less fluid deposited would also increase the number of spottings possible under these conditions. Albeit, it is an empirical process, one in which the pin, buffer, and substrate properties must be matched to produce the desired microarray (see Figure 4.1).

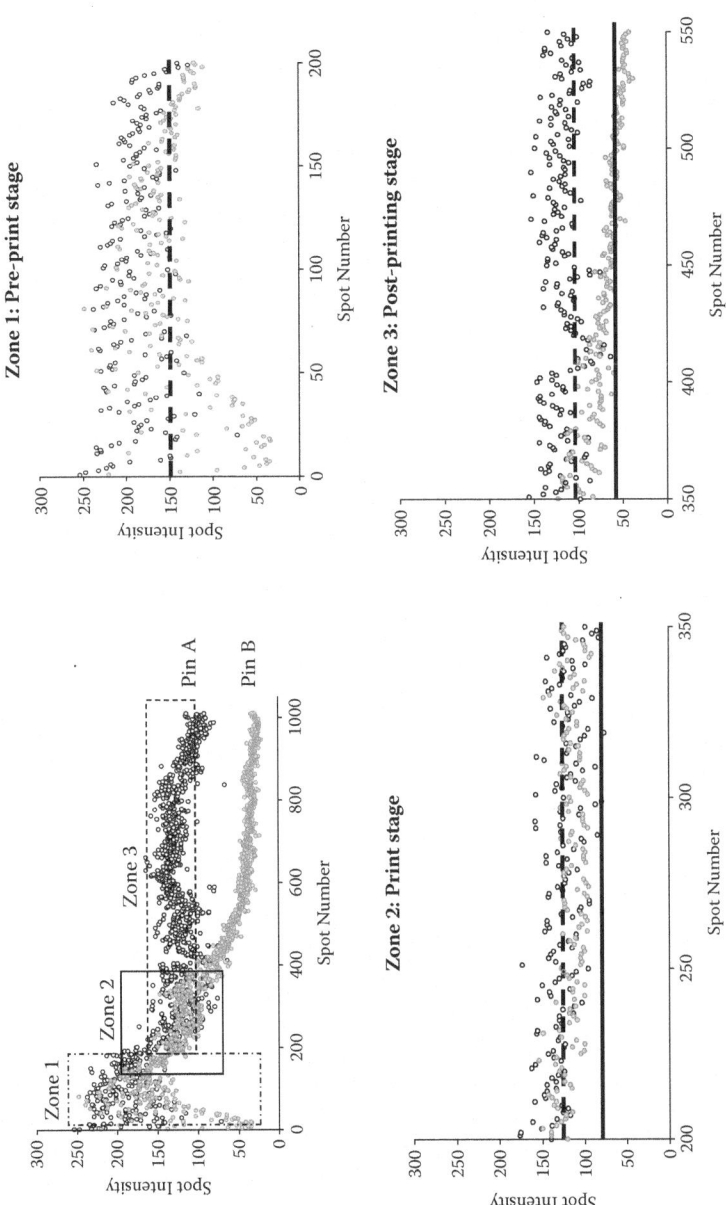

FIGURE 4.18 Quill pin printing profiles.

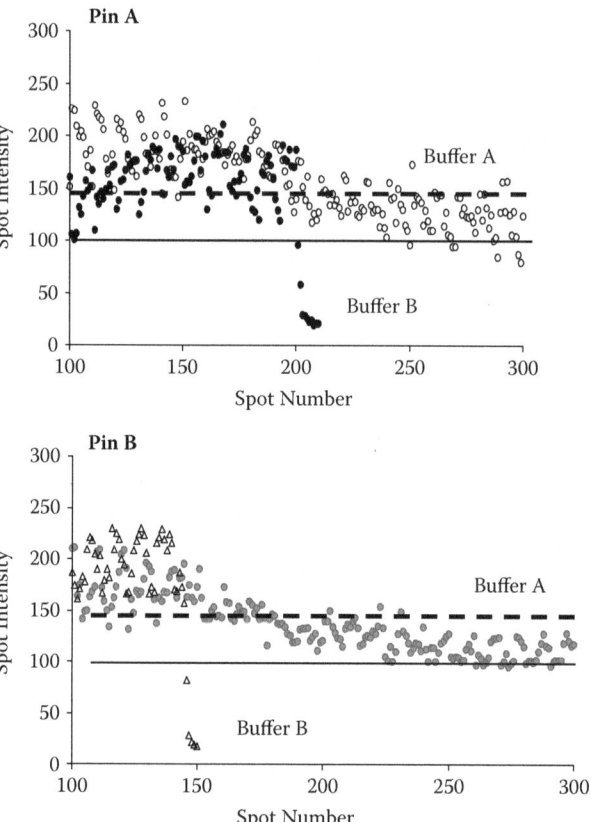

FIGURE 4.19   Buffer effect on pin performance.

One cannot overemphasize the importance of the buffer composition. A case in point is provided in Figure 4.19 in which our two pins (A and B) previously filled with Buffer A are now filled with Buffer B. The initial printing zone characteristics of these pins remain essentially the same, but the number of spots produced by either pin is dramatically curtailed. How could this happen? In this case, Buffer B was composed of a rather volatile component while Buffer A was nonvolatile. The pins dipped in Buffer B simply dried out, leaving salt deposits to block the tip of the pin preventing it from any further printing.

In another example, shown in Figure 4.20, a detergent SDS is used to control spot diameter of oligonucleotide probes immobilized onto a plastic (hydrophobic) surface. First, at zero sodium dodecyl sulfate (SDS) content a very small spot is produced (~100 micron diameter) even though the same quill pin on a glass surface is reported by the manufacturer to produce a spot diameter approaching 300 to 330 microns (TeleChem Inc., Chipmaker™ CMP-10). SDS wetting action overcomes the surface tension on the substrate allowing the probe to spread out from the tip of the quill, eventually filling in around the flat tip, leaving a characteristic rounded square

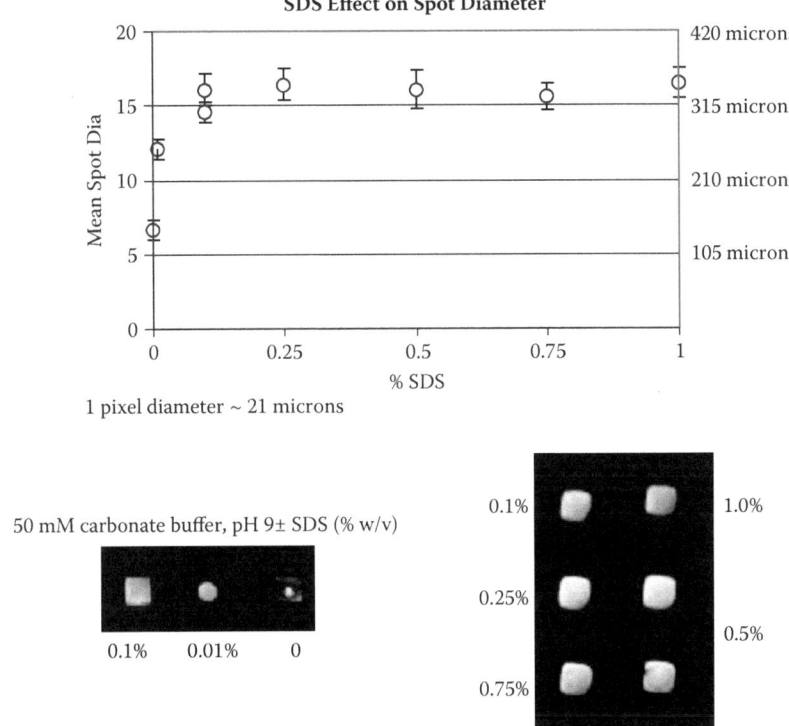

FIGURE 4.20 Detergent effect on spotting.

spot pattern. As expected, printing of these oligonucleotide probes in SDS buffer on a glass surface results in the deposition of very large spots by the CMP-10 quill.

MICROARRAY DESIGN

The design of the array is not a trivial task. In addition to the probes of interest, one must place an appropriate number of control elements. These should include both positive and negative control elements, such as housekeeping genes, as well as a number of controls that can be used to monitor the efficiency of important steps within the process. For example, you may wish to spike in internal standards that track recovery or labeling efficiency among different samples. It is also important to consider how you will print. How many replicates? And, should these replicates be placed next to each other or be randomly distributed throughout the array? If there is concern about print uniformity then replicates should be distributed across the array.

The pin configuration and print order may also have an impact. Balázsi et al. (2003) have shown that the printing process can lead to significant biasing of gene expression data. That is, how you arrange the pins in the print-head and the probes in the source plate can affect the outcome of your experiment. Balázsi and co-workers examined microarray gene expression data collected on the growth cycle

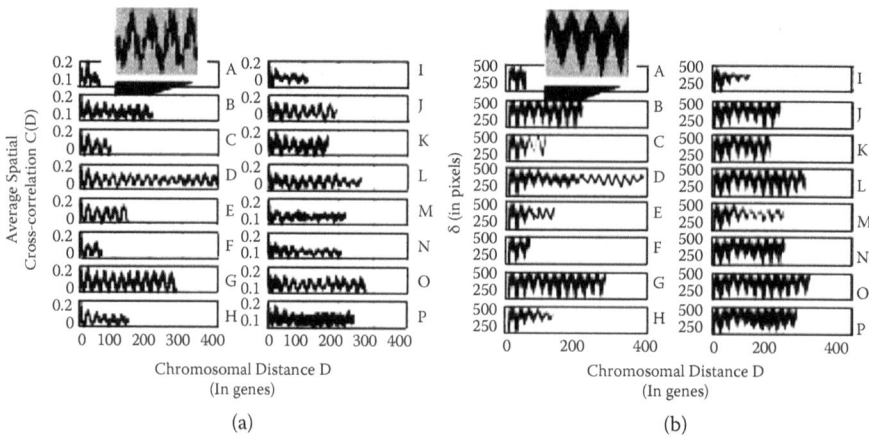

FIGURE 4.21 Gene expression periodicity effects due to pin tip bias. (From Balázsi G, Kay KA, Barabási A-L, et al. Spurious spatial periodicity of co-expression in microarray data due to printing design. *Nucleic Acids Res.* 31(15): 4425–4433, 2003. With permission.)

for *Saccharomyces cerevisiae* from the work of Spellman et al. (1998) at Stanford University. In this case, cDNA probes were arranged in 64 (96-well) microtiter plates according to chromosomal order (centromer to left telomere, then right telomere). The print pattern revealed that the array had been printed using four pins in a commonly employed 2 × 2 pattern yielding blocks of 44 × 44 spots or 1936 probes printed per probe. Each pin can be tracked through the printing process, and the genes were grouped according to print position. When the corresponding gene expression values were arranged according to chromosomal distance, a spatial periodicity was observed: a two gene period superimposed on a 24 gene periodicity (Figure 4.21). This periodicity was traced back to tip-specific biases introduced by the 2 × 2 printhead configuration (Yang et al., 2002; Balázsi et al., 2003). Applying algorithms to filter out the above periodicity from the gene expression data, a new 176 gene periodicity appeared. The 176 gene period was attributed to print location bias—that is, 176 spots (or four pins printing 44 spots per print cycle) are printed before the printer returns to a given position on the slide (Figure 4.22). Corrections for such systematic error due to printing bias were shown to improve the gene clustering (Figure 4.23) for the printed arrays with the average minimum distances between functional classes decreased from 13.3369 to 12.4067.

   The take home message here is to be aware of the performance characteristics of individual printing pins (or ink-jets) and how multiple pins match up. Re-inking of pins can lead to systematic differences in spot diameters or spot intensity, dividing the array into regions. This can lead to periodicity effects (Figure 4.24). Individual pins may also exhibit different printing patterns, delivering slightly different volumes or spot diameters. If there are significant differences (e.g., based upon signal mean ± standard deviation) obtained from the same probe pairs printed by different pins, then the data will need to be adjusted to normalize these differences.

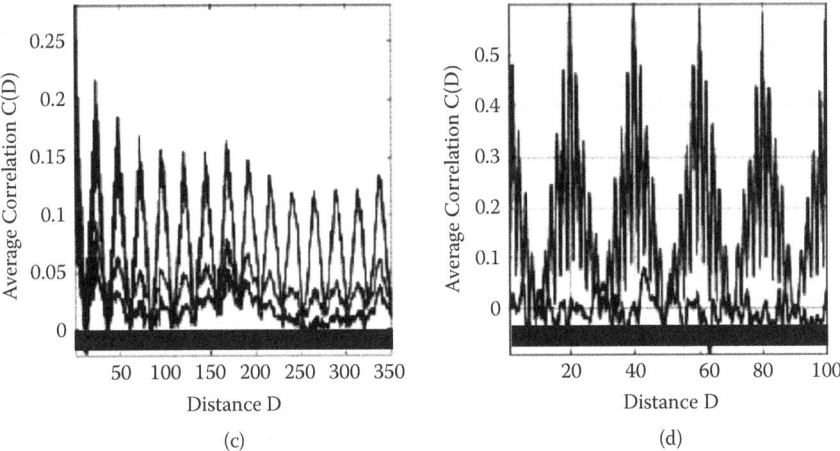

FIGURE 4.22 Higher-order print periodicity caused by print location. (From Balázsi G, Kay KA, Barabási A-L, et al. Spurious spatial periodicity of co-expression in microarray data due to printing design. *Nucleic Acids Res.* 31(15): 4425–4433, 2003. With permission.)

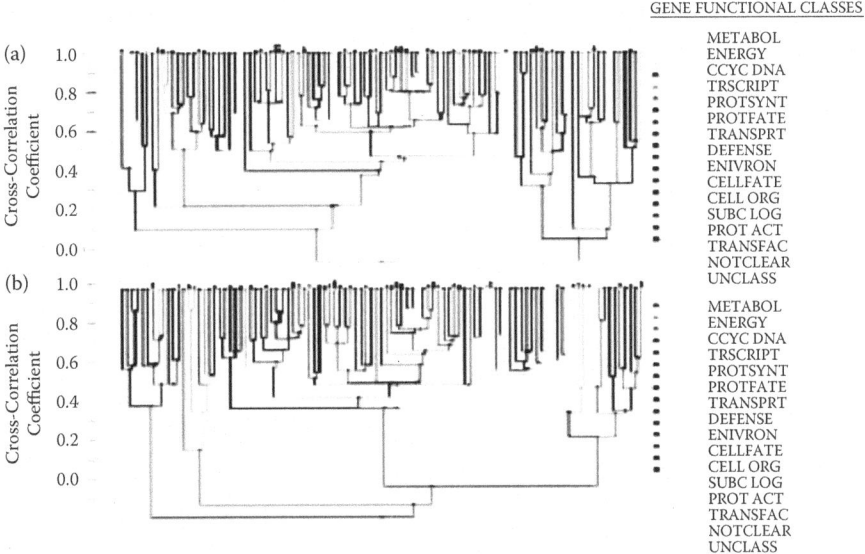

FIGURE 4.23 Corrections to gene clustering for pin printing bias. (From Balázsi G, Kay KA, Barabási A-L, et al. Spurious spatial periodicity of co-expression in microarray data due to printing design. *Nucleic Acids Res.* 31(15): 4425–4433, 2003. With permission.)

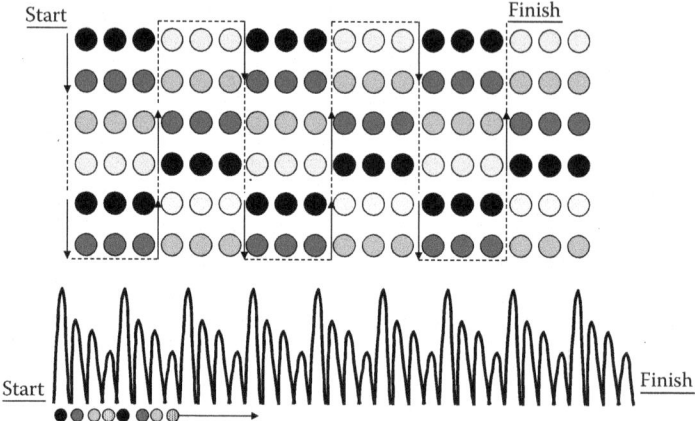

FIGURE 4.24   Periodicity effects due to re-inking.

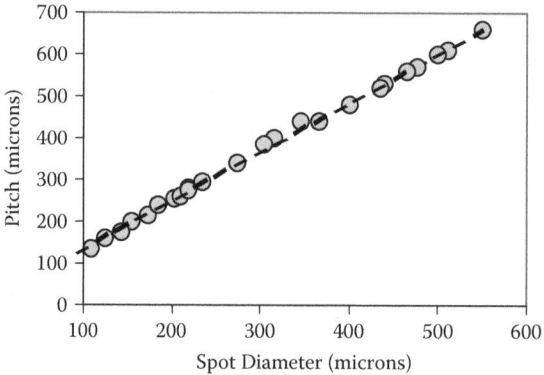

FIGURE 4.25   Relationship between spot size and pitch. (From Table 1, ChipMaker™ Micro Spotting Pin Matrix, http://arrayit.com/Products. With permission.)

SETTING UP THE PRINT RUN

Let's assume that you are now ready to create your first array on a previously selected substrate. How many elements (spots) do you wish to print? This will determine what kind of pin you will need, and although it appears to be a rather fundamental question to ask, it may not be operationally as simple to answer. As we have noted spot density is directly related to spot size and pitch (Figure 4.25). The pitch will determine how many spots you can actually print on the slide (Figure 4.26). The pin will deliver a specific droplet volume that will spread to a certain diameter largely based upon the tip's diameter and the print buffer used (Figure 4.27). So, the larger the spot diameter the less spots you can print (Figure 4.28). The selected pin will only fit into that manufacturer's print-head (i.e., you cannot interchange different manufacturer's pins and print-heads without hardware modifications). Obviously, the number of pins

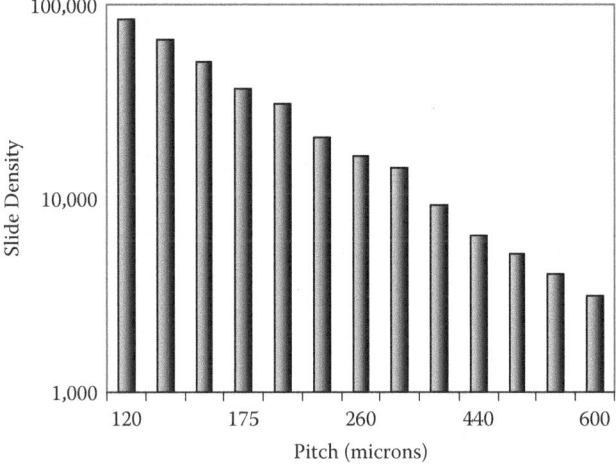

FIGURE 4.26 Spot density versus pitch. (From Table 1. ChipMaker™ Micro Spotting Pin Matrix, http://arrayit.com/Products. With permission.)

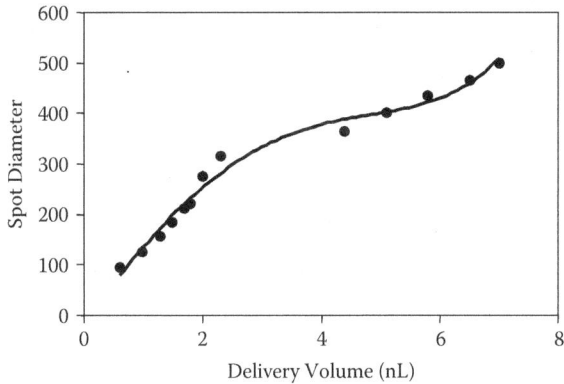

FIGURE 4.27 Pin delivery volume versus spot diameter. (From Table 1. ChipMaker™ Micro Spotting Pin Matrix, http://arrayit.com/Products. With permission.)

and their configuration in the print-head are determined by the print-head design, and this can also vary among manufacturers. Pins and print-heads are expensive, so choose wisely.

Now that you have the correct pin and print-head, you may wish to consider the number of pins and their arrangement (configuration) in the print-head in terms of throughput. That is, if you are time limited then the more pins you configure the faster the printing will be accomplished. However, also consider that printing with a larger number of pins will mean that you will need to use more duplicate probe source wells in the source plate for that purpose. This may also complicate the preparation of the source plate. At that point it is most desirable to enlist the aid of a

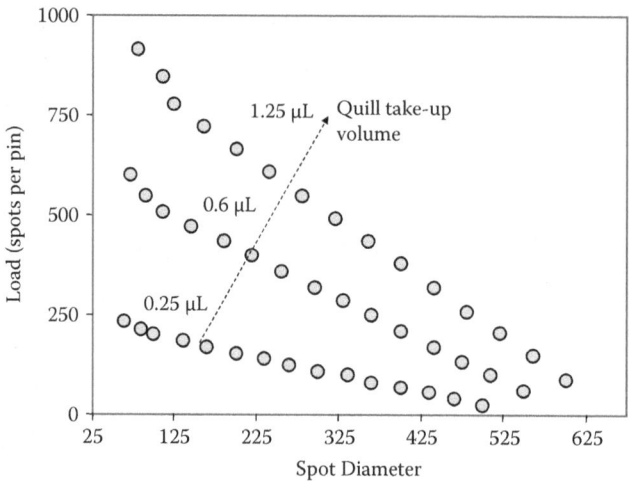

FIGURE 4.28   Relationship between pin load, delivery, and spot diameter. (From Table 1. ChipMaker™ Micro Spotting Pin Matrix, http://arrayit.com/Products. With permission.)

robotic liquid handler to fill the source plate from the probe stocks. This will reduce errors in filling and save considerable time by allowing the preparation of many source plates at once.

PRINTING PARAMETERS

There are a number of printer-defined parameters that must be taken into consideration when setting up the run (Figure 4.29). These are briefly described below:

> *Inking Depth*—Depending upon the quill type (or solid pin) selected there are certain precautions in filling that need to be followed. Generally, quill pins should not be dipped too far into the source plate well because the printing ink will deposit on the outside of the quill. This can have two effects. First, surface wetting means that the quill will eventually drain off the excess upon first striking the substrate, leading to very large spots of varying spot diameters. Thus, more preprints will be required to achieve uniform spot diameters. The other issue with improper depth filling is that capillary action can be slowed or prevented by fluid coming in from the side. Upon removal of the quill from the well, an air gap can be formed which essentially blocks delivery of the fluid from the tip. Quill manufacturers can provide the recommended filling depth for their pins.
>
> *Inking Time*—Likewise, pins may have different capillary fill and wetting rates depending upon the pin's surface characteristics and geometry. For instances, inking times on quills are in the range of a few seconds, while it may not be necessary to keep solid pins in contact with the source plate for more than a second to allow uptake.

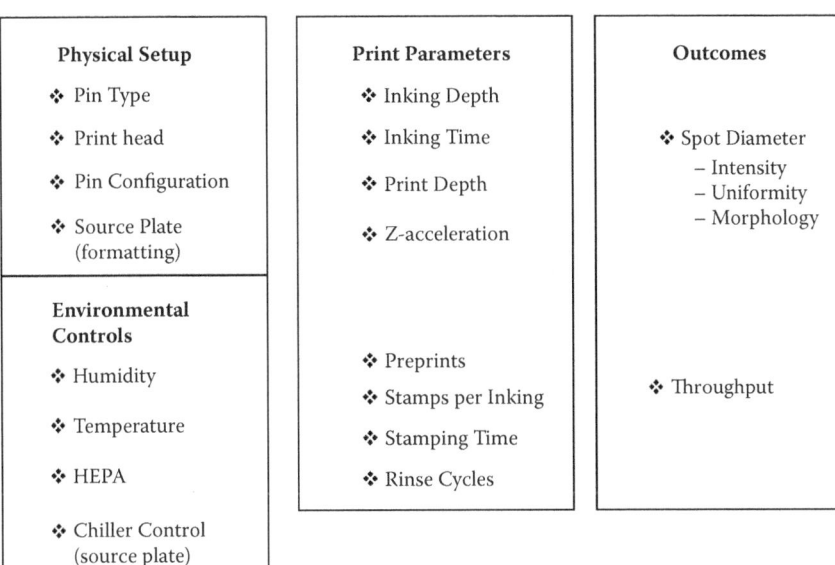

FIGURE 4.29   Printer setup parameters.

Print Depth—This is a critical parameter effecting pin performance and lifetime. Quill pins striking a hard surface (such as the glass slide) have a limited lifetime. Manufacturers rate their pin's lifetimes according to the number of spottings. However, misuse can dramatically reduce this number. In particular, there is the risk of overdriving the pin. Such excessive force at the tip of the pin will cause damage and reduce the pin's performance leading to poor-quality array production. Avoid overdriving by adjusting the print depth. This can be accomplished by adjusting the Z-height of the print-head assembly and in some cases the deck height of the substrate. Essentially, the pin should be seen to barely float up upon striking the surface. For example, TeleChem quills are generally held to a print depth of about 50 microns relative to the slide surface. If your arrayer has Z-acceleration control as well, then the striking process can be slowed by reducing the final acceleration of the print-head just prior to striking the surface.

Preprints—Quill pins for the reasons described above require a number of preprints. This is largely an empirical exercise. For that reason the best approach is to conduct a print test and measure the number of printings necessary to achieve a consistent spot size. For example, it may require 10 spottings before that condition is reached. Therefore, the number of preprints would be about 15 to assure a high-quality array production.

Stamps per Inking—Takes into account how many printings can be obtained before the print quality deteriorates. This determines how many slides could be printed per inking. There is a limit to the number of acceptable spots printed upon a single inking. For example, spot quality or diameter may begin to fall around 80 prints. Thus, the useful range for printing with a single dip into the source plate would be, for example, 75 prints. If 15 preprints were required, then the pin with a single inking would effectively deliver a single spot to each of 60 slides. The more time involved in re-inking, the longer the print run will take because most of the time for printing is accumulated in rinsing the pin and replacing the ink in the quill.

Stamping Time—This is the amount of time that the pin resides on the substrate surface. The longer the time spent on the surface, the greater is the ink volume deposited on the substrate. This will have a tendency to cause spreading of the fluid, thereby increasing the spot diameter. However, other factors such as the substrate's contact angle and the capillary hydrostatic head also influence the size and spread of the droplet.

Z Acceleration—As previously mentioned certain printing mechanisms involve striking the surface with the quill pin in order to dispense a droplet onto the substrate. Control of the acceleration rate can be useful in ejecting the droplet without crashing the pin into the substrate (sometimes referred to as "overdriving") (Figure 4.30). Not all arrayers have this feature.

Finally, when setting up a print routine for the first time, it is advisable to use dummy or break-away pins (Figure 4.31). That way in the event of a programming error you have reduced your liability in replacing damaged pins. Certain arrayers also offer features such as running in a "slow motion" mode which allows the user to more easily follow in real time the printing steps, such as pin positioning within the reservoir well or stamping depth.

FIGURE 4.30   Damaged pin due to excessive overdriving into substrate.

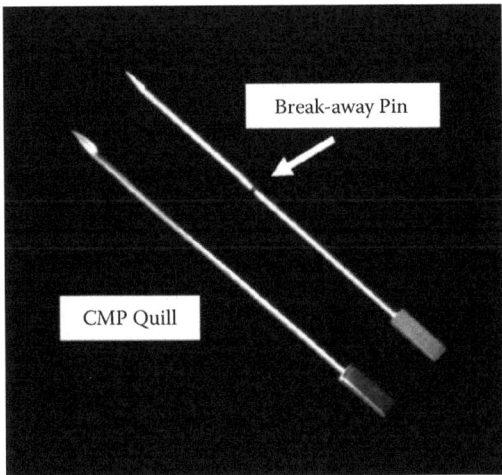

FIGURE 4.31 Break-away "dummy" pins.

PREPARING THE PROBE INK

There are many approaches by which to prepare the probe for immobilization to the substrate by printing. The exact nature of the buffer (ink) composition will depend upon the substrate's surface characteristics and the surface chemistry employed for immobilization. However, there are some fundamental precautions. First, the print buffer needs to be compatible with the probe. Certain buffers and salts will cause precipitation or probe aggregation. It is highly recommended that the physical-chemical stability of the probe be well understood prior to printing. All printing solutions should be filtered to remove aggregates or debris that might occlude the capillary tip. Stock solutions should be prepared and aliquots frozen and stored until needed. The source plate may be prepared in advance and stored frozen depending upon the probe ink stability. In any case, the source plate should be brought to ambient temperature to assure that all components have re-dissolved. It is recommended that the source plate be centrifuged to remove any entrapped air bubbles from the bottom of the wells. While manual preparation of the source plate is possible, the use of a robotic dispensing system is highly recommended in order to avoid mistakes in placement of the probe inks in the wells and to avoid cross-contamination. The use of such a device will also lead to a more uniform preparation of subsequent source plates.

OPTIMIZATION OF PROBE CONCENTRATION

What concentration of probe should be used for printing the array? While one can theoretically estimate the monolayer surface coverage for a particular biomolecule, this question is best answered by empirical determination. There are some practical reasons for performing probe loading versus hybridization efficiency. First, there may be differences in the probe stocks due to variation in their production. It is well known that the synthesis and purification of oligonucleotide probes can vary considerably among vendors or even within lots from the same vendor. It is not uncommon to obtain probes with varying amounts of salt or other materials remaining after high performance liquid chromatograph (HPLC) purification. As a result, differences in the chemical compositions of the various probe stocks can lead to significant differences in the immobilization efficiency. Second, certain secondary structures within oligonucleotides can lead to concatenation and aggregation, thereby reducing the coupling efficiency to the surface. Proteins can also undergo aggregation or become denatured by adsorption to the surface, thereby reducing the overall binding efficiency.

Here are some general starting ranges for optimized loading. Obviously, this may vary depending upon molecular weight and size of the biomolecule:

Oligonucleotides: 5 µM to 40 µM
cDNA: 1 nM to 1 µM
Proteins: 0.1 to 1 mg/mL

PROTOCOLS FOR PRINTING NUCLEIC ACIDS

cDNA MICROARRAY

Slide-based microarray technology as we have described it was first introduced to the world by Schena et al. (1995). The processes and equipment for preparing (the arrayer) and analyzing (the laser scanner) microarray slides composed a portion of Dari Shalon's thesis work at Stanford University. polymerase chain reaction (PCR) products (cDNA probes) were attached to poly-L-lysine coated glass microscope slides. The mechanism of attachment most likely involves electrostatic interaction through adsorption of the negatively charged nucleic acid to the positively charged lysine residues coating the glass substrate. The probes (48 cDNAs) were laid out in duplicate in an array format using a custom-built arrayer that was equipped with a single quill pin. The following is a summary of that early printing protocol:

1. Load quill tip with 1 µL PCR product (cDNA probe) from a 96-well microtiter plate.
2. Print probe onto 40 slides depositing ~5 nL per slide at 500 micron center-to-center spacing.
3. Rehydrate slide for 2 hours in a humid chamber.
4. Snap dry at 100°C for 1 minute.
5. Rinse slide with 0.1% SDS.
6. Block (cap) residual lysine residues with a carboxylic acid by acylation using succinic anhydride prepared in 1-methyl-2-pyrrolidinone, boric acid buffer.
7. Just prior to hybridization heat denature cDNA probes by 2 minute soak in distilled water held at 90°C.

The fundamental processes described above have not changed significantly. In fact, many researchers continue to follow this simple protocol, while a few have introduced some slight modifications for improvement in spotting consistency or reduced backgrounds (e.g., Hedge et al., 2000; Diel, 2001; Hessner, 2003). Aminosilane coated slides (such as Corning's CMT-GAPS™ slides) are reported to offer a more uniform surface coating and a reduction in fluorescent background over the poly-L-lysine slides. Problematic spot morphologies such as the occurrence of donuts (also known as ring spots) have been reported to be substantially reduced using the aminosilane surface (Hedge et al., 2000). While the earlier printing inks employed the use of high salts such as 3X SSC (saline sodium citrate) for depositing spots of cDNA, there remained issues regarding spot homogeneity even on the aminosilane surface.

OLIGONUCLEOTIDES

The method of printing single-stranded oligonucleotide probes is similar to that used for the printing of cDNA. The input concentration to achieve optimal surface loading is usually 10-fold higher. In our laboratory we typically print cDNA (e.g., >300 bp to ~1000 bp) at 1 nM and 5′ amino-oligonucleotides (15 to 30mer) at 20 µM. However, we have found that certain oligonucleotides are optimal at higher or lower concentrations and recommend performing a loading study (e.g., from 5 to 40 µM) in most instances. For synthetic oligonucleotides we highly recommend using HPLC purified and desalted stocks, especially for amino-oligonucleotides. Print buffer composition should remain simple. For most printing applications a sodium phosphate buffer, pH ~8 to 9, works well. In our laboratory, we employ a carbonate buffer system (50 to 150 mM sodium carbonate-bicarbonate, pH 9) which works very well for the immobilization of oligonucleotides (or proteins) to activated supports. We can easily change spot size by the addition of SDS to this buffer (0 to 0.25%).

For aldehyde slides print oligonucleotides at 10 to 50 pmole/µL (10 to 50 µM) and cDNAs at 0.2 to 1 µg/µL (~0.3 to 1.5 µM based upon 1 µg of 1000 bp cDNA = 1.52 pmoles). For aminosilane slides, resuspend DNA to a maximum of 0.25 mg/mL (~0.4 µM) in phosphate buffer (150 mM, pH 8.5).

Accelr8 Technology offered a polymer coated glass slide product under the trade name of OptiPlate™-DNA. This plate was later acquired by Schott. A printing protocol for amine-oligonucleotides differentiated between long and short print runs: for short runs—50% relative humidity for contact printing at the following buffer: 20 µM oligonucleotide in 300 mM sodium phosphate, pH 8.5, 0.005% Tween-20, 0.001% sarkosyl; for long runs—30% relative humidity in a print buffer made up of 150 mM sodium phosphate, pH 8.5, and 0.001% Tween 20. Sarcosyl is not included in this buffer. Dimethyl sulfoxide (DMSO) was not recommended for use.

U-Vision Biotech (www.u-vision-biotech.com) marketed an epoxy-activated slide product under the trade name EasySpot™ Oligo for the immobilization of oligonucleotides (20 to 70mers). Unmodified oligos, PCR products, and RNA are preferred over amine-modified forms. The company stated, "Our experiments showed that the amine modification slightly lowers the attachment efficiency of oligonucleotides." Their protocol suggested resuspending DNA to 2 µM ~16 µM in 50% DMSO-ddH20 only and cautioned not to use any salts such as SSC (sodium citrate-saline) and sodium bicarbonate. The presence of salts on this slide chemistry affected the spot morphology and the efficiency of immobilization.

Longer oligonucleotides (50 to 70mer) have been employed. Kane et al. (2000) described the covalent attachment of 5′-amino 50mer oligonucleotides to 3D-Link slides (Surmodics, Eden Prairie, MN). The 50mers were prepared at 20 μM in 150 mM sodium phosphate buffer, pH 8.5. From the now defunct gene-arrays listserv (gene-arrays@itssrv1.ucsf.edu) there had been considerable discussion regarding the printing of longer oligonucleotides onto different substrates. For example,

70 mers (Operon), 40 μM, 50% DMSO, Corning UltraGAPS—"you can go down to 20 μM but the spots look a lot better with 40," Gregory Khitrov, 7/31/2003

50mers (Operon), 10 μM, 150 mM sodium phosphate, pH 8.5 + 0.01% SDS, epoxy slides—"We also found 40 μM to be excessive. I did a dilution series and could see no change in signal until I reduced the oligo concentration below 5 μM. So we print oligos at 10 μM," Patty Holman, 7/30/2003

Most likely all of these observations are valid depending upon substrate, probe, and printing conditions.

Use of DMSO

DMSO (Figure 4.32) was introduced as an additive with demonstrated improvements in spot morphology and an increase in hybridization efficiency (Hedge et al., 2000). This solvent serves as a denaturant for nucleic acids and presumably permits more efficient tethering of single-stranded probes. Its hygroscopic property is responsible for slowing of evaporation of printed spots as well as probes remaining in the source plate. Hedge et al. (2000) examined spot morphology from cDNA spotted down on aminosilane slides in 3X SSC containing 50% DMSO as a function of relative humidity (RH) and temperature and concluded that optimal conditions were obtained at 22.2°C and 45% RH.

Use of Betaine

Even though Diehl et al. (2001) agreed that the addition of DMSO to print buffer improves spot uniformity, they argue that DMSO is also toxic and a good solvent for other materials. As a result, they explored alternative chemistries to replace DMSO and also to improve upon post-print blocking conditions. In the latter case, this was an effort to find a replacement for borate –NMP (1-methyl-2-pyrrolidinone) buffer used for preparing solutions of succinic anhydride for capping of residual amine groups.

The Heidelberg group (Diehl et al., 2001) chose (carboxymethyl) trimethylammonium hydroxide, commonly known as betaine (Figure 4.33) as a substitute for DMSO. Why these researchers selected betaine and did not consider other additives

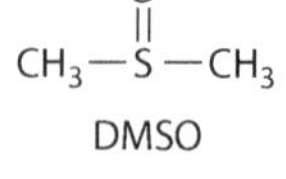

$$CH_3 - \overset{\overset{\displaystyle O}{\|}}{S} - CH_3$$

DMSO

FIGURE 4.32   Dimethyl sulfoxide (DMSO) structure.

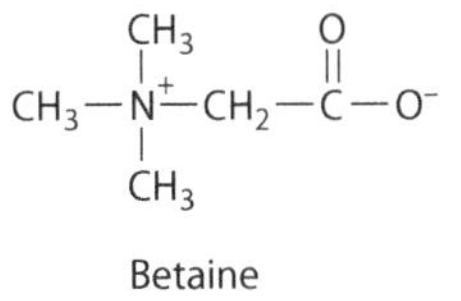

$$CH_3 - \overset{\overset{\displaystyle CH_3}{|}}{\underset{\underset{\displaystyle CH_3}{|}}{N^+}} - CH_2 - \overset{\overset{\displaystyle O}{||}}{C} - O^-$$

Betaine

FIGURE 4.33 Betaine structure.

was not discussed. However, we do know from other works (see cited references) that this compound is effective in reducing stability differences between A:T and G:C base-pairing during hybridization, much like that of tetramethylammonium chloride or formamide (Rees et al., 1993). This provides a simple means to denature DNA and maintain such probes in a single-stranded state. Single-stranded probes are preferred for attaching the probes to a support. In addition, higher concentrations of betaine are rather viscous and slow down evaporation. Thus, betaine would most likely improve spot size and uniformity by slowing evaporation and thereby impeding the spread of the DNA spot out into the familiar donut shape. It also would slow evaporation in the wells of the source plate, preventing the unwanted and nonuniform plate edge evaporation effect on probe concentration. Thus, 3X SSC + 1.5 M betaine was used to print down a 500 bp cDNA onto both poly-L-lysine and aminosilane glass slides. In addition, a commercial print buffer ArrayIt™ micro-spotting solution or MSS (TeleChem International, Inc.) was compared. Curiously, 3X SSC + 50% DMSO was not included in this study. While MSS may contain DMSO, it would have been a better designed set of experiments to have also included 3X SSC + 50% DMSO as a control. Nevertheless, it is clear from the work that the inclusion of betaine into the print buffer was an improvement over SSC or the MSS on several fronts. First, the SSC-betaine spotting was found to increase hybridization efficiency as measured by a 2.5-fold higher hybridization signal intensity for probe-target hybrids relative to those probes spotted in SSC or MSS alone (Figure 4.34). Spot morphology was

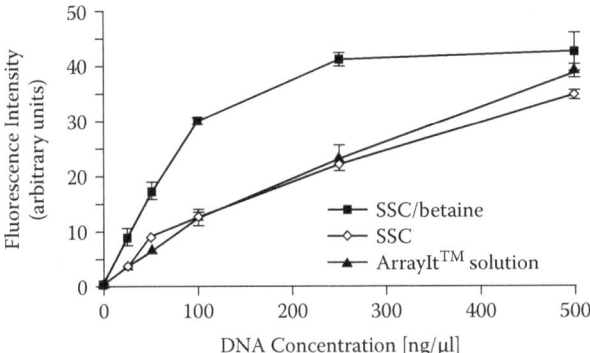

FIGURE 4.34 Efficiency in the delivery of DNA onto a substrate using various additives. (From Diehl F, Grahlmann S, Beier M, Hoheisel JD. Manufacturing DNA microarrays of high spot homogeneity and reduced background signal. *Nucleic Acid Res.* 29(7) e38: 1–5, 2001. With permission.)

thought to have improved based upon the observed levels of signal intensity variation within a spot. The addition of betaine reduced the spot pixel intensity CVs from 14% down to 5% when DNA was spotted at 100 ng/mL. Finally, betaine contributed the lowest on-spot background (the buffer's residual background) relative to SSC or MSS. In summary, the inclusion of betaine in the print buffer improves printing.

The other issue addressed by Diehl et al. (2001) is that of a more refined control of the capping process to prevent the occurrence of an elevated inter-spot background. The dissolution of various print buffer components back into solution can lead to comet tailing effects as well as re-adsorption back onto the slide resulting in a buildup of the background. Slides may appear blotchy with uneven signal distribution. In other instances, a decrease in spot uniformity is evident or there may be a significant reduction in signal intensity. The most severe case occurs when the inter-spot background intensities are much greater than the spot intensities, leading to the appearance of "black" spots surrounded by bright inter-spot backgrounds. These authors reasoned that the aqueous succinic anhydride capping buffer (composed of 96% NMP and 4% sodium borate) may lead to the re-dissolving of probe DNA that is subsequently randomly re-deposited over the entire slide leading to an elevated background. As a result, a reformulation of succinic anhydride into a nonaqueous media of dichloroethane (DCE) solvent containing N-methylimidazol (acylatin catalyst) was undertaken. Significant improvements in inter-spot backgrounds were evident.

USE OF POLYVINYL ALCOHOL

Wu et al. (2006) compared various hydroxylated additives for their effect on the performance of antibody arrays. The compounds included polyvinyl alcohol (PVA), glycerol, glucose, sucrose, and trehalose, all of which have been used for protein stabilization and as components in print buffer inks. PVA 9000 (0.5%) provided exceptional performance relative to the other compounds in terms of spot morphology, consistency in spot diameter, and antibody activity. Moreover, shelf-life (1 day versus 1 month storage at 4°C under nitrogen) for the antibody array was improved by the addition of PVA in print buffer compared to the other hydroxylated additives, including trehalose.

EVAPORATION

McQauin et al. (2003) undertook a detailed study on the effects of relative humidity and the direct comparison between DMSO versus betaine in print buffer on the overall performance of quill pin printing. A video microscope was employed to visualize and track the drying behavior of the various printing inks. A Cy5-labeled 466 bp dsDNA probe was used to monitor the printing process. Drop drying behavior, bulk evaporation from the quill reservoir, surface tension changes, and spotting characterstics (spot diameter, spread, and number deposited) were examined at different RH levels. Print buffers 3X SSC, 3X SSC + 50% DMSO and 3X SSC + 1.5 M betaine were evaluated at 40%, 60%, and 80% RH for spot intensity, spot diameter, and intra-spot variation, CV (Figure 4.35). The reductions in quill drop volumes and droplet drying times were measured by video microscope, and the quill reservoir

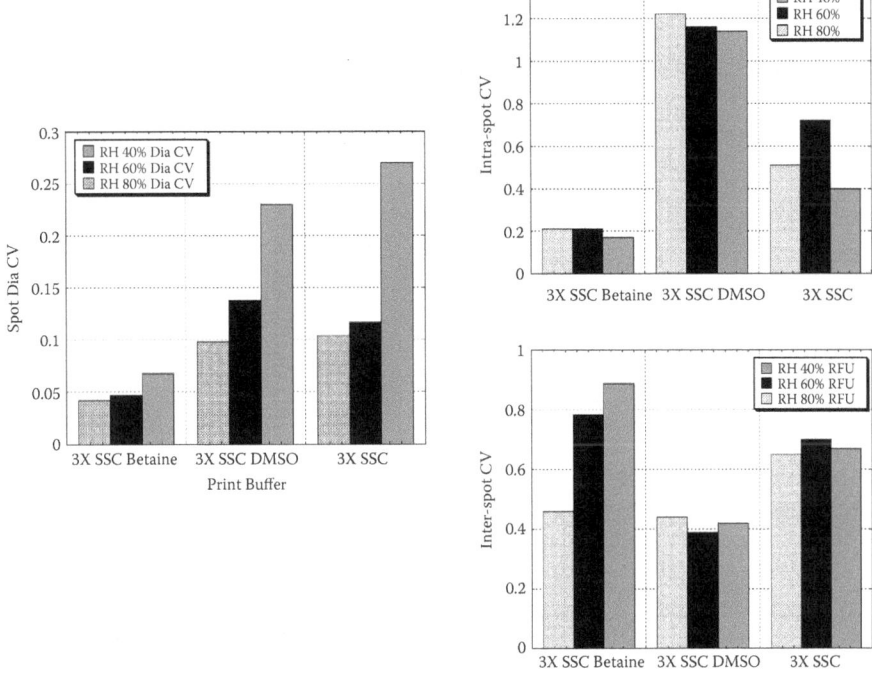

FIGURE 4.35 Spot variation upon printing in various buffers at different relative humidity. (From McQuain MK, Seale K, Peek J, Levy S, Haselton FR. Effects of relative humidity and buffer additives on the contact printing of microarrays with quill pins. *Anal. Biochem.* 320: 281–291, 2003. With permission.)

volume changes were determined by weight. The results of the study can be summarized as quoted, "solvent evaporation from the print buffer reservoir is the major factor responsible for the variations in the transfer of fluid to the slide surface" (p. 281).

So, in order to control spot deposition and ultimately the spot's diameter and morphology, you must first control the rate of evaporation from the quill's reservoir. The change in surface tension causes variation in the spot characteristics. Evaporation of water from the bulk solution held in the quill's reservoir increases the salt and probe concentration which in turn increases the solution's surface tension. As a result, less fluid volume is transferred from the pin, and the smaller droplet that is transferred to the substrate surface is at a higher surface tension. This results in a higher surface contact angle. Opposing surface tension force is the pinning force. Essentially, pinning forces that are promoted by surface features tend to fix the contact line of the droplet and drive the DNA toward the contact line (solvent perimeter). Solutes such as the salts and probe spread to the perimeter by convection as the droplet evaporates (Figure 4.36). Uneven evaporation leads to differences in spot uniformity. But, how effective were DMSO and betaine at improving print performance? This really depends upon what is most important. From McQuain et al. (2003) the following rankings can be summarized (Table 4.2) based upon variation in inter- and intra-spot

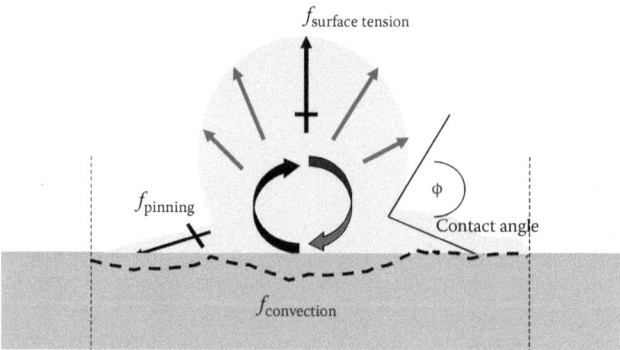

FIGURE 4.36   Droplet drying mechanisms.

signal intensity as well as spot diameter. The addition of DMSO provided the best spot-to-spot signal reproducibility, while betaine appeared to produce the most consistent spot diameters and spot homogeneity. Interestingly, betaine produced spots with greater signal intensity variation than either 3X SSC + DMSO or 3X SSC alone, while intra-spot CVs were the highest when DMSO was included in the buffer. So, if you are most concerned about spot-to-spot variations use 3X SSC + 50% DMSO, but if spot homogeneity is more important then betaine may be a better choice as the print buffer additive.

Of course, one must ask whether or not a combination of DMSO and betaine would be of even greater benefit for spot quality? Hessner et al. (2003a) in their efforts to develop a tracking system to measure microarray performance found that 1.5 M betaine in 3% DMSO provided the greatest probe retention on poly-L-lysine slides. Their major concern was that when cDNA probes were printed at low concentration (<100 ng/μL) one could not accurately distinguish expression differences. With a reduced probe population on the surface it would be possible to saturate the probe-target binding sites (i.e., target excess). As Hessner et al. point out, under such a condition the transcript ratios would be compressed. To achieve quantitative

TABLE 4.2
Ranking Print Buffer Additives

Buffer Composition	Inter-Spot Spot to Spot	Intra-Spot Within Spot	Spot Diameter
3X SSC	2	2	3
3X SSC + DMSO	1	3	2
3X SSC + Betaine	3	1	1

Source: From McQuain MK, Seale K, Peek J, Levy S, Haselton FR. Effects of relative humidity and buffer additives on the contact printing of microarrays with quill pins. *Anal. Biochem.* 320: 281–291, 2003. With permission.

Note: From McQuain, 2003, rankings based on calculated coefficients of variation from reported mean intensity, standard deviation 1 = lowest %CV (least variation).

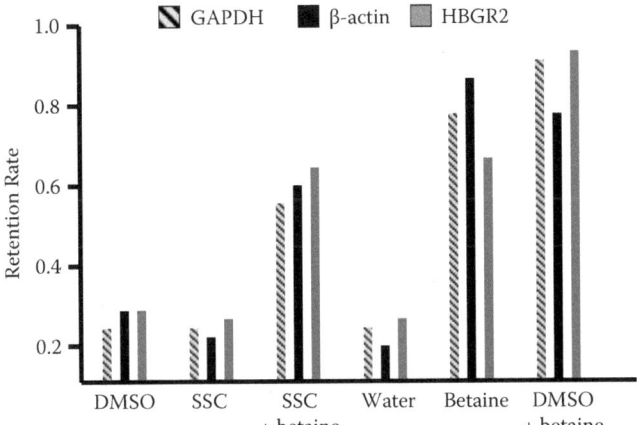

FIGURE 4.37 cDNA retention on glass subtrates after printing in various buffers. (From Hessner MJ, Wang A, Hulse K, et al. Three color cDNA microarrays: Quantitative assessment through the use of fluorescein-labeled probes. *Nucleic Acid Res.* 31(4) e14: 1–6, 2003. With permission.)

information from microarrays (i.e., more than just knowing whether or not the target is present), it is desirable that probes remain in "binding" excess over targets. The problem encountered was that when probes were printed down on the glass slide only a portion of the cDNA remained attached to the surface (or remained available for hybridization) following processing. This was found to be print buffer dependent (Figure 4.37). Both 50% DMSO and 3X SSC were about as effective as water as a print buffer in terms of cDNA probe retention (~20 to 30%), while 3X SSC + 1.5 M betaine retained ~60 to 70% of hybridizable probes on the surface. Interestingly, betaine alone appeared to provide better retention than in combination with 3X SSC, although the data scatter permits us only to suggest a trend. However, the combination of DMSO and betaine also provided a level of retention comparable to that of betaine (~70 to +100%) with the added benefit of being able to titrate with DMSO to control spot diameter. Thus, Hessner et al. (2003a) determined that 1.5 M betaine in 3% DMSO provided the optimal print buffer for their studies involving immobilization of cDNA probes onto poly-L-lysine slides.

PRINT QUALITY ASSESSMENT

The more important outcome of the Hessner et al. studies was development of a method useful for determining microarray spot quality (Hessner, 2003a, 2003b). One of the drawbacks to using microarrays has been in fact how to address quality control (QA/QC) issues regarding the printed product prior to use. Once you have printed the array, how do you best determine spotting consistency from slide to slide or batch to batch? There have been several approaches used such as Syber Green II staining (Battaglia, 2000) or the hybridization of fluorescently labeled sets of randomly synthesized short oligonucleotides. While these methods do work, it is necessary to melt-off or dissociate the short oligonucleotides from the support by chemical means. Baattaglia et al. (2000) claimed a reversible staining for Syber Green; however, in our experience we have found the need for extensive destaining to reduce the fluorescence background to an acceptable level. We had also noted certain sequence bias in staining with SYBR Green such as polyA strings (R. Matson, unpublished). Later Zipper et al. (2004) undertook an extensive study of SYBR binding interaction with ds DNA. They noted sequence-specific binding as well. In either case, there remains the lingering question as to what effect such reagents have on the microarray's performance. It is similar to the question of how many times one can strip and reuse the microarray before performance deteriorates.

An alternative approach is provided by Hessner et al. (2003a) in which the cDNA probes are permanently labeled using fluorescein-labeled primers to the clone's vector insert region. Fluorescein is excited at 488 nm and emits at 508 nm, while Cy3 may be excited at 543 nm to reduce any spectral overlap with fluorescein. Thus, fluorescein-labeled cDNA probes may be printed down and the slide scanned for QC/QA purposes prior to hybridization. Because the same region is primer labeled in each cDNA, a direct comparison between the relative fluorescence units (RFUs) and the amount of cDNA probe can be determined. One can now evaluate the microarray slide for a variety of parameters: spot diameter, intensity, morphology, and retention upon processing. Hessner et al. examined 50 pairs of slides varying in spot quality and found a significant ($P < 0.001$) difference in hybridization performance based upon the fluorescein probe quality (Figure 4.38). Those probes with low signal to noise are most likely to produce hybridized arrays having low signal-to-noise values, and such sets of microarrays do not show good inter-slide correlation.

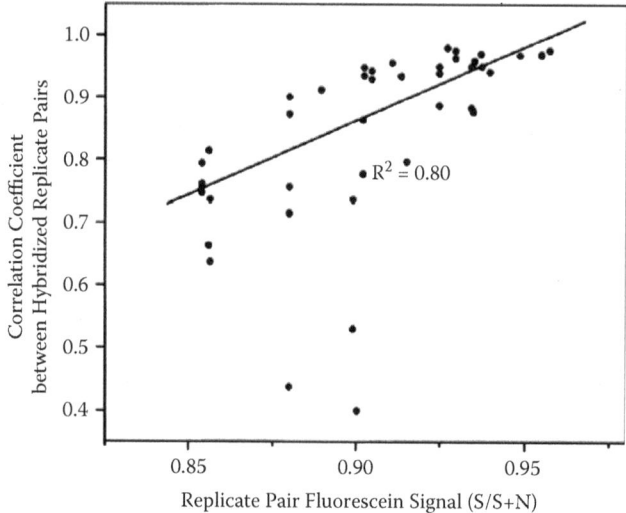

FIGURE 4.38 Spot quality versus probe quality. (From Hessner MJ, Wang A, Hulse K, et al. Three color cDNA microarrays: Quantitative assessment through the use of fluorescein-labeled probes. *Nucleic Acid Res.* 31(4) e14: 1–6, 2003. With permission.)

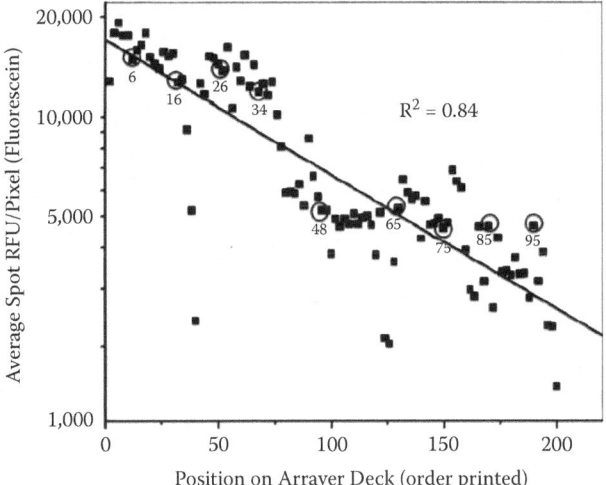

FIGURE 4.39 Spot quality versus print order. (From Hessner MJ, Wang X, Khan S, et al. Use of a three-color cDNA microarray platform to measure and control support-bound probe for improved data quality and reproducibility. *Nucleic Acid Res.* 31(11) e60: 1–9, 2003. With permission.)

In a further examination of printing behavior on slide performance, Hessner et al. (2003b) looked at the effect of print order on spot quality (Figure 4.39). There is an obvious trend in that the first arrays printed have higher overall spot intensity. A spot quality metric called a *quality composite score* (qcomp) derived from the weighting of five spot-related criteria (size, S/N, background level, background uniformity, and saturation) was developed based upon Matarray software (Wang, 2001, 2003). Using qcomp to screen for spot quality on microarrays, it was possible to remove spots with low scores from the analysis. In doing so Hessner et al. (2003b) concluded that limiting probe concentration leads to greater hybridization signal variability, while probes with high fluorescence intensity (higher probe concentration) are less variable. In their study they observed that a fluorescein intensity threshold at 5000 RFU per pixel was necessary for good hybridization performance. A slide acceptance criterion was suggested to achieve reproducible results allowing differential expression change to 1.5-fold:

1. Array mean element intensity >5000 RFU/pixel
2. Signal intensity CV <10%
3. Mean S/S+N score >0.85
4. Spot size CV <20%

The work by Hessner et al. and Wang et al. (Max McGee National Research Center for Juvenile Diabetes, Milwaukee, WI) provides a quantitative and systematic approach toward cDNA probe microarray quality assessment. Its utility for the assessment of oligonucleotide arrays is less certain. Oligonucleotide probes labeled with fluorescein may not be as sensitive due to weaker signal strength at lower probe concentration with the added potential for quenching the fluorescence signal at high probe concentration.

A simpler method has been described by Shearstone and co-workers (2002). Instead of employing dye-labeled oligonucleotides they spiked into the print buffer (containing unlabeled oligonucleotide probes) either Cy3- or Cy5-dCTP. Spiked Cy3-dCTP was more sensitive than Cy5-dCTP. The Cy3-dCTP could be reliably detected from 10 µM down to at least 20 nM. The latter concentration was chosen in order to assure that there would be no carryover. The potential for cross-contamination was recognized due to difficulty in removing the dye from the quills at higher concentrations. It was noted that higher concentrations of the spiked Cy3-dCTP (e.g., 500 nM) may have interfered slightly with oligonucleotide attachment. The effect was minor at about 87% of signal strength compared to control hybridization to probes without spike. This may be within the range of assay variation. However, what is particularly attractive about the dye-labeled dNTP route is that it is relatively inexpensive compared to primer labeling. Moreover, spot characterization (e.g., spot morphology, signal intensity, spot diameter) could be easily determined from the scans (Figure 4.40).

BACKGROUNDS

An understanding of what contributes to background and how best to correct microarray data for background remains an important issue for both DNA and protein microarray analysis. Fundamentally, we can divide background into two basic categories: off-spot contributions (i.e., nonspecific signal surrounding a spot) and on-spot contributions (i.e., nonspecific signal within the specific signal region). For example, consider a fluorescence pixel intensity scan for a portion of a microarray as described in Figure 4.41. The positive signal (a) is easily distinguished, while the much weaker signal at (b) could be identified as either a nonspecific signal such as fluorescence arising from the print buffer, or as a true signal, in which case by our example it would be a false-positive signal. Nevertheless, by means of a background subtract (e.g., b − d) and establishing threshold ratios (e.g., a/b > n, value), it would be possible to score the signal at (b) as desired in order to improve the data. Some of the image analysis software examines the pixel intensity histograms for the on-spot and off-spot signals and then defines a local background (e.g., at a radius r from the spot) from which to subtract background. This is an acceptable practice provided that the backgrounds are similar within the spot and outside of the spot. However, this background assumption may not always be valid. As depicted in Figure 4.41 the nonspecific signal (c) could contribute to the overall intensity within the spot. An example of this type of background would be fluorescent residue from components in the print buffer (e.g., at spot b) or rinsing solutions. Another approach would be to background subtract (a − b) assuming an equivalent background contribution to the on-spot intensity.

However, what if the background was primarily associated with the spot and that the background varied from spot to spot? Martinez et al. (2003) examined on-spot contaminating fluorescence backgrounds for a number of commercial and in-house printed slides. Using the Axon 4000B scanner as well as a hyperspectral-imaging

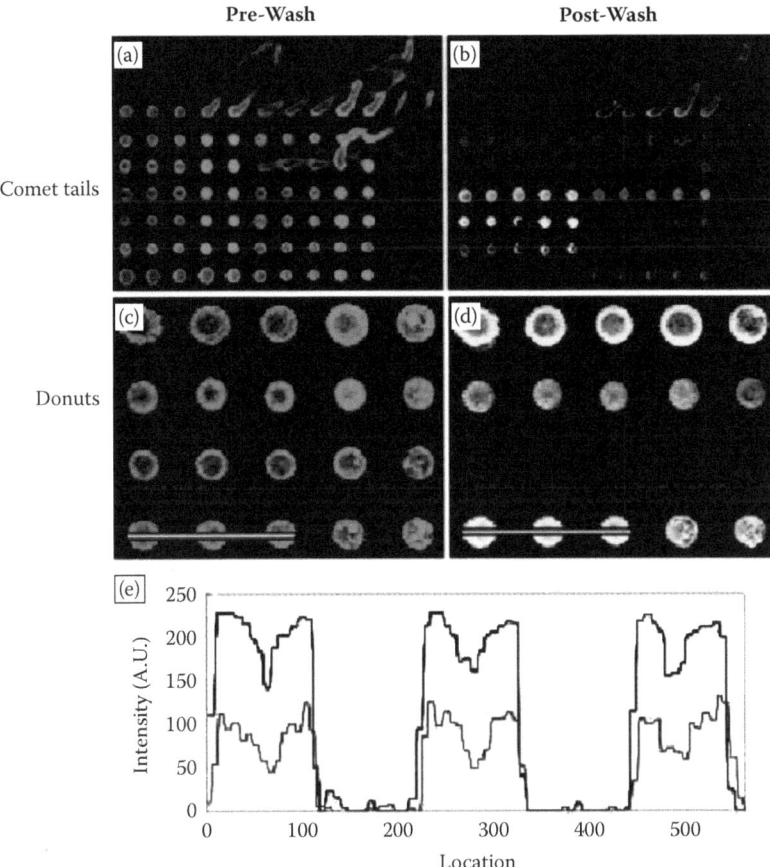

FIGURE 4.40 Assessment of spot quality using dye-labeled dNTPs. (From Shearstone JR, Allaire NE, Getman ME, Perrin S. Nondestructive quality control for microarray production. *BioTechniques* 32(5): 1051–1057, 2002. With permission.)

scanner (developed at Sandia National Laboratories, Albuquerque, NM) that is capable of looking very closely at intra-spot intensity profile, significant spot-localized backgrounds were detectable in the green channel in the absence of Cy3 (Figure 4.42). Such contaminations discovered in mock hybridizations were found to be highly variable with average spot intensities ranging from 840 ± 689 for Corning preprinted CMT to 682 ± 382 for Operon's OpArray preprinted yeast array slides. The printing of 70mer oligonucleotide probes onto a series of commercial slides also exhibited on-spot backgrounds, especially when the print buffer included TeleChem's MSS where backgrounds were, for example, 3116 ± 1405 on Corning's GAPS I aminosilane slides. A post-printing treatment involving the following sequential rinsing protocol reduced these backgrounds to ~200 relative intensity units:

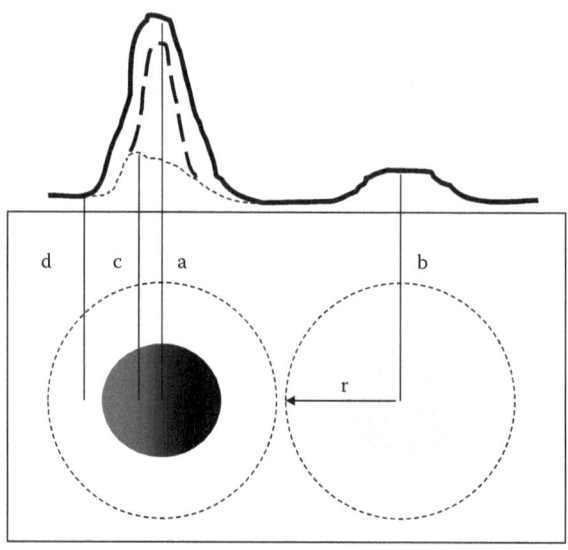

a = array pixel intensity profile
b = non-specific bkg on-spot (false positive)
c = hidden non-specific bkg on-spot
d = off-spot background

FIGURE 4.41   Background profiles.

1. 0.1% SDS, 10 min.
2. 2X SSC, 2 min.
3. boiling water, 3 min.
4. 100% ethanol, ice cold, 5 min.

Aging of certain slides (e.g., Corning GAPS) by exposure to humidified air for several hours prior to printing also reduced the background. It is not known whether or not this would apply to all substrates.

Utilizing the hyperspectral imaging scanner together with MCR (multivariate curve resolution) algorithm analysis, it was possible to evaluate the contribution of on-spot backgrounds to errors in the Cy5/Cy3 gene expression ratios. Martinez et al. (2003) reported that for the green channel (Cy3) intensities ~75% of spots were off by a factor of 2, while 50% were off by a factor of 3, and at least 25% of all spots exhibited errors more than a factor of 4.5 (Figure 4.43). Such variation has an obvious impact on the Cy5/Cy3 expression ratio and upon the interpretation of the data set in terms of the biology being studied.

PROTOCOLS FOR PRINTING PROTEINS

The printing of proteins is not difficult as long as the proteins are antibodies. While this is not obviously a completely accurate statement, there is much more danger

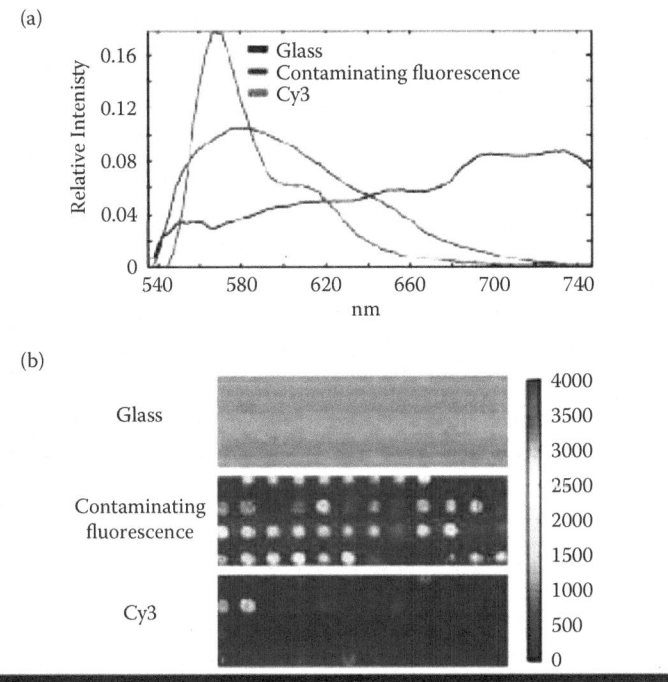

FIGURE 4.42 Examining microarray backgrounds using hyperspectral scanning. (From Martinez MJ, Aragon AD, Rodriguez AL, et al. Identification and removal of contaminating fluorescence from commercial and in-house printed DNA microarrays. *Nucleic Acid Res.* 31(4) e18: 1–8, 2003. With permission.)

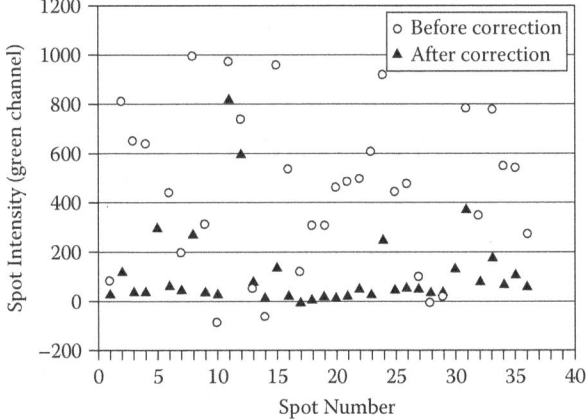

FIGURE 4.43 Background correction based upon multivariate curve resolution following hyperspectral scans. (From Martinez MJ, Aragon AD, Rodriguez AL, et al. Identification and removal of contaminating fluorescence from commercial and in-house printed DNA microarrays. *Nucleic Acid Res.* 31(4) e18: 1–8, 2003. With permission.)

in generalizing about printing proteins than printing down oligonucleotides. While the physical-chemical properties of nucleic acids are very well understood and we have had over a decade to learn how to print them, our knowledge base for creating protein microarrays is essentially derived from work on antibody arrays. While some may disagree, I would contend that the antibody array represents the easiest of examples to have followed. The complexity of the proteome will provide ample challenge to those interested in using microarrays.

ANTIBODY ARRAYS

There are many protocols now available for printing antibodies onto various substrates. However, most protocols will involve the following fundamental processing steps: exchange of protein from storage buffer into print buffer, adjustment of protein concentration, arraying, post-printing rinse to remove unbound excess protein, and a blocking step. In some cases where the protein is covalently immobilized, there may also be a capping step in order to inactivate residual reactive groups. Perhaps one of the best examples was that provided by Haab et al. (2001) in which 94 antibody-antigen pairs were printed onto a poly-L-lysine slide. This work originated from Pat Brown's lab at Stanford University where the DNA microarray was invented based upon the same slide chemistry. The protocol follows:

1. Transfer antibodies and antigens from glycerol buffer into glycerol-free PBS solution using spin columns (BioSpin P6, Bio-Rad Laboratories).
2. Prepare proteins at 0.1 to 0.3 mg/mL.
3. Transfer 4 μL into 384-well source plate.
4. Array solutions from source plate onto the poly-L-lysine coated glass slide.
5. Rinse microarrays briefly in 3% nonfat milk in PBS, pH 7.4 containing 0.1% Tween-20. Note that this step is to remove unbound protein.
6. Soak slides overnight at 4°C in 3% nonfat milk in PBS, pH 7.4 containing 0.02% sodium azide as preservative. Note that this step is used to block the slide surface in order to reduce nonspecific adsorption of analytes (or other interfering substances in the sample) in subsequent analysis steps.
7. Just prior to use, rinse slides at room temperature with PBS (three times, 1 minute each soak). Maintain slides in PBS buffer up until incubation with sample.

At the forefront of arraying proteins has been the demonstration by MacBeath and Schreiber (2000) in which proteins were immobilized to a glass slide coated with BSA and subsequently derivatized with NHS (*N*-hydroxysuccinimide) groups. The BSA-NHS slide served two roles. First, the BSA as a common blocking protein masked out any regions on the slide that might otherwise contribute to nonspecific binding. Second, the NHS-modified BSA served as a convenient scaffold for covalent immobilization of the protein probe. The adsorbed BSA layer most likely prevented the probe from any surface interactions with the substrate, while keeping it oriented out into the surrounding media for efficient capture of analyte.

The protocol of MacBeath and Schrieber (2000):

1. Prepare proteins for spotting in PBS, pH 7.5 containing glycerol (40% v/v) at 100 µg/mL.
2. Array proteins onto NHS-BSA coated slide.
3. Incubate protein microarrays in a humidified chamber at room temperature.
4. After 3 hours remove slides and drop slides face down into PBS, pH 8.0 containing 500 mM glycine. Note that glycine is commonly used as a capping reagent, in this case, to inactivate residual NHS esters.
5. Soak for 1 minute and then turn slides upright and immerse in PBS-glycine. Incubate with gentle agitation for 1 hour at room temperature with a final rinse in PBS prior to use.

In the previous examples the substrate of choice was the glass microscope slide. This format allowed these researchers to rapidly move ahead with experiments. This was largely due to the accessibility of existing equipment and technologies used in DNA microarray analysis such as arrayers, scanners, and image analysis software programs. At about the same time a higher-throughput microarray format was under development. This new format was based upon the microtiter plate, a standard format for automated liquid handling and target screening assays.

Genometrix (Mendoza et al., 1999) first introduced the use of optically flat glass plates that were masked with Teflon to create a pattern of 96 wells. The surface was coated with aminosilane and subsequently reacted with bis-sulfo-succinimidyl suberate to create an NHS surface for covalent attachment of proteins. Various immunoglobulins (i.e., creation of an antigen array) were printed down into individual wells in 50 mM carbonate buffer, pH 8.3 at 100 µg/mL using a capillary contact printer to deposit droplets (200 pL) of protein ~275 micron diameter spots on 300 micron centers. Arrays were first rinsed with Tris buffered saline (TBS) containing 0.1% Tween-20 to remove unbound protein. Blocking was accomplished using Blocker Casein (Pierce) in PBS for 1 hour at room temperature. A semi-automated microarray ELISA was performed in a well volume of 25 to 35 µL depending upon the incubation step using a HYDRA® 96 Liquid Pipettor (Robbins Scientific).

Matson and co-workers from Beckman Coulter first described the use of a low form 96-well plastic microplate for automated micro-ELISA immunoassays (Matson et al., 2001). The polypropylene plate was first modified by a radiofrequency plasma amination process (Matson et al., 1995) followed by conversion to an acyl fluoride surface chemistry for rapid covalent attachment of biomolecules. Proteins (1 to 2 mg/mL) were prepared in 50 mM carbonate buffer, pH 9 containing 4% sodium sulfate (to improve spot uniformity) and printed using a conventional arrayer system. For example, approximately 200 pL droplets of monoclonal antibodies (anti-cytokine) were deposited into the bottom of the micro-well using a Cartesian PS7200 system equipped with Majer Precision quill pins (Figure 4.44). Following a drying step the wells were soaked in a carbonate-casein buffer to quench residual reactive groups and block the wells in order to reduce nonspecific protein adsorption.

Moody et al. (2001) immobilized anti-cytokine MAbs in a 3 × 3 pattern to the bottom of a Maxisorp™ polystyrene 96-well plate (Nalge Nunc) using a Biochip

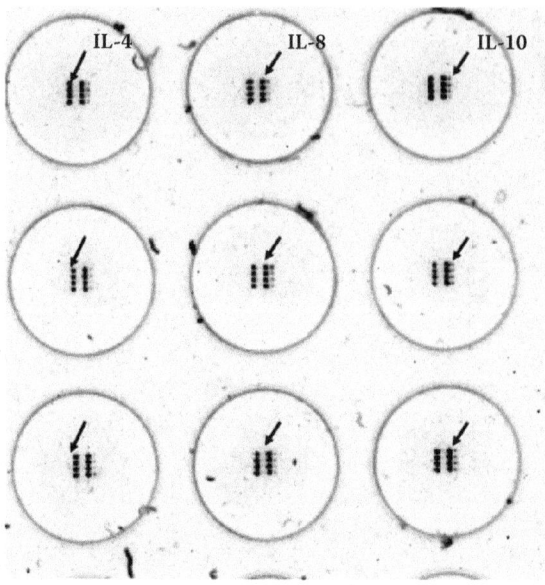

FIGURE 4.44   Cytokine antibody array.

Arrayer (Packard Instruments). The MAbs were prepared at 50 μg/mL in Dulbeco's PBS without calcium or magnesium (Life Technologies) and deposited at 20 nL per spot with a resulting spot diameter of about 400 microns. Well microarrays were then blocked with SuperBlock buffer (Pierce Endogen) for use in a micro-ELISA. Pierce's SearchLight™ microplate antibody microarray products (now part of Aushon's product line) are based upon this technology.

Angenendt et al. (2003) evaluated several slide surface chemistries for use as protein and antibody microarrays. They reasoned that because proteins vary greatly in their surface charge and relative hydrophobicity that a careful selection of surface chemistry may be important in order to obtain optimal performance for a particular protein. Thus, several commercially available slide surface chemistries were evaluated for performance in model arrays composing a protein dilution series including anti-fibrinogen antibody (antibody array) and human serum albumin (antigen array). The surfaces included poly-L-lysine, polystyrene, epoxy terminated polyethylene glycol (PEG) or dendrimer slides, various amine derivatized surfaces, and nitrocellulose coated slides. All proteins were printed in PBS, rinsed in TBS, and then blocked in 3% nonfat dry milk powder dissolved in TBS-0.1% Tween-20. A final rinse in TBS was performed prior to incubation. While there was no attempt to optimize print buffer or blocking conditions for each of the selected surfaces, it was apparent that with the exception of activated polystyrene most chemistries performed at about the same level (i.e., within two- to threefold at saturation). Detection limits ranged for the antigen array (~60 to ~90 amol) and the antibody array (~100 to <400 amol). Dendrimer coated slides appeared to offer some advantage for immobilization of antibodies providing the lowest detection limit and highest mean signal intensity. However, with CVs ranging from 16 to 43% it would be difficult to predict an optimal substrate for proteins. The study does show that diverse surface chemistries can be used to create protein microarrays at similar levels of performance by applying a simple printing protocol based upon PBS, a buffer commonly employed in working with proteins. What was not addressed is protein stability on these supports.

In summary, we have examined a number of studies concerned with the printing of proteins onto various substrates possessing different coating chemistries. The origin of most employed classic methodologies rests upon the decades of research on the development of the ELISA or in the coupling of proteins to affinity matrices. Table 4.3 provides an overview.

Where the protein microarray differs from the classic ELISA is in the much smaller quantities of proteins that are deposited on the substrate. So while micrograms (μg ~10 to 6 grams) of protein are employed in the coating of an ELISA well, microarray spots may contain only picograms (pg ~10 to 12 grams) or less of protein. The accurate delivery of small volumes (pL-nL) containing small amounts of protein can be problematic. A certain amount of protein is likely to adsorb onto the quills or capillaries used for printing. The degree at which adsorption takes place will be dependent upon a number of factors such as buffer composition, the quill's surface features, and more importantly, the physical-chemical nature of the protein.

Delehanty and Ligler (2003) from the Naval Research Laboratory (Washington, DC) described glass capillary piezoelectric dispensing of antibodies and the effects of buffer composition on printing efficiency. Here, the printing ink is aspirated into a glass capillary surrounded by the piezoelectric collar. The application of a voltage to this element causes the aspirate solution to compress and expel a small droplet from the capillary orifice. Typically, picoliter volumes ($0.1 < x < 1$ nL) are dispensed to the surface. The observed problem is that of nonspecific adsorption of protein within the borosilicate capillary tube. In this study the effect of ionic strength and carrier protein (BSA) were examined in terms of the outcome for microarray printing. As we have pointed out, there are many factors that can unduly influence printing performance. Proteins are notoriously bad when it comes to nonspecific adsorption. It should not be much of a surprise that some portion of the protein probe will adsorb to the printing device whether it is a stainless steel quill or a glass capillary.

In the case of the piezo system, biotinylated antibodies were dispensed at 1 nL to produce spot diameters of 230 microns on avidin coated glass slides (NeutrAvidin™, Pierce Chemical). The concentration of biotinylated antibody (Cy5-labeled mouse IgG) was varied from 2.5 to 20 μg/mL, while the ionic strength varied by dilution of PBS (10 mM to 150 mM, where 150 mM = 137 mM NaCl, 3 mM KCl, 10 mM phosphate, pH 7.4) in distilled water. Bovine serum albumin (BSA) was added as a carrier protein at 0.1% (w/v). After printing the slides were rinsed in 150 mM PBS containing 0.05% Tween-20 (PBST), followed by distilled water, and were then dried under a nitrogen stream.

Two effects were observed. First, in the absence of carrier BSA the amount of IgG deposited onto the slide was reduced at high ionic strength. Presumably more IgG remained adsorbed to the borosilicate glass capillary under high salt conditions. As the input concentration of IgG increased, there was less of an ionic strength buffer effect. Most likely at the higher IgG (20 μg/mL protein), concentration sufficient coating of the capillary was achieved. However, the addition to the print buffer of

TABLE 4.3
Conditions for Printing Protein Microarrays

Study	Print Buffer	Additives	Protein Conc.	Substrate	Rinses	Blocking Buffer
1999 Mendoza et al.	50 mM carbonate, pH 8	None	100 µg/mL	NHS-glass	TBST	PBS Casein
2000 MacBeath and Schreiber	PBS	40% glycerol	100 µg/mL	NHS-BSA glass Aldehyde-glass	PBS-glycine, pH 8 PBS, pH 7.5	None PBS 1% BSA
2000 Joos	PBS	10% glycerol 0.1% SDS 5 µg/mL BSA	0.2–1.2 mg/mL	Aldehyde-glass	PBST	PBS 1.5% BSA 5% milk
2001 Haab et al.	PBS	None	0.1–0.3 mg/mL	PolyLysine glass	PBST, 3% milk	PBS 3% milk 0.02% NaN$_3$
2001 Matson et al.	50 mM carbonate, pH 9	4% sodium sulfate	0.2–1 mg/mL	ACF-plastic (PP)	TBST	TBST 0.1% casein
2001 Moody et al.	PBS	None	50 µg/mL	Nunc Maxisorb plastic (PSt)	None	SuperBlock
2003 Angenendt et al.	PBS	0.1% NaN$_3$ preservative	0.3 µg–1.2 mg/mL	Various: glass coated; membrane coated; plastic w/epoxy, poly-L-lysine, amine, NC	TBS	TBST 3% milk
2003 Delehanty and Ligler	PBS	0.1% BSA	Biotinylated IgG 2.5–20 µg/mL	Avidin-glass	PBST	PBS 1% BSA 0.01% NaN$_3$ (pre-spotted slide storage)

Note: BSA, bovine serum albumin; TBS, Tris buffered saline; PBS, phosphate buffered saline; PBST, phosphate buffered saline-tween; TBST, tris buffered saline-tween.

0.1% BSA (1 mg/mL) was more effective in coating the capillary, thereby significantly reducing the nonspecific adsorption of IgG. Thus, more IgG was deposited onto the slide. An added benefit with the addition of BSA was the finding that spot morphology improved. There was a reduction in the "donut" spot morphology most likely due to less peripheral drying occurring with the higher protein content within the spot.

NEWER METHODS FOR PRINTING

Acoustic Dispensing

Wong and Diamond (2009) describe the use of acoustic dispensing for microarray printing. Ultrasonic waves are focused under a source plate resulting in the ejection of a droplet upward. A coupling fluid, such as water, is required between the acoustic transducer and the bottom of the source plate to propagate the energy. Positioning a slide or inverted microplate above the source plate permits collection of the droplet at a precise location. The droplet volume transferred from the source plate well can be controlled by adjusting the ultrasonic transducer energy level dissipated (Figure 4.45). Thus, 1 to 2 nL droplets can be transferred to the targeted slide or microplate well.

Conventional contact split pin printing of microarrays (OmniGrid Accent, GeneMachines; SMP4 Stealth microarray pins, Telechem Int'l Inc.) was compared

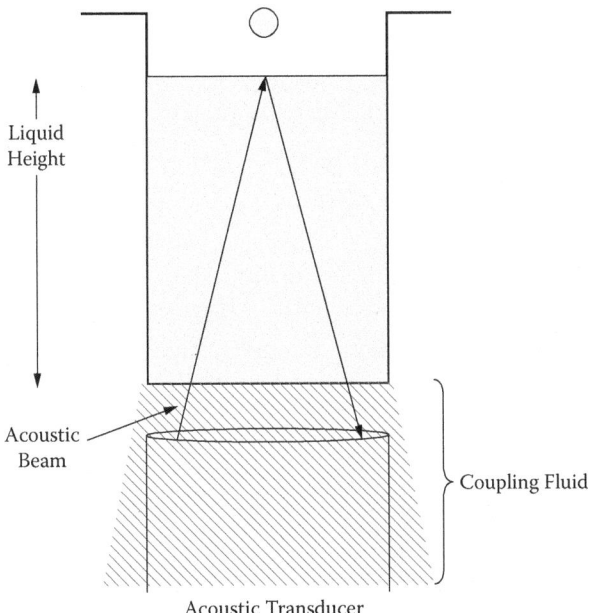

FIGURE 4.45 Acoustic dispensing. (From Wong EY, Diamond SL. Advancing microarray assembly with acoustic dispensing technology. *Anal. Chem.* 81(1), 509–514, 2009. With permission.)

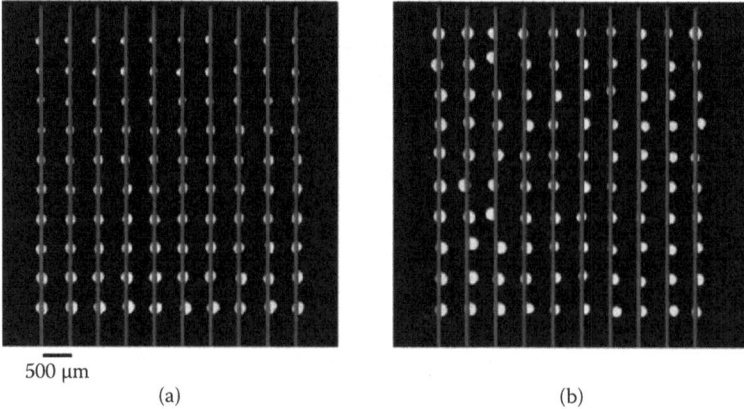

500 µm

(a)   (b)

FIGURE 4.46 Comparison of pin- and acoustic dispense–generated microarrays. (From Wong EY, Diamond SL. Advancing microarray assembly with acoustic dispensing technology. *Anal. Chem.* 81(1), 509–514, 2009. With permission.)

with that of acoustic printing using the ATS-100 acoustic dispenser (EDC Biosystems). In this demonstration, 2 nL droplets of a fluorescent dye in 10% glycerol were printed in an array format at 500 µm × 500 µm spacing using both systems (Figure 4.46). The arrayed droplet features were then compared:

1. *The conventional microarray spots were of uniform spacing but varied in spot diameter.* This is a well-known phenomenon of contact printing which can be lessened by optimizing the pre-print and pin acceleration parameters. Uniform spot diameters with CV < 5% are very achievable from contact printing.
2. *The acoustic dispenser produced spots of uniform spot diameter, but the spots were slightly misaligned.* The alignment becomes less of an issue provided that the image analysis software algorithm is flexible enough to correctly identify and measure the misaligned spots. However, if the spot diameters are small but the array set out at a relatively larger center-to-center spacing, there is the risk of flagging badly misaligned spots as debris or satellites.

There are certain advantages that acoustical dispensing provides:

Drop-on-Drop: In this study, droplets of the fluorescent dye AMC (7-amino-4-methylcoumarin) in water were dispensed into a previously dispensed array of 40 nL droplets of 10% glycerol-water. This was followed by a stepwise dispensing of fluorescein isothiocyanate (FITC) (in DMSO) to each spot. Fluorescent scanning revealed that a uniform signal for each dye was achieved at CV < 5%. Thus, acoustical dispensing permits drop-on-drop additions. This is not practical with contact printing. It is possible with other noncontact printing technologies such as solenoid driven systems (e.g., BioDot's Biojet technology).

Variable Droplet Volume: Spots of varying diameter were demonstrated by dispensing multiple 2 nL droplets at fixed locations. For example, a 2 nL droplet of 10% glycerol-water yields a 200 μm spot diameter, while 20 dispenses at the same location produced a 750 μm spot. Again, pins do not permit transfer of variable volumes on demand, but piezo and solenoid driven dispensers have this capability.

True Noncontact Dispensing: Acoustical dispensing is accomplished free of contact with the sample (albeit, delivery involves a source plate) and the microarray substrate. This reduces the risk of cross-contamination and carryover that can plague contact pin printing due to incomplete removal of the sample from pins between dispenses. Other noncontact dispensers are subject to the same because their nozzles and valves are in contact with the sample.

Capability for Bead Dispensing: The EDC acoustical dispenser was found to quantitatively deliver 1 μm diameter fluorescent beads from suspensions containing 107 to 108 beads per milliliter. This was an impressive demonstration in creating a 11 × 20 array of microsphere gradients at CVs ranging from 8 to 14%.

FLEXOGRAPHIC PRINTING

The large-scale and high-throughput manufacture of antibody arrays on planar substrates is problematic. Most antibody arrays are prepared using arrayer-based technologies utilizing contact pin printing, piezo ink-jet, or solenoid driven dispensers. Generally, antibodies are dispensed onto substrates such as glass slides or microplates. The dispense times are on the order of one drop or feature per second depending upon the complexity of the microarray pattern. For example, a Biojet noncontact dispenser system (BioDot, Inc.) was able to deliver a 10 × 10 array in 109 seconds (Sparks et al., 2006). However, this is somewhat deceiving because considerable time is lost in pre-printing (contact pin printing), rinsing, and drying times. A general "rule of thumb" would be about 1 to 2 minutes per cycle per unique feature. For example, to print one substrate with 384 unique features requires 6.4 hours of run time using a single split pin (Rose, 2000). Of course, the more dispensers or pins used, the less time is required. Thus, using a print head with eight dispensers would accomplish the task in under 1 hour. Yet, these times are relatively slow for large-scale production purposes.

Phillips (2012) describe the application of flexographic printing of antibodies. Flexographic printing is traditionally used in the production of graphic labels for packaging and electronics. It is a high-throughput contact inking reel-to-reel process involving a photopolymer relief plate that transfers the image onto the substrate. The relief plate is first "inked" by the anilox (engraved cylinder) and then pressed up against the substrate with the aid of an impression cylinder (Figure 4.47). This is a continuous process that can print at a rate of hundreds of meters per minute using web widths greater than 1 meter. In contrast, reel-to-reel dispensers and conveyor systems commonly used in the production of lateral flow devices can approach 6 to 12 meters/minute but are limited by web widths of 98 mm (BioDot, Inc.).

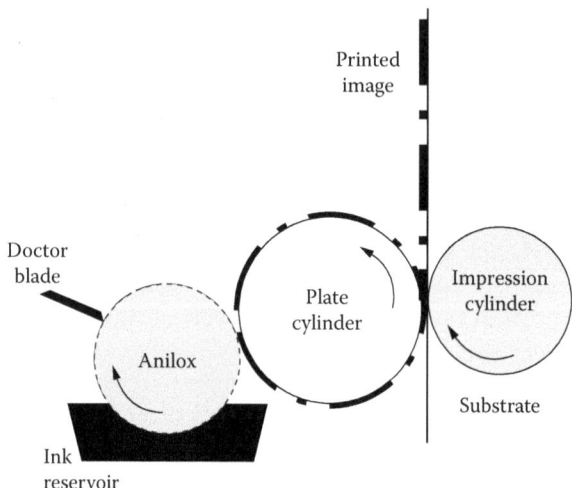

FIGURE 4.47   Flexigraphic printing process. (From Phillips, 2012. With permission.)

In this study, optimization of the flexographic printing parameters were undertaken that eventually lead to the creation of 6 × 6 array patterning of a single antibody. Unfortunately, the images provided are not sharp and appear to be of low print quality. Certainly, these results are not of a quality comparable to that of current microarray product offerings. Likewise, the representative antibody arrays are not of uniform spot morphology (Figure 4.48). This is most likely due to the choice of substrate coatings. Diffusion on nitrocellulose leading to nonuniform spotting is well known and requires considerable optimization of both binding and drying conditions. Prolonged generation of the chemiluminescent signal can also lead to "signal bleed" that further exaggerates the nonuniformity.

FIGURE 4.48   Flexigraphic printed antibody array. (From Phillips, 2010. With permission.)

This study does an excellent accounting of the optimization processes necessary to develop a flexographic printing of antibodies. Further improvements in substrate materials, binding conditions, and post-print processing are needed.

Microarray-to-Microarray Slide Snapping Transfer

Li et al. (2012) have devised a method to perform multiplex immunoassays while avoiding issues of cross-reactivity from secondary (signal) antibody interactions (see Chapter 5). Two slides are used in the process. A nitrocellulose "assay" slide is spotted with capture antibodies. The corresponding secondary detection antibodies are printed down on an aminosilane "transfer" slide at specific locations that permit a precise alignment with capture antibodies on the assay slide. Following incubation of the assay slide with sample, the two slides are sandwiched together using a handheld alignment tool.

The tool is composed of two block parts, each with a vacuum chuck. The assay slide is nestled into a recess on one part, while the transfer slide is seated into a similar recess on the opposing piece. Alignment pins on the transfer block are guided manually into holes on the assay block. The blocks are clamped together using a C-clamp. Vacuum is applied delivering the detection antibodies in droplets aligned over the corresponding capture antibody on the assay slide. The authors refer to this as "snapping" of the slides. Vacuum is then released and the two slides separated. The assay slide is subsequently developed and signal detected using a microarray slide scanner.

There are two design features worthy of further comment. First, the opposing faces of the assay and transfer slides represent mirror images. This presents an alignment problem. The opposing slides are likely to not be dimensionally exact (i.e., within microns), making it very difficult to simply align the slide edges. To solve this problem an alignment fiduciary was printed on the back of the transfer slide matching the top rightmost spot on the assay slide. This enabled precise alignment of the two slides. Experiments were performed demonstrating the ability to precisely overlap a sandwich immunoassay in this manner (Figure 4.49).

The second aspect concerns the control of the spot size on the two slides. For the assay slide, 300 micron diameter spots of capture antibody (1.2 nL droplet dispense) were arrayed. On the transfer slide the detection antibodies were dispensed in 8 nL droplets. Upon "snapping" the detection antibody droplet produced a 700 micron diameter overlay of the capture antibody spot. This further reduced the alignment stringency. The center-to-center spacing of the arrays was set at 800 microns leaving a 100 micron spacing between spots following transfer.

A 10-plex sandwich immunoassay was performed using the snap slide method. Analysis of protein antigen standards in PBS buffer revealed that pg/mL levels of sensitivity were achievable (Figure 4.50). Limits of detection (LODs) were established at either 2 SD or 3 SD relative to the background values. These were compared with standard ELISA for these proteins reported in the literature by the supplier (R&D Systems). The LODs for the single-plex standard ELISA were found to be lower. The multiplex ELISA averaged sixfold higher in LOD (2 SD) than that of the single-plex in the comparison of 8/9 antigens that were assayed by both methods. A single antigen was found to be fivefold higher in LOD with the standard ELISA.

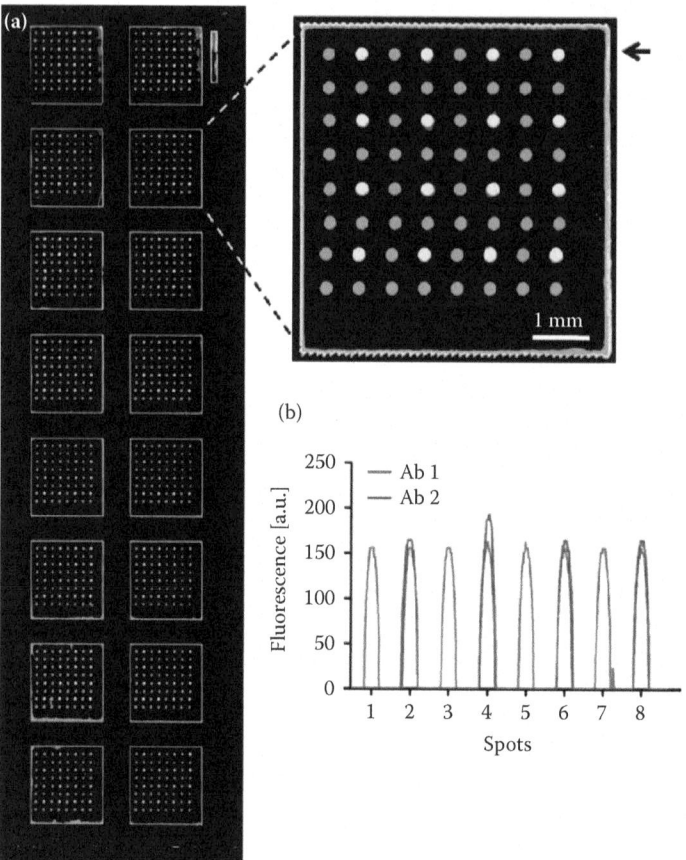

FIGURE 4.49 Microarray-to-microarray transfer by snapping leads to precision overlay. (From Li H, Bergeron S, Juncker D. Microarray-to-microarray transfer of reagents by snapping of two chips for cross-reactivity-free multiplex immunoassays. *Anal. Chem.* 84(11): 4776–4783, 2012. With permission.)

Antigens spiked into 10% serum were reported to be higher due to the presence of endogenous levels. Overall, the quality of printing was very good, and with respectable binding curves from 16 replicates. Slides could be stored dry for up to 1 month at –20°C without loss in performance. At 3 months storage a loss in activity was indicated by a rise in the LOD by four- to fivefold.

MICRO-CONTACT (μCP) PRINTING

The micro-patterning of substrates is often accomplished by transfer of chemicals, macromolecules, or cells adsorbed to a poly(dimethylsiloxane) (PDMS) stamp. The PDMS soft lithography and replication technologies were largely developed by George Whitesides' group at Harvard in the 1990s. In particular, the micro-patterning of self-assembled monolayers (SAMs) consisting of long-chain alkanes,

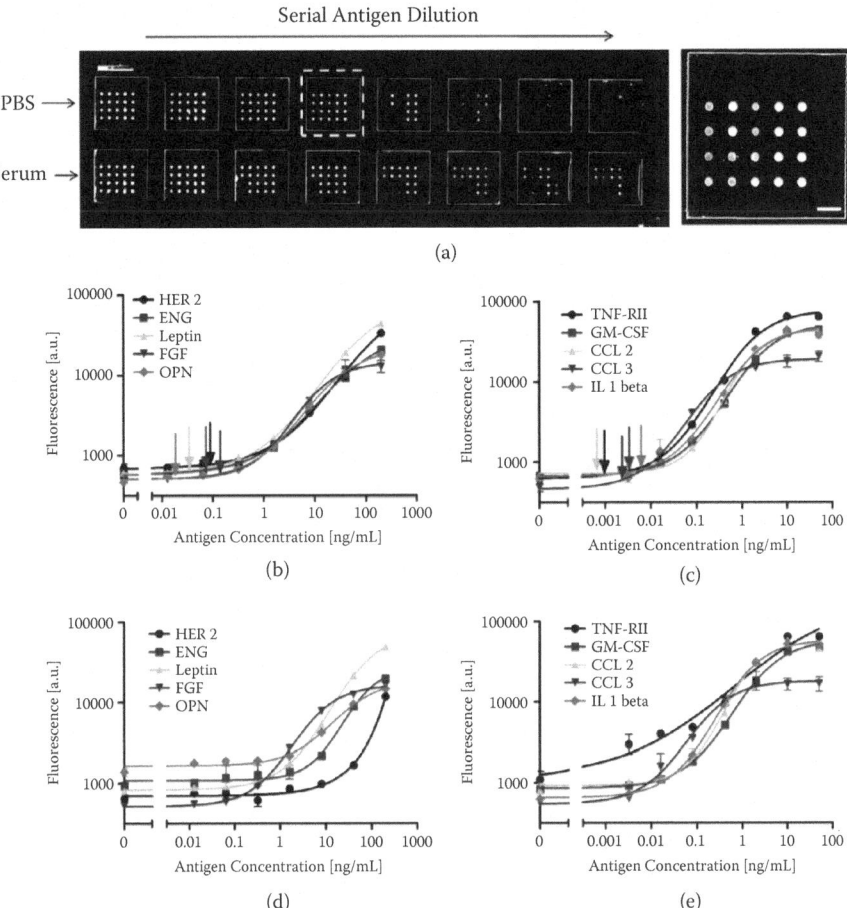

FIGURE 4.50 Binding curves for sandwich immunoassays on microarrays created by snapping transfer. (From Li H, Bergeron S, Juncker D. Microarray-to-microarray transfer of reagents by snapping of two chips for cross-reactivity-free multiplex immunoassays. *Anal. Chem.* 84(11): 4776–4783, 2012. With permission.)

especially thiolates and PEGs on gold surfaces are popular. However, Chien et al. (2012) chose to construct PDMS stamps in order to instead print patterns of polydopamine (PDA) onto various substrate surfaces. The resulting PDA layer can then be modified for immobilization of proteins, cells, or metal nanoparticles.

PDA is an interesting polymer whose origin is derived from studies of adhesive proteins found in marine mussels. The adhesive nature of these proteins is attributed to the presence of repeat units of dihydroxy phenylalanine-lysine residues, referred to as the DOPA-K motif. Dopamine, 2-(3,4-dihydroxyphenyl)ethylamine (Figure 4.51), has been reported to undergo oxidative polymerization producing a similar adherent film. The exact polymeric structure has been debated. Dreyer et al. (2012) have proposed a new structure for poly(dopamine) that suggests it

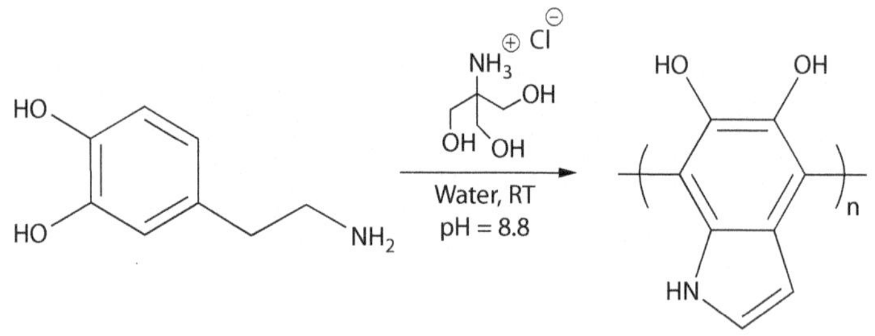

FIGURE 4.51 Oxidation of dopamine. (From Dreyer DR, Miller DJ, Freeman BD, Paul DR, Bielawski CW. Elucidating the structure of poly(dopamine). *Langmuir* 28: 6428–6435, 2012. With permission.)

actually is composed of aggregates of monomers held together by a combination of noncovalent forces, especially charge transfer, π-stacking, and hydrogen bonds (Figure 4.52).

While SAMs are generally restricted to metalized substrates, PDA is able to adhere to a variety of substrate materials including plastics and glass. PDA can also be effectively transferred from PDMS stamps providing a rather simple and straightforward means to replicate patterns. Once a PDMS stamp is constructed, the structure is wetted with dopamine hydrochloride in Tris buffer, pH 8.5. The stamp is then pressed down on the recipient substrate and held for 10 minutes to complete the pattern transfer.

In this study, bovine serum albumin (BSA) (FITC labeled) was immobilized on micro-patterned glass slides derived by the deposition of a mixture of dopamine and PEI-g-PEG (Figure 4.53). Because no cross-linking agent was included, it is assumed that the protein remained adsorbed to the surface by electrostatic interaction. However, Lee et al. (2009) have presented evidence suggesting that covalent attachment of proteins is possible by nucleophilic addition of available protein lysine residues with that of reactive catachol-quinone in PDA.

This study demonstrated the ability to pattern PDA on glass substrates and subsequently immobilize cells, protein, and gold and silver nanoparticles. It is conceivable

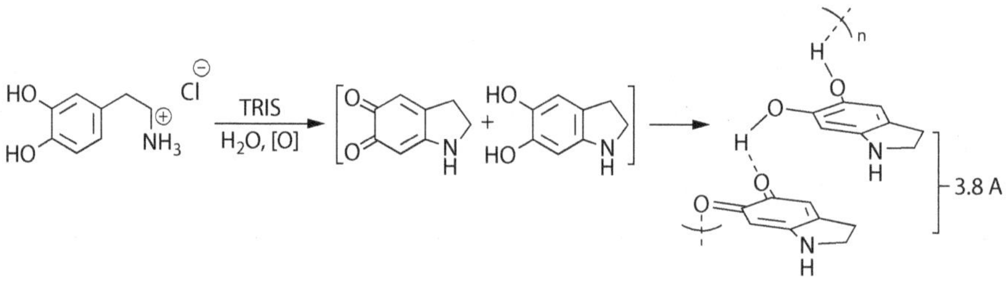

FIGURE 4.52 New structural model for poly(dopamine). (From Dreyer DR, Miller DJ, Freeman BD, Paul DR, Bielawski CW. Elucidating the structure of poly(dopamine). *Langmuir* 28: 6428–6435, 2012. With permission.)

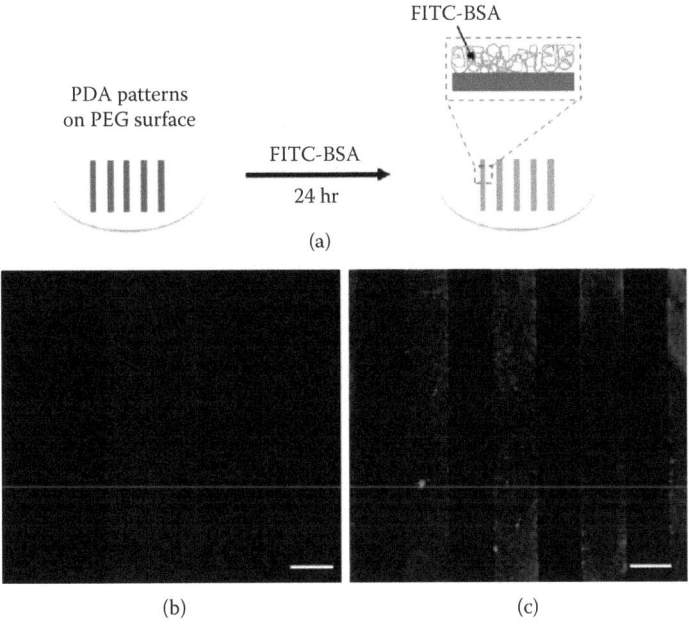

PDA patterns
on PEG surface

FITC-BSA

FITC-BSA
24 hr

(a)

(b)   (c)

FIGURE 4.53 Bovine serum albumin immobilization on polydopamine (PDA)-imprinted substrate. (From Chien H-W, Kuo W-H, Wang M-J, Tsai S-W, Tsai W-B. Tunable micropatterned substrates based on poly(dopamine) deposition via microcontact printing. *Langmuir* 28: 5775–5782, 2012. With permission.)

that automation of the process would provide a means to manufacture complex microarrays incorporating regions for, for example, electrochemical detection.

Creation of Peptide Nucleic Acid Microarrays by µCP

Calabretta et al. (2011) have demonstrated the micro-patterning of peptide nucleic acids (PNAs) arrays by micro-contact printing. Glass slide substrates were functionalized with aldehyde groups, while PNAs were synthesized with a terminal amine. This permits the tethering of the PNA by covalent attachment of the PNA-amine to the glass surface through Schiff's base formation (Figure 4.54). A piezo noncontact printer (Scienion AG, Berlin, Germany) was used to dispense droplets of various PNA-amines in an array pattern on flat PDMS stamps. Up to three separate stampings of the array were possible without a decrease in the hybridization efficiency with complementary DNA (Figure 4.55). The average spot size was reported to be 200 microns in diameter.

Experiments were undertaken to characterize the performance of the PNA array. For printing of PNA-amine on the PDMS stamp the optimal ink concentration was 20 µM. The PDMS stamp was held against the aldehyde activated slide for 15 minutes. A reduction in concentration from 10 to 1 µM resulted in reduced intensity of stamped images as determined from hybridization of fluorescently (TAMRA)

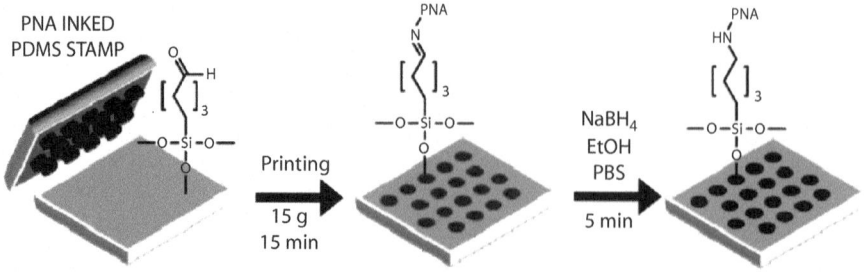

FIGURE 4.54 Patterning via reactive microcontact printing. (From Calabretta A, Wasserber D, Posthuma-Trumpie GA, et al. Patterning of peptide nucleic acids using a reactive micro-contact printing. *Langmuir* 27(4): 1536–1542, 2011. With permission.)

labeled complementary DNA. An increase in concentration from 20 to 100 µM did not lead to increased signal intensity.

The specificity of the imprinted PNA array was accessed by hybridization of complementary and mismatched DNA (11- to 13mer oligonucleotides) at T_m ranging from 48 to 75°C for the perfect matches. As predicted, perfect matches were found to be more stable than the mismatch based upon melt curve analysis. However, when PNA/DNA perfect match and single-base mismatch stabilities were compared both in the solution phase as well as by solid-phase hybridization, the effect of mismatch

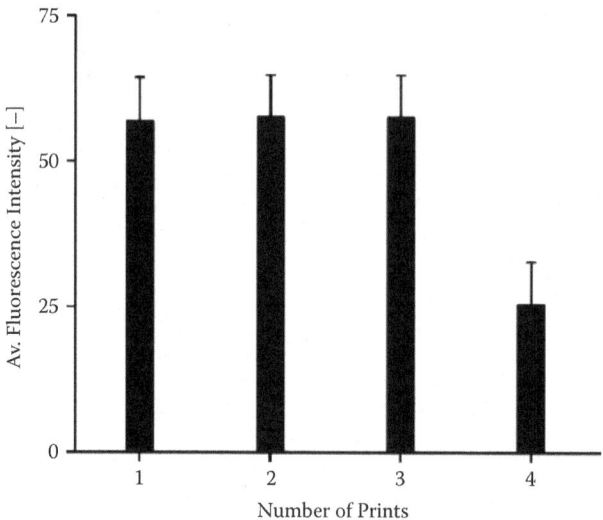

FIGURE 4.55 Number of replicate prints achieved by flat stamping. (From Calabretta A, Wasserber D, Posthuma-Trumpie GA, et al. Patterning of peptide nucleic acids using a reactive microcontact printing. *Langmuir* 27(4): 1536–1542, 2011. With permission.)

was found to be significantly different. Solution phase hybridization showed a difference in $\Delta T_m = T_m$ match $- T_m$ mismatch of 15°C, while the surface bound match versus mismatch resulted in only a $\Delta T_m = 5$°C. The reason for the added stability of the surface bound hybrid is unknown.

There were several advantages cited for undertaking the micro-patterning approach to array printing:

1. Reduction in processing time—The print and transfer was accomplished in 15 minutes resulting in covalent attachment of the PNA. The previous method involved an overnight incubation.
2. Improved spot morphology—Micro-contact printing resulted in spots having a higher degree of surface homogeneity than by conventional microarray printing processes.
3. Simple stamping process—Removes the operational constraints associated with arrayer-based printing such as the precise control of print buffer composition, temperature, and humidity.
4. Replicate transfers can be accomplished without re-inking.

Covalent Subtractive Print by µCP

Coyer et al. (2011) describe a micro-patterning process called *covalent subtractive printing* for the deposition of mixed SAMs of alkanethiols in which features are represented at both the micrometer and nanometer scales. First, nano-templates were produced by electron-beam lithography of silicon wafers to produce the desired micro-patterns (reverse image). Then, flat PDMS stamps were inked with a fibronectin solution and applied to the nano-template. Release of the PDMS stamp from the nano-template resulted in the removal (subtraction) of excess protein from the stamp creating the desired micro-pattern.

The inked stamps holding the fibronectin micro-pattern were then applied to an alkanethiol SAM-gold substrate. The SAMs were made up of alkanethiols doped with carboxy terminated alkanethiols representing approximately 1% of the mixture. Carboxyl groups were converted to active esters using EDC-NHS chemistry. Upon stamping, the fibronectin covalently coupled to the active ester thereby creating the micro-pattern (Figure 4.56). The pattern could then be detected using fluorescent dye labeled anti-fibronectin antibodies.

A micro-pattern of fibronectin was developed to control the formation of focal adhesions. These are structures that are required by cells to adhere to each other during cell spreading. Focal adhesions form junctions between one cell and the extracellular matrix of another cell. Vinculin is a structural protein of the focal adhesion that regulates cell adhesion strength. Thus, the patterning of fibronectin was designed to limit the localization of vinculin, thereby controlling cell spreading (Figure 4.57).

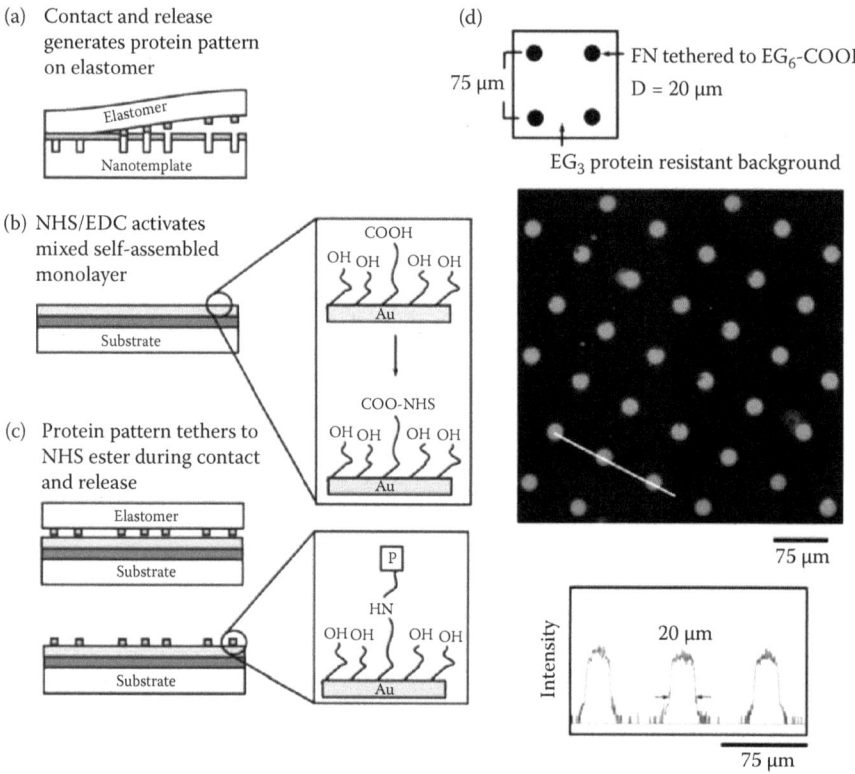

FIGURE 4.56   Subtractive contact printing of fibronectin. (From Coyer SR, Delamarche E, Garcia AJ. Protein tethering into multiscale geometries by covalent subtractive printing. *Adv. Mater.* 23(13): 1550–1553, 2011. With permission.)

REFERENCES

Angenendt P, Glokler J, Sobek J, Lehrach H, Cahill DJ. Next generation of protein microarray support materials: Evaluation for protein and antibody microarray applications. *J. Chromatogr. A* 1009: 97–104, 2003.

Balázsi G, Kay KA, Barabási A-L, Oltvai ZN. Spurious spatial periodicity of co-expression in microarray data due to printing design. *Nucl. Acids Res.* 31(15): 4425–4433, 2003.

Battaglia C, Salani G, Consolandi C, et al. Analysis of DNA microarrays by non-destructive fluorescent staining using SYBR green II. *Biotechniques* 29(1): 78–81, 2000.

Calabretta A, Wasserber D, Posthuma-Trumpie GA, et al. Patterning of peptide nucleic acids using a reactive microcontact printing. *Langmuir* 27(4): 1536–1542, 2011.

Chien H-W, Kuo W-H, Wang M-J, Tsai S-W, Tsai W-B. Tunable micropatterned substrates based on poly(dopamine) deposition via microcontact printing. *Langmuir* 28: 5775–5782, 2012.

Coyer SR, Delamarche E, Garcia AJ. Protein tethering into multiscale geometries by covalent subtractive printing. *Adv. Mater.* 23(13): 1550–1553, 2011.

Daub M. TopSpot Technology: Highly parallel dispensing for production of microarrays. *EuroBiochips Conference*, Berlin, Germany, 2002.

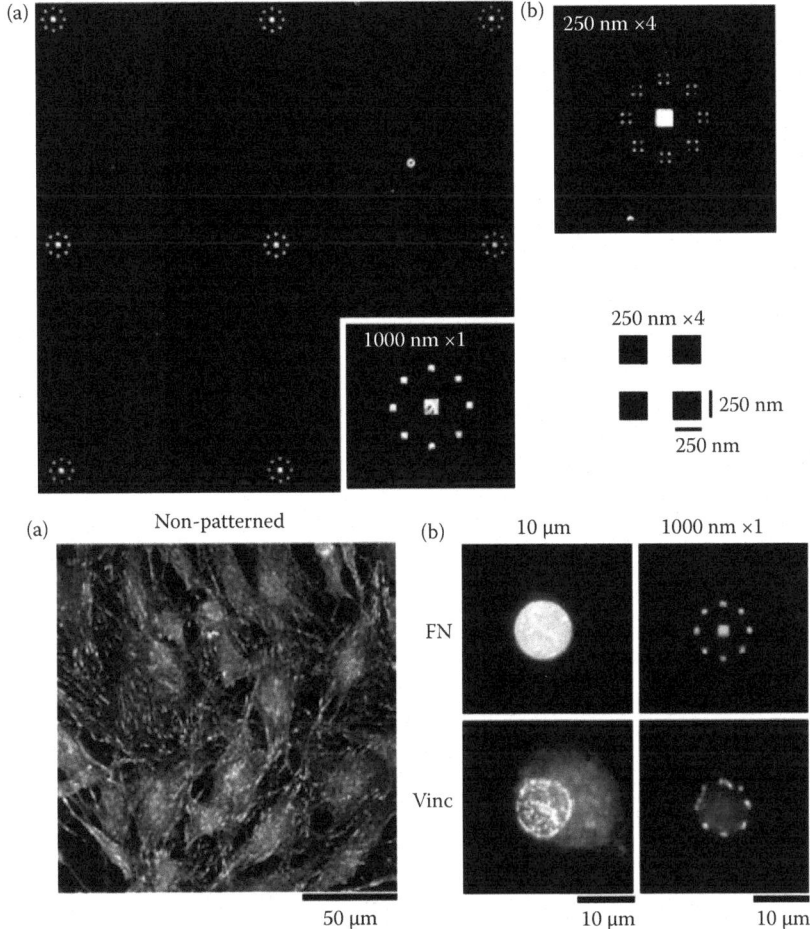

FIGURE 4.57 Localization of cell adhesion protein by covalent subtractive printing. (From Coyer SR, Delamarche E, Garcia AJ. Protein tethering into multiscale geometries by covalent subtractive printing. *Adv. Mater.* 23(13): 1550–1553, 2011. With permission.)

Delehanty JB, Ligler FS. Method for printing functional protein microarrays. *BioTechniques* 34(2): 380–385, 2003.

Diehl F, Grahlmann S, Beier M, Hoheisel JD. Manufacturing DNA microarrays of high spot homogeneity and reduced background signal. *Nucl. Acid Res.* 29(7) e38: 1–5, 2001.

Dreyer DR, Miller DJ, Freeman BD, Paul DR, Bielawski CW. Elucidating the structure of poly(dopamine). *Langmuir* 28: 6428–6435, 2012.

George RA, Woolley JP, Spellman PT. Ceramic capillaries for use in microarray fabrication. *Genome Res.* 11: 1780–1783, 2001.

Haab BB, Dunham MJ, Brown PO. Protein microarrays for highly parallel detection and quantitation of specific proteins and antibodies in complex solutions. *Genome Biol.* 2(2): research0004.1–0004.13, 2001.

Hedge P, Qi R, Abernathy C, et al. A concise guide to cDNA microarray analysis. *BioTechniques* 29(3): 548–562, 2000.

Hessner MJ, Wang A, Hulse K, et al. Three color cDNA microarrays: Quantitative assessment through the use of fluorescein-labeled probes. *Nucl. Acid Res.* 31(4) e14: 1–6, 2003a.

Hessner MJ, Wang X, Khan S, et al. Use of a three-color cDNA microarray platform to measure and control support-bound probe for improved data quality and reproducibility. *Nucl. Acid Res.* 31(11) e60: 1–9, 2003b.

Joos TO, Schrenk M, Hopfl P, et al. A microarray enzyme-linked immunosorbent assay for autoimmune diagnostics. *Electrophoresis* 21: 2641–2650, 2000.

Kane MD, Jatkoe TA, Stumpf CR, Lu J, Thomas JD, Madore SJ. Assessment of the sensitivity and specificity of oligonucleotide (50mer) microarrays. *Nucl. Acid Res.* 28(22): 4552–4557, 2000.

Lane S, Birse C, Zhou S, Matson R, Liu H. DNA array studies demonstrate convergent regulation of virulence factors by Cph1, Cph2, and Efg1 in *Candida albicans. J. Biol. Chem.* 276(52), 48988–48996, 2001.

Li H, Bergeron S, Juncker D. Microarray-to-microarray transfer of reagents by snapping of two chips for cross-reactivity-free multiplex immunoassays. *Anal. Chem.* 84(11): 4776–4783, 2012.

Macas J, Nouzova M, Galbraith DW. Adapting the Biomek®2000 Laboratory Automated Workstation for printing DNA microarrays. *BioTechiques* 25(1): 106–109, 1998.

MacBeath G, Schreiber SL. Printing proteins as microarrays for high-throughput function determination. *Science* 289: 1760–1763, 2000.

Martinez MJ, Aragon AD, Rodriguez AL, et al. Identification and removal of contaminating fluorescence from commercial and in-house printed DNA microarrays. *Nucl. Acid Res.* 31(4) e18: 1–8, 2003.

Matson RS, Rampal J, Pentoney SL, Anderson PD, Coassin P. Biopolymer synthesis on polypropylene supports: Oligonucleotide arrays. *Anal. Biochem.* 224: 110–116, 1995.

Matson RS, Milton RC, Cress, MC, Rampal JB. Microarray-based cytokine immunosorbent assay. *Oak Ridge Conference*, Poster No. 20, Seattle, WA, 2001.

McQuain MK, Seale K, Peek J, Levy S, Haselton FR. Effects of relative humidity and buffer additives on the contact printing of microarrays with quill pins. *Anal. Biochem.* 320: 281–291, 2003.

Mendoza LG, McQary P, Mongan A, Gangadharan R, Brignac S, Eggers M. High-throughput microarray-based enzyme-linked immunosorbent assay (ELISA*). BioTechiques* 24(4): 778–788, 1999.

Moody MD, Van Arsdell SW, Murphy KP, Orencole SF, Burns C. Array-based ELISAs for high-throughput analysis. *BioTechniques* 31(1): 1–7, 2001.

Okamoto T, Suzuki T, Yamamoto N. Microarray fabrication with covalent attachment of DNA using Bubble Jet technology. *Nat. Biotechnol.* 18: 438–441, 2000.

Phillips CO, Govindarajan S, Hamblyn SM, et al. Patterning of antibodies using flexographic printing. *Langmuir* 28(25): 9878–9884, 2012.

Rees WA, Yager TD, Korte J, Von Hippel PH. Betaine can eliminate the base pair composition dependence of DNA melting. *Biochemistry* 32(1): 137–144, 1993.

Reese MO, van Dam RM, Scherer A, Quake SR. Microfabricated fountain pens for high-density DNA arrays. *Genome Res.* 13: 2348–2352, 2003.

Rose D. Chapter 2: Microfluidic technologies and instrumentation for printing DNA microarrays in Microarray Printing Technologies in *Microarray Biochip Technology.* Schena M. ed., Natick MA, Eaton Publishing: 19–38, 2000.

Schena M, Shalon D, Davis RW, Brown PO. Quantitative monitoring of gene expression patterns with a complementary DNA microarray. *Science* 270: 467–470, 1995.

Shearstone JR, Allaire NE, Getman ME, Perrin S. Nondestructive quality control for microarray production. *BioTechniques* 32(5): 1051–1057, 2002.

Silzel JW, Cerecek B, Dodson C, Tsong T, Obremski RJ. Mass-sensing, multianalyte microarray immunoassay with imaging detection. *Clin. Chem.* 44(9): 2036–2043, 1998.

Sparks CM, Gondran CH, Havrilla GJ, Hastings EP. Automated nanoliter solution deposition for total reflection X-ray fluorescence analysis of semiconductor samples. *Spectrochimica Acta* Part B, 61: 1091–1097, 2006.

Spellman PT, Sherlock G, Zhang MQ, et al. Comprehensive identification of cell cycle-regulated genes of the yeast *Saccharomyces cerevisiae* by microarray hybridization. *Mol. Biol. Cell* 9: 3273–3297, 1998.

Stimpson DI, Cooley PW, Knepper SM, Wallace DB. Parallel production of oligonucleotide arrays using membranes and reagent jet printing. *BioTechniques* 25(5): 886–890, 1998.

Wang X, Ghosh S, Guo S-W. Quantitative quality control in microarray image processing and data acquisition. *Nucl. Acid Res.* 29(15) e75: 1–8, 2001.

Wang X, Hessner MJ, Wu Y, Pati N, Ghosh S. Quantitative quality control in microarray experiments and the application in data filtering, normalization and false positive rate prediction. *Bioinformatics* 19(11): 1341–1347, 2003.

Wong EY, Diamond SL. Advancing microarray assembly with acoustic dispensing technology. *Anal. Chem.* 81(1): 509–514, 2009.

Wu P, Grainger DW. Comparison of hydroxylated print additives on antibody microarray performance. *J. Proteome Res.* 5(11): 2956–2965, 2006.

Yang H, Dudoit S, Luu P, et al. Normalization of cDNA microarray data: A robust composite method addressing single and multiple slide systematic variation. *Nucl. Acids Res.* 30(4) e15: 1–10, 2002.

Zipper H, Brunner H, Bernhagen J, Vitzthum F. Investigations on DNA intercalation and surface binding by SYBR Green I, its structure determination and methodological implications. *Nucl. Acids Res.* 32(12): e103, 2004.

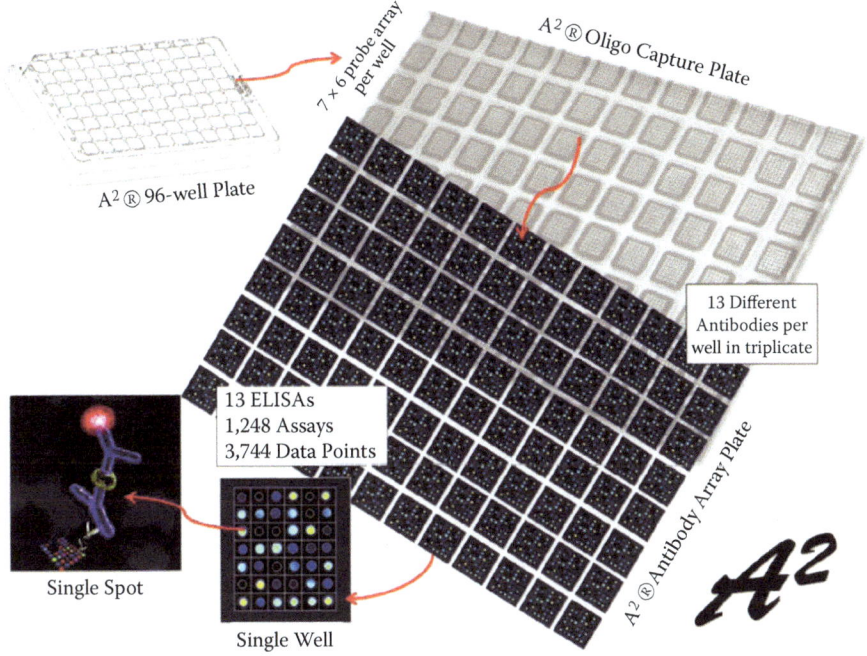

7 × 6 probe array per well

A^2 ® Oligo Capture Plate

A^2 ® 96-well Plate

13 Different
Antibodies per
well in triplicate

13 ELISAs
1,248 Assays
3,744 Data Points

A^2 ® Antibody Array Plate

A^2

Single Spot

Single Well

FIGURE 2.11 Anatomy of the A^2 plate assay.

Multiplexing

Introduction of more than one unique analyte-specific sensor into a single sample compartment that results in the detection of one or more analytes within the sample.

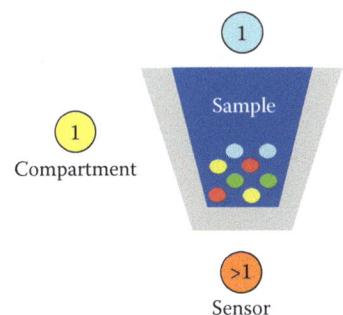

1

1
Compartment

Sample

>1
Sensor

FIGURE 7.1 Multiplex assay defined.

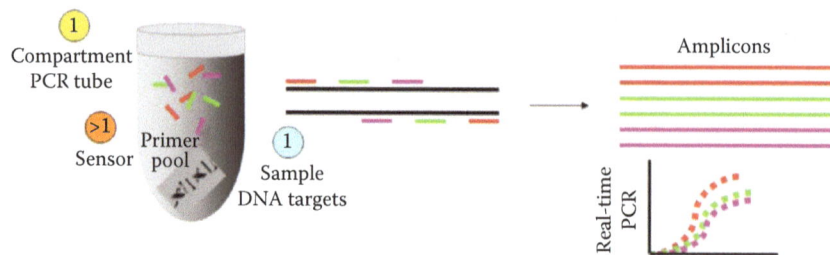

FIGURE 7.2   Multiplex polymerase chain reaction (PCR) assay.

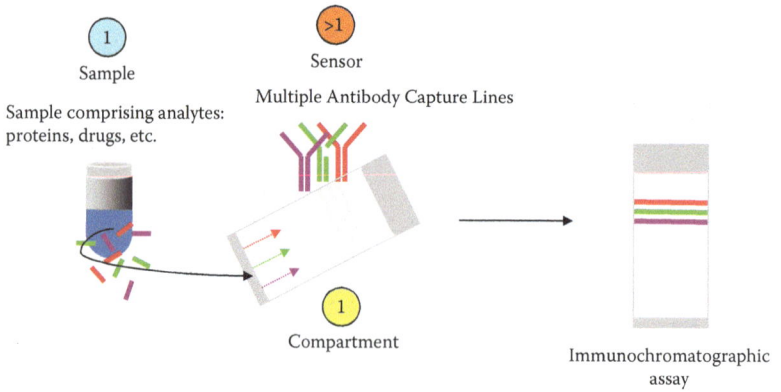

FIGURE 7.4   Multiplex lateral flow (LF) assay.

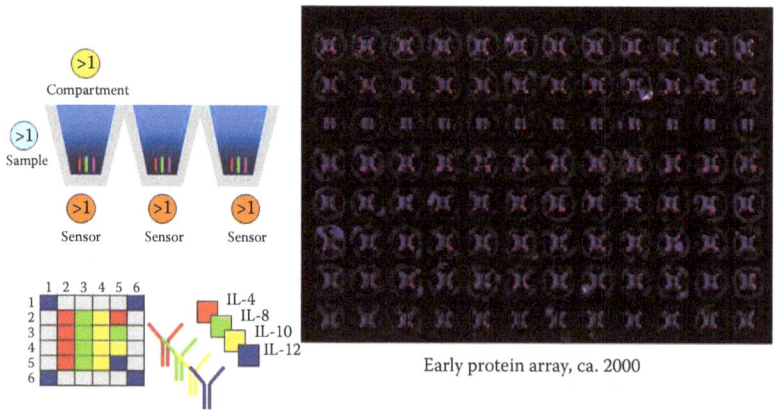

FIGURE 7.6   Multiplex microtiter plate (MTP) assay.

5 Gene Expression
Microarray-Based Applications

INTRODUCTION

We owe much of our present understanding of gene expression to microarray studies. Two enabling technologies—the Affymetrix biochip and the glass microarray slide—are responsible. The commercialization of Affymetrix's GeneChip provided perhaps the first standardized tool for genomics-based research. Aside from cost considerations, the GeneChip represented a "closed" system. Researchers at the time were limited by the oligonucleotide probe content provided on these biochips. This made it difficult for biologists who were not studying human genomics or genetic mutation. The glass slide microarray changed all of that in an instance.

In this chapter we trace the development of the gene expression microarray and discuss representative applications.

The microarray slide originated from the work by the Brown and Davis groups at Stanford University (Schena et al., 1995). It was an adaptation from filter grid arraying of cDNA clones to examine gene expression on a global scale. With miniaturization several benefits were realized: First, higher-density arrays could be produced allowing the monitoring of many more genes. Second, the sample volume, reagent consumption, and processing steps were greatly reduced. Third, fluorescence labeling and detection could be used in place of radiolabeling and autoradiography. With the development of the slide scanner based upon fluorescence detection, the throughput from sample to result was greatly increased. With dual-label competitive hybridization analysis (e.g., Cy3 labeled control gene population compared to Cy5 labeled sample population), the slide microarray became an enabling technology for genomic studies.

APPLICATIONS DEMONSTRATING DNA MICROARRAY UTILITY

GENE EXPRESSION

"The temporal, developmental, topographical, histological, and physiological patterns in which a gene is expressed provide clues to its biological role" (Schena et al., 1995, p. 467). With this introductory statement by Schena et al., the Stanford groups of Davis and Brown heralded what would soon become a new paradigm for biological investigation. At the center of this global examination of gene expression was the glass slide cDNA microarray. The Stanford paper in *Science* not only showed the power of the microarray in determining gene expression patterns but also provided much of the detail for others to construct their own microarrays. This is perhaps the more important contribution made by this landmark paper. Based upon Dari Shalon's Ph.D. thesis and methods later published on Ron Brown's Web site, the biologist was quickly enabled into the building of a robotic pin printer, preparation of slide arrays, the utility of two-color labeling, and the construction of a laser scanner (Shalon et al., 1996). So while others fiddled to the tune of sequencing by hybridization (SBH) or mutation detection in hopes of moving DNA arrays rapidly into diagnostics, the field of genomics was borne. SBH by now has essentially been abandoned, and while the potential for DNA arrays in diagnostics remains, the slide microarray continues to advance as an important genomics tool.

Armed with this new tool Schena et al. (1996) created a microarray of 1046 human cDNAs of unknown sequence. These were derived from human peripheral blood lymphocytes that had been transformed with Epstein-Barr virus. Suitably sized inserts (>600 bp) were cloned into a lambda vector, subsequently infected into an *Escherichia coli* strain, and finally amplified by polymerase chain reaction (PCR) using 5'-amino modified primers. The resulting 5'-amino modified cDNA amplicons were then arrayed onto silylated microscope slides. Next, the expression levels in human Jurkat cells undergoing heat shock or phorbol ester induction were examined. Total mRNA from control and induced cells were labeled using reverse transcriptase with the incorporation of fluorescene-dCTP (control; green label) or Cy5-dCTP (induced; red label). In this particular case, the two populations were hybridized to separate arrays. However, the labels were also swapped to verify that any differences in labeling efficiency did not affect the result.

Greater than 95% of the arrayed cDNA probes showed hybridization at signal intensities ranging over three logs, and only a few "genes" displayed significant differences in expression. In fact, only 17/1046 (1.6%) had changes between twofold to approximately sixfold. While the absolute expression levels of these genes varied considerably between microarray hybridization and RNA blots, the relative fold-changes showed a good correlation (R2 ~0.8) between that of the microarray and blot results (Figure 5.1). As we see from other work, most genes are not a priori expressed in response to a particular metabolic or environmental change, but rather a smaller number are significantly induced or repressed (Figure 5.2). Moreover, the response of these particular genes is important in the drug discovery process. Here is a case in point: 4 out of 17 (23.5%) of the twofold expressed genes were discovered to be novel upon sequencing of the cDNA clones. This also points to the power of the cDNA microarray in that the monitoring of gene expression patterns and the discovery of novel genes interactions can be done so without knowledge of gene sequence.

While the Schena et al. papers (1995, 1996) served as first demonstrations of cDNA microarray technology, it was clear that further refinements would be necessary in order for the full potential of the microarray to be realized. Arraying technology was at its infancy and suffered from inconsistency in uniform spotting, making it difficult to compare two slides. Refinements in labeling and detection were also needed. In spite of these shortcomings, relative expression levels could be monitored with some degree of confidence in the data by employing comparative hybridization (DeRisi, 1996). In this approach the labeled control and test mRNA populations are

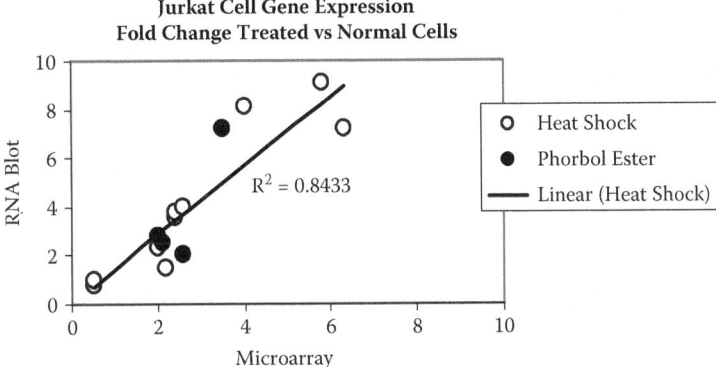

FIGURE 5.1   A comparison of gene expression levels: Northern (mRNA) versus microarray. (From Schena M, Shalon D, Heller R, Chai A, Brown PO, Davis RW. Parallel human genome analysis: Microarray-based expression monitoring of 1000 genes. *PNAS* 93: 10614–10619, 1996. With permission.)

The Comfort Zone

Sensitivity: Ability to discriminate important genetic events

Fold-Discrimination
≥ 2.0

Study	Diff. Expressed/Total Genes	% Diff. Expressed
Lock et al	1080/7026	15.4
Moustafa et al	213/12530	1.7
Lapteva et al	9/1081	0.8
Nishizuka et al	29/7000	0.4
Reilly et al	332/11000	3.0
Maeda et al	8/2304	0.3
Katsuma et al	82/4224	1.3

How significant are lower levels of expression?

FIGURE 5.2   The comfort zone.

mixed and then applied to a single slide array, thus avoiding problems associated with differences in slide-to-slide uniformity. Also, well-characterized synthetic targets could be doped into the samples to serve as internal standards and permit quantitative estimates of relative abundance levels between the two expressed gene populations. The two-color, comparative hybridization approach also allowed for a visual interpretation of the microarray results. DeRisi et al. (1996) labeled tumorigenic cell mRNA with a green dye and nontumorigenic control cell mRNA with a red dye. Thus genes overexpressed in tumor cells would show up as green spots, those preferentially expressed in control cells would have red spots, while equivalent levels of gene-specific expression between the cell lines would show an equal mix of red and green resulting in the appearance of yellow spots. Of course, if no genes were expressed then the spots would remain dark. In the DeRisi (1996) study an additional level of control was provided by examining 90 "housekeeping" genes that would remain invariant between control and test mRNA populations. In fact, some differential expression will occur, and therefore user-defined cutoffs in relative gene expression are necessary. For example, in this study the red/green (R/G) ratio for housekeeping gene expression was 1.13 while for internal standards it was $R/G = 0.97$. Setting minimal cutoffs at 3 SD, then ratios <0.52 and >2.4 are required to measure statistically significant levels of differential expression. At least a twofold change is necessary (i.e., the comfort zone) in order to compensate for variance in constitutive biological levels and assay performance. As a result, DeRisi et al. (1996) found that only about 9% of the 870 genes (15/870 down-regulated; 63/870 up-regulated) on the microarray were observed to be differentially expressed. Attempts to utilize data from lower levels of differential expression (i.e., less than twofold) have to date been controversial, although with careful experimental control this should be possible.

One of the most elegant early experiments with cDNA microarrays was the assessment of gene expression during the classic metabolic (diauxic) shift in *Saccharomyces cerevisiae* in going from anaerobic to aerobic states (DeRisi et al., 1997). Thus, superimposed upon the metabolic pathways characterizing the flow of metabolites during glycolysis and gluconeogenesis was now the temporal relationship of the pathway genes as the shift from fermentation to respiration takes place (Figure 5.3). What was most impressive about this study was that the microarray contained essentially the entire yeast genome of 6400 ORFs (open reading frames) providing a global genomic overview. Gene expression "snapshots" could be taken during the diauxic shift using microarrays and the entire metabolic process reviewed frame by frame. The mRNA from cells grown in glucose-rich media (fermentation state) were labeled with Cy3 (green) and served as the reference, while mRNA from cells transitioning into glucose-depleted media (shifting to aerobic respiration) were labeled with Cy5 (red). Note that the introduction of the Cy5/Cy3 ratio has been largely adopted as standard practice for microarray gene expression analysis although newer dyes such as the ALEXA series (Molecular Probes, Inc.) are now also in use.

The diauxic shift experiment turned out to be an extremely important demonstration. While *S. cerevisiae* exhibited very little in the way of differential expression activity (19/6400 genes showing twofold expression) during exponential growth in glucose-rich media, this was certainly not the case during glucose depletion. At expression levels measured at more than or equal to twofold, greater than 25% of the yeast genome had undergone induction (710/6400) or repression (1030/6400) in response to the anaerobic to aerobic shift. The experiments also resulted in the discovery of at least 400 genes that at the time were found to have no known function and did not appear on public databases. Those genes coding for enzymes associated with the metabolic pathways permitted a dynamic view of shifts in metabolite flow. For example, ALD2 gene encoding for aldehyde dehydrogenase and ACS1 gene for acetyl-CoA synthase undergo 12.4- and 13-fold inductions, respectively. These enzymes are responsible for moving acetyl-CoA into the tricarboxylic acid (TCA) and glyoxylate cycles. At the same time it was found that similar inductions of PCK1 gene (encoding phosphoendpyruvate (PEP) carboxykinase) and FBP1 gene (encoding fructose 1,6-biphosphatase) were responsible for reversing the flow of

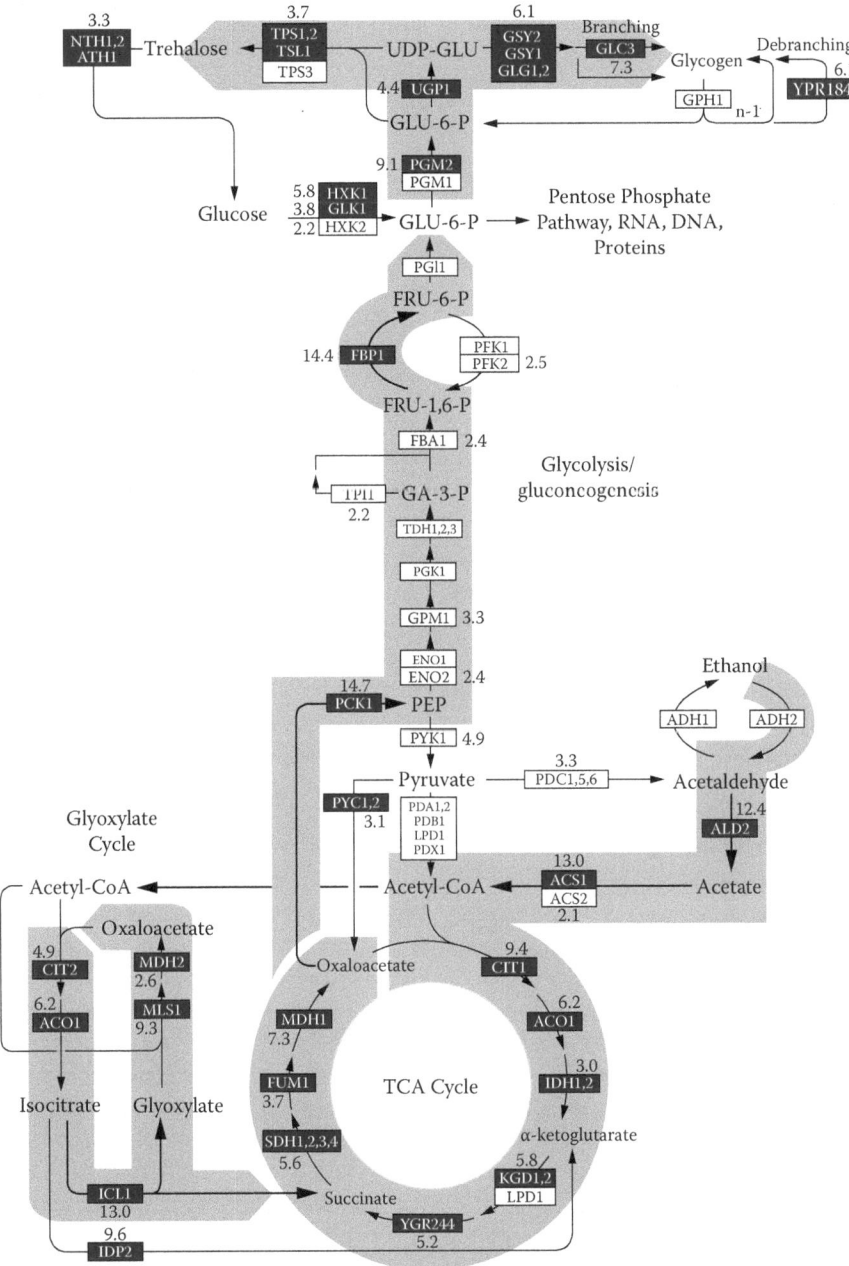

FIGURE 5.3 The diauxic shift in *Saccharomyces cerevisiae*—Monitoring metabolic pathways using the gene expression microarray. (From DeRisi J, Iyer VR, Brown PO. Exploring the metabolic and genetic control of gene expression on a genomic scale. *Science* 278: 680–686, 1997. With permission.)

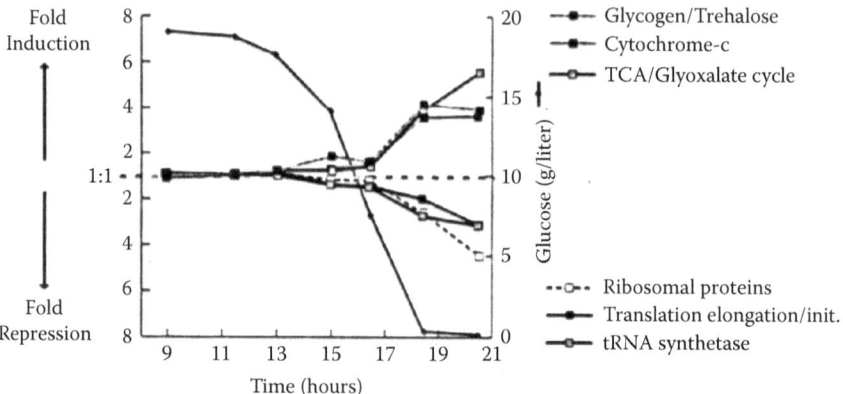

FIGURE 5.4   Global assessment of gene expression among various metabolic states. (From DeRisi J, Iyer VR, Brown PO. Exploring the metabolic and genetic control of gene expression on a genomic scale. *Science* 278: 680–686, 1997. With permission.)

oxaloacetate ultimately into glucose-6-phosphate for glycogen storage. Obviously, with changes in 25% of the yeast genome occurring at some point during the diauxic shift, a large number of regulator events remain to be mapped in this manner. One way to address this is to group gene families rather than monitor individual genes (see Eisen et al., 1998—cluster analysis). For example, genes associated with a particular pathway such as glycolysis or protein synthesis can be monitored collectively for average fold induction or repression during the time course of glucose starvation (Figure 5.4). For additional details regarding the strategy for constructing specific yeast ORF arrays for gene expression analysis and yeast strain comparisons, see Lashkari et al. (1997). In a related application of this approach, Ferea et al. (1999) examined variation in gene expression of progeny during adaptive evolution. Here, the yeast genomes were monitored in evolving strains subjected to growth under a glucose-limited chemostat for 250 generations. All of the evolved strains studied appeared to have similar changes in their gene expression profiles relative to the parental strain. Essentially, those genes involved in respiration were up-regulated while genes encoding enzymes of the glycolytic pathway (fermentation of excess glucose) were repressed relative to the parental strain (Figure 5.5). This is consistent with the yeast's physiological response in shifting from fermentation to respiration. Each new generation adapts to be more efficient in the utilization of glucose (via oxidation) under limiting concentrations.

Iyer et al. (1999) extended microarray pathway expression profiling to examine the effects of serum on fibroblast growth. It is well known that certain growth factors are required in order to propagate mammalian cells in culture. Serum sources such as fetal calf serum contain these growth factors. Using a cDNA microarray made up of 8613 human genes, the changes in mRNA levels were monitored from 15 minutes to 24 hours following the introduction of fibroblasts into serum. Cluster analysis on 517 (6% of the cDNA microarray) genes showing significant change (Δ 2.20 fold-expression change) was conducted. These were grouped into 10 cluster families on the basis of the similarity of their expression profiles (Figure 5.6). Genes

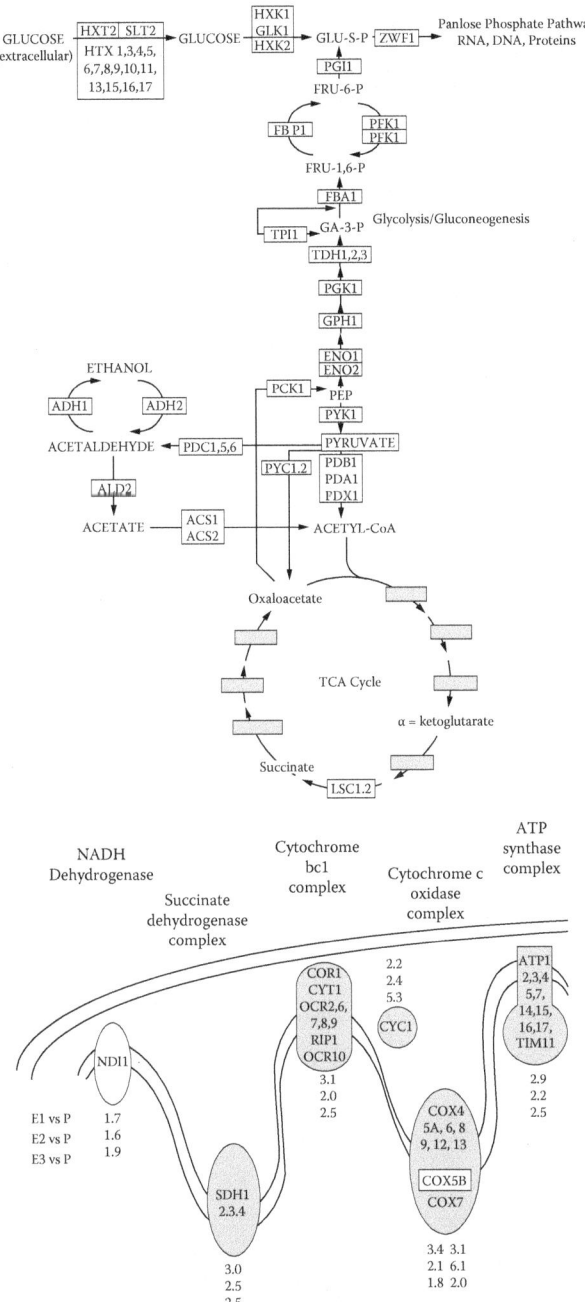

FIGURE 5.5 Examining the regulation of yeast genes from evolving strains. (From Ferea TL, Botstein D, Brown PO, Rosenzwig RF. Systematic changes in gene expression patterns following adaptive evolution in yeast. *PNAS* 96: 9721–9726, 1999. With permission.)

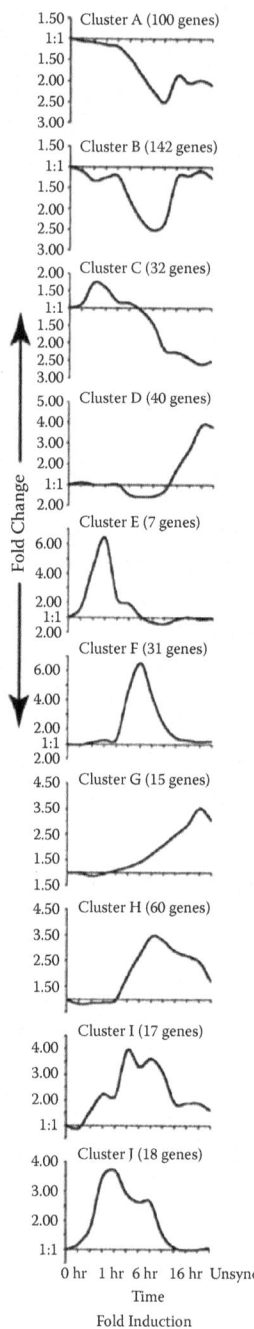

FIGURE 5.6 Gene cluster profiling. (From Iyer VR, Eisen MB, Ross DT, et al. The transcriptional program in the response of human fibroblasts to serum. *Science* 283: 83–87, 1999. With permission.)

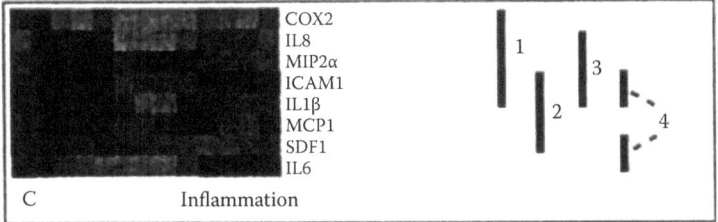

FIGURE 5.7 Human fibroblast responsive to serum-monitoring genes associated with inflammation. (From Iyer VR, Eisen MB, Ross DT, et al. The transcriptional program in the response of human fibroblasts to serum. *Science* 283: 83–87, 1999. With permission.)

involved in the encoding of transcription factors and signal transduction proteins were induced within 15 minutes following fibroblast transfer from serum-poor into serum-rich media. Those genes involved in cell cycle progression begin to appear (induced) about 16 hours following passage into serum. For example, induction of mRNA encoding various subunits of the RNA/DNA polymerases and cyclins that are regulators of the growth phase appear in this cluster. Of particular interest was the discovery that genes involved in wound healing are induced within hours following serum stimulation. For example, COX2 (chemotaxis, neutrophil activation), MCP1 (macrophage recruitment), IL-8 (T lymphocyte activation), ICAM-1 (B lymphocyte activation), and VEGF (angiogenesis) among others are all induced within a few hours (Figure 5.7). In addition, at least 200 previously unknown genes were identified as participating in the fibroblast growth and cell cycle progression.

The cDNA microarray format provides the ability to monitor gene expression without prior knowledge of the probe cDNA sequence. However, because cDNAs are typically 200 bp to 600 bp in size, there is considerable potential for the occurrence of cross-hybridization. When the gene sequence is known, oligonucleotide probes can be carefully designed to avoid this problem. However, the selection of unique gene-specific probe sequence of an appropriate size is also problematic because the presence of partial but nevertheless hybridizable sequence copies may be present at random within the genome. Under low stringency conditions it is possible to hybridize 6- to 8mer oligonucleotides (Drmanac et al., 1990).

In order to reduce the effect of cross-hybridization (leading to false positives), researchers at Affymetrix created a series of gene-specific but closely related complementary probes (PM, perfect match) and then introduced corresponding mismatched probes (MM, mismatch) in which the single-base mismatch was placed at a central location in the sequence. For short oligonucleotide probes (e.g., 25mers in the case of the GeneChip array) this central location provides the optimal instability relative to the perfect match. As a result, the ratio of PM/MM provides a convenient means to improve both specificity and sensitivity on the array. Wodicka et al. (1997) created such an array for measuring yeast gene expression based upon 25mer oligonucleotides covering the 6200 ORFs. Each ORF was represented by 20 PM and 20 MM probes. Why were there so many probes? Simply put, not all probes hybridize in a predictable manner, and thus averaging across a number of probes improves the outcome. Thus, the yeast expression chip included over 65,000 probe features and required a set of four chip subarrays.

The Wodicka et al. (1997) paper also defined the performance of the Affymetrix chip. First, semi-quantitative measurement of the absolute abundance of mRNA species was possible. Hybridization of total yeast genomic DNA to the chips revealed the mean hybridization signal across 6049 probe sets to vary by 25% CV. The use of gDNA serves to normalize because most genes are represented only once in the population. In fact, the majority (98%) of the intensities were found to cluster well within two standard deviations. Thus, the concentration of a given mRNA could be estimated at >95% probability to reside within twofold of its actual concentration. Measurement at widely different total gDNA concentrations did not appreciably affect this outcome.

Finally, the Affymetrix chip further corroborated the fact that only a relatively small number of genes show significant levels of differential expression in response to particular stimuli. In this case, a comparison between rich and minimal media revealed that 36 mRNA (genes) were more abundant (5- to 10-fold higher) in rich media, while 140 genes were more abundant in minimal media. This collectively represents less than 3% of the yeast genome. In terms of "absolute" concentration or copy number per cell, the distribution of mRNA was similar between cells grown in rich or minimal media (Figure 5.8). An estimated 50% of the mRNA population (~15,000 total copies per yeast cell) are present between 0.1 and 1 copy per cell; 26% (1 to 10 copies per cell); 5% (>10 copies per cell), and 19% (<0.1 copies per cell).

An interesting application of the Affymetrix chip yeast genome array is direct allelic variation scannning (Winzeler et al., 1998). The yeast expression chip provided 21.8% coverage of nonrepetitive regions of the yeast genome. Winzeler et al. reasoned that this was enough to capture a small but significant portion of the genetic variation found between strains. The hope was that markers could be found that would map out phenotypic differences. Two strains of S. cerevisiae that were phenotypically distinguishable were tested. When labeled genomic DNA was hybridized to the high-density array, 3714 (contributing ~4.7% of the estimated strain-to-strain variation) marker candidates were found that had a greater than 99% probability of differentiating the two strains. These biallelic markers were spaced at about 3500 bp. Cycloheximide sensitivity (phenotype) was mapped to the PDR5 gene (a multi-drug resistance pump) in the one strain exhibiting hypersensitivity to cycloheximide. Thus, phenotypic variation between the yeast strains was accurately mapped directly based upon differential hybridization of genomic DNA.

de Saizieu et al. (1998) examined the transcriptional activity of bacterial genomes (influenza, pneumonia) using an Affymetrix chip of 64,000 probes that were complementary to +100 genes for each genome or about 150 probe pairs per gene. This represents about 5% coverage for the Streptococcus pneumoniae genome. Because bacterial mRNA lacks the 3′-poly(A) tailing, enrichment of mRNA from total RNA by affinity purification is not possible. Instead, labeling of total RNA must be undertaken. In this case, the most efficient labeling was accomplished using psoralen-biotin providing incorporation of about one biotin per 120 nucleotides. Fragmented, biotinylated RNA prepared in this manner was hybridized to the array and signal developed using streptavidin-R-phycoerytherin. A confocal laser scanner was used

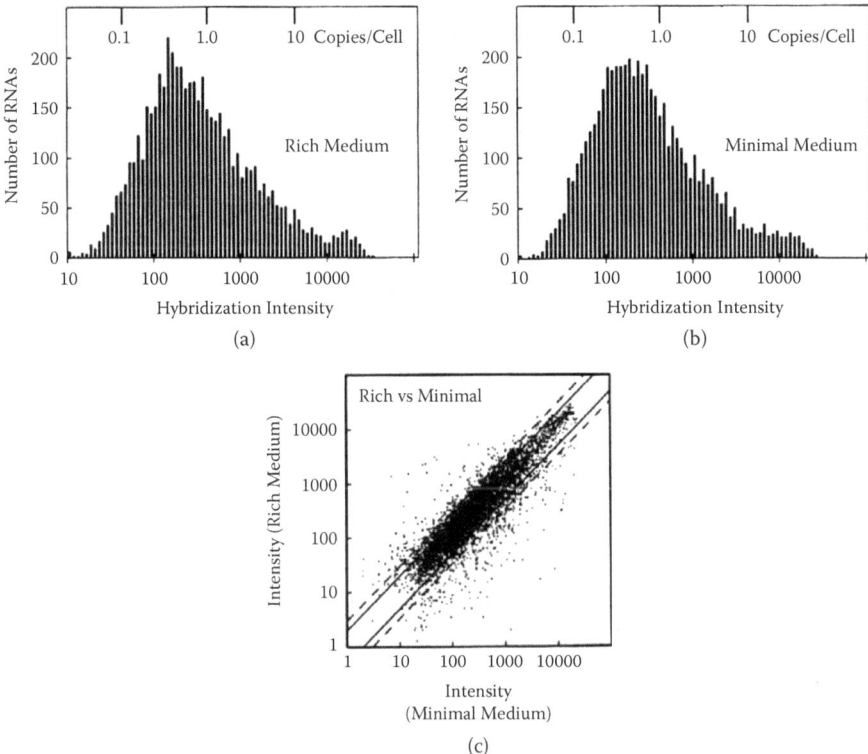

FIGURE 5.8 Distribution of expression levels for cells grown in rich or minimal media. (From Wodicka L, Dong H, Mittmann M, Ho M-H, Lockhart DJ. Genome-wide expression monitoring in *Saccharomyces cerevisiae. Nat. Biotechnol.* 15: 1359–1367, 1997. With permission.)

for detection. The researchers estimated that they could detect two transcripts per cell based upon labeling efficiency and an estimated 4% mRNA content in total bacterial RNA. Thus, chip detection of labeled transcripts was found to be more sensitive than from Northern blots. Specific genes (e.g., basal levels of cinA) undetectable on Northerns were quantifiable on the microarray. In addition it was possible to monitor gene expression in moving from exponential to stationary growth phase (Figure 5.9).

The issue of whether or not to use the enriched poly(A) RNA or total RNA was addressed in a later paper by Mahadeveppa and Warrington (1999). Using human adenocarcinoma cells, the recovery of detectable transcripts from varying numbers of cells was examined for both protocols. Of the ~1800 genes represented on the chip about 35% were observed to be detectable from either preparation. Of these "detectable" transcripts (number of copies per cell basis) about 86% of the same transcripts were reported across several levels of relative abundance. For example, at levels over 10 copies per cell, poly(A) RNA derived transcripts ranged from 102 to 134, while total RNA ranged from 119 to 141; and at levels less than two copies per cell, poly(A)

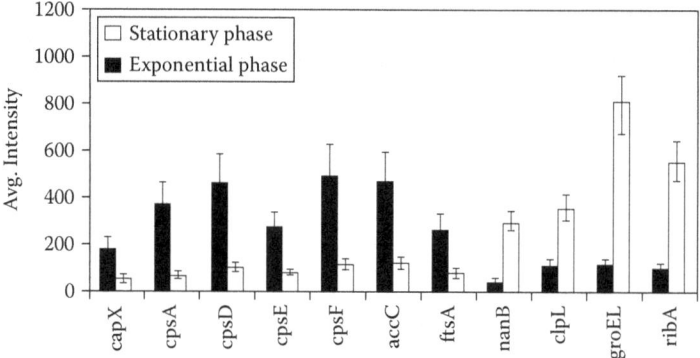

FIGURE 5.9   Gene expression microarray monitoring of bacterial growth. (From de Saizieu A, Certa U, Warrington J, Gray C, Keck W, Mous J. Bacterial transcript imaging by hybridization of total RNA to oligonucleotide arrays. *Nat. Biotechnol.* 16: 45–48, 1998. With permission.)

RNA ranged from 141 to 165 transcripts, while total RNA ranged from 152 to 162 transcripts. Intermediate abundance levels were also similar. Added benefits from total RNA labeling include the use of less starting material and improved yield of higher-quality material (i.e., less degradation by elimination of the extraction process).

In this section we have described some of the important earlier work on the DNA microarray. What we have discovered from these examples is that the microarray format can be used quite effectively to view differences (or similarities) in gene expression between a control and test population of cells. It is also evident that such temporal changes do not represent genome-wide levels of response but rather involve a relatively small number of genes (1 to 10%) that are up- or down-regulated. Because of the large number of genes represented on the microarray and the level of biological variability, it is difficult to draw a conclusion from examining single-gene events. A more meaningful approach for large gene expression data sets is to monitor these events by grouping genes of similar expression patterns together in clusters. As we have seen, clusters tend to bring together genes that participate in related functions or those that are involved in the juncture of metabolic pathways. In the next few sections we look at representative applications in the biomedical research field for DNA microarrays.

BIOMEDICAL RESEARCH APPLICATIONS

DRUG DISCOVERY

As we examine various applications for microarrays, it is important to understand the use of the terms *pharmacogenomics* and *pharmacogenetics*. Microarrays are used for studies in both fields but for slightly different purposes. Pharmacogenomics refers to the application of genomics to the drug discovery and development process. Here we use microarrays to see which targeted genes are turned off or turned on in response to candidate drugs. In primary screening, large numbers (1000s) of candidate drugs (compound libraries) are directed against a few targets (e.g., receptors). Generally this involves a biological assay using the putative target and measuring direct drug binding and response. Microarrays are usually not used in primary screening. They are most suitable for analyses involving the secondary screening of a few lead drug candidates that are produced from the primary screen. In this case we have identified the intended target (gene product) and are attempting to determine which drugs interact. As we have also discussed, microarrays can be constructed of gene-specific probes (i.e., cDNA clone) without prior knowledge of the gene's function. The ability to examine alterations in gene expression patterns in response to a candidate drug may lead to the identification of new (gene) targets. For example, a gene of unknown function may associate or cluster with known genes in response to a specific drug, giving us important clues to the unknown gene's function and suitability as a target (Ivanov et al., 2000).

Pharmacogenetics involves the understanding of an individual's genetic makeup relative to drug action (i.e., given a particular genotype we ask which drug is most effective for treatment without adverse or off-target side-effects). Microarrays are used here to monitor for the up- or down-regulation of genes involved in various off-target pathways in order to access potential drug toxicity. This is often referred to as the field of *toxicogenomics*. In this case, a single drug whose target has been identified is examined for off-target responses (e.g., activation or shut-down of important metabolic pathways). For additional review on the role of the DNA micorarray in the drug discovery process see the collection of reviews by Jain (2000), Zanders (2000), and Ivanov et al. (2000). In this section we will examine studies aimed at adopting DNA microarrays for the drug discovery process.

The fundamental approach in using microarrays for gene expression analysis is really to seek out differences between a control cell gene population and that of the gene population from the test cells in our experiment. We then examine those genes whose expression level did change and attempt to discover what caused the change. In reality, we observe many changes and as we have seen this leads to a rather interesting but complicated story. In drug discovery we attempt to find compounds that interact with a specific target of interest. The ideal candidate drug is to be specific for the target and free of unwanted side-effects. However, traditional approaches to determining "off-target" drug responses are time consuming, expensive, and not always precise.

Because of the DNA microarray's ability to examine global changes in gene expression, it has become an important new tool for measuring off-target drug interactions. However, with so many changes how do we determine whether or not the drug's mechanism of action is through interaction with the presumed target? One clever way is to "knock out" a particular gene (i.e., the putative target) and access whether or not the drug is still effective. For example, Marton et al. (1998) used mutant yeast strains and looked at the gene expression patterns in the presence or absence of a specific drug. Comparison of these "signatures" provided clues to the mechanism of action of specific drugs. They chose to study the calcineurin signaling pathway in yeast (Figure 5.10). In yeast, calcineurin is involved in the regulation of a number of key cellular functions such as the onset of mitosis. In mammals this calcium-activated protein phosphatase has been implicated in a wide range of

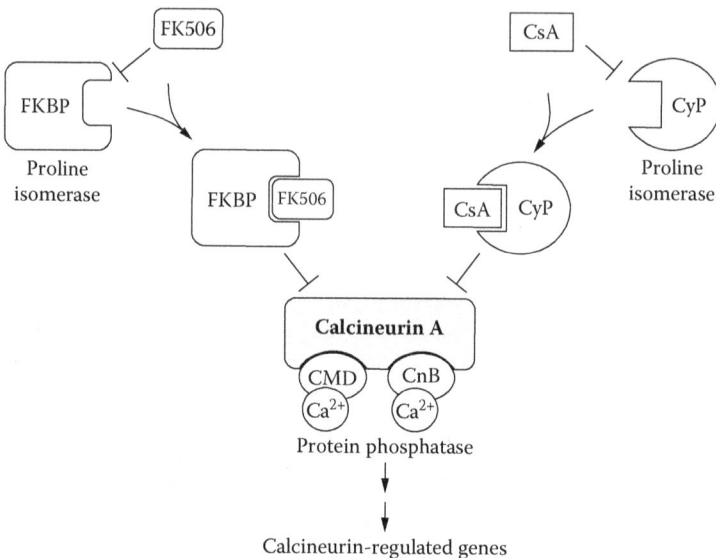

FIGURE 5.10   Calcineurin signaling pathway in yeast. (From Marton MJ, et al., *Nature Med.* 4(11): 1293–1301, 1998. With permission.)

biological functions from T-cell activation to being an effector in short- to long-term memory transition. Calcineurin activity is inhibited by two immunosuppressant drugs: tachrolimus (FK506) and cyclosporin A (CsA). Do these drugs interact in the same manner? In order to answer this question the gene expression microarray signatures for each of the drugs in wild-type yeast were obtained. Next, signatures were obtained in mutant strains in which the putative target was deleted. It was the hypothesis that if the mutated gene encoded a protein in the pathway affected by the drug, that the drug's mutant signature would be different or absent from that of the wild-type signature.

Drug Toxicity

Cleary et al. (2001) studied amphotericin B toxicity to human mononuclear cells using a commercially available cDNA nylon membrane-based microarray (Clontech). Amphotericin B, a fungicide, elicits immune responses such as activation of IL-1β and TNF-α. The IL-1β gene expression is related to the accumulation of intracellular calcium that is known to be mediated by amphotericin B (see Cleary et al., references 4,5). Because calcium plays a role as a second messenger, it was recognized that additional genes may undergo regulation in response to the fungicide.

The Clontech microarray used in this study contained 588 cDNA fragments arranged in several relevant functional categories: oncogene, tumor suppressor genes; ion channels, transducers; apoptosis; transcription factors; cell receptors such as for IL, hormones, chemokines; and extracellular cell signaling related genes. Of the 588 genes represented on the microarray, 16 genes were up-regulated and 4 genes down-regulated in the THP-1 cell line after 6 hours following administration of the drug. The majority of these genes (75%) exhibited greater than 10-fold differences relative to control cells. These genes (encoding the listed proteins) were identified as having important regulatory function implicating potential mechanisms of action for amphotericin B:

1. Transcription factor AP-1 is up-regulated. AP-1 is involved in the induction of genes encoding inflammatory response (e.g., IL-1).
2. MAL protein is up-regulated. This protein is involved in cell signaling and protein trafficking.
3. Caspace 4 is down-regulated and represents a potential block to activation of the apoptosis pathway.
4. Cell adhesion protein ICAM-1 is up-regulated. This protein plays a role in cell-cell adhesion and leukocyte migration.
5. IL-8 is up-regulated. This chemokine is responsible for neutrophil activation.

Thus, the application of the cDNA microarray revealed that the fungicide appears to affect a number of cellular processes. While the content of the Clontech microarray was rather limited, it did provide a substantial amount of new information regarding amphotericin B-mediated cellular toxicity. Now that we have identified most of the human genome, a higher-density array could be easily applied here to refine the study. By applying this global look at gene expression, we can greatly enhance our understanding of drug toxicity.

Reilly et al. (2001) using a mouse model studied the global gene expression profile associated with drug-induced liver toxicity brought about by the analgesic drug acetaminophen (APAP). The mechanism of APAP-drug toxicity is not well understood because the drug is implicated in a variety of biochemical events leading to cellular damage such as oxidative stress, disruption of calcium and mitochondrial hemostasis, alterations to transcription, inflammation, and programmed cell death pathways. For this reason a global examination of gene expression events using the microarray was undertaken in the hope that additional information regarding the mechanisms of toxicity would be found.

APAP was administered to mice and the progression of hepatotoxicity monitored by histochemical means as well as by microarray analysis. The Affymetrix oligonucleotide array (Mul1K sub A, sub B) was used to access gene expression activity across more than 11,000 genes and ESTs. Monitoring began 6 hours after administration in order to establish the early onset of the hepatotoxicity. Fold-change greater than 2 was used as the threshold level. Of the 11,000 genes on the array, 332 or ~3% of the genes were scored as up- or down-regulated. Of these, >50% exhibited fold-change between 2 and 2.9, and >90% were represented between 2- and 10-fold change. APAP toxicity as revealed by the gene expression analysis was found to be even more extensive than described in a similar proteomic study conducted by Fountoulakis et al. (2000). The results of the oligonucleotide microarray were reported to be consistent with RT-PCR estimates of fold-change for selected genes within several of the assigned functional clusters. These clusters included, among others, stress-responsive genes ($n = 22$), cell cycling and growth inhibition ($n = 14$), inflammation ($n = 14$), cell signaling ($n = 18$), and cell metabolism ($n = 13$).

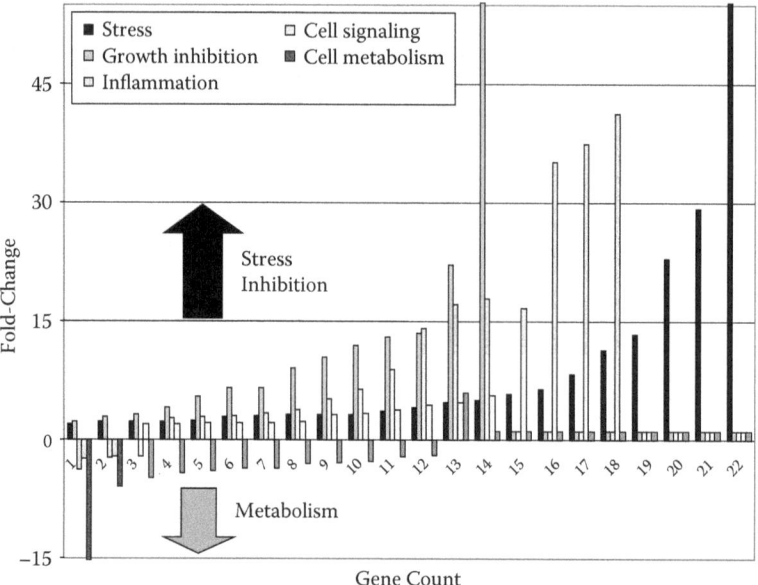

FIGURE 5.11   Gene expression changes associated with acetaminophen toxicity in liver. (From Reilly TP, Bourdi M, Brady JN, et al. Expression profiling of acetaminophen liver toxicity in mice using microarray technology. *Biochem. Biophys. Res. Commun.* 282: 321–328, 2001. With permission.)

As depicted in Figure 5.11, the drug-induced liver toxicity results in the induction of genes involved in stress and inflammation, while genes involved in cellular metabolism are down-regulated. What is missing in this report is information regarding the timing of expression. It would be of interest to see which genes responded first to APAP and how that temporal expression varied during the course of the hepatotoxicity. These time course snapshots may have provided useful information regarding APAP's off-target mechanism of action.

Katsuma et al. (2001) undertook the time-course approach to help elucidate factors involved in drug-induced lung fibrosis. They followed histopathological changes associated with bleomycin-induced pulmonary fibrosis in a mouse model and correlated these with the gene expression profiles obtained using a cDNA microarray. A "lung chip" was prepared with spotting of 4224 cDNAs obtained from a normalized lung cDNA library. Mice were subjected to bleomycin and sacrificed 2, 5, 7, and 14 days following intratracheal instillation. Lungs were fixed in formalin and embedded in paraffin, and thin sections were examined for pathology. While the left lung was reserved for pathology, the right lung was used for RNA preparation. Two micrograms of poly (A)+ RNA were used to prepare labeled cDNA by incorporation of Cy3- or Cy5-dUTPs into control and test populations, respectively. Approximately 80% of the array was found to hybridize to the labeled cDNA targets from the lung. A twofold change in expression level was scored as being significant.

Of the 4224 cDNA clones, 159 clones (82 nonredundant genes) were observed to be differentially expressed over the 14-day period. Most genes were found to be up-regulated by day 5 with very few genes down-regulated. The differential expressed genes were divided into four clusters for the purpose of analysis. Cluster 1 genes up-regulated at 5 days then returned to a basal level; Cluster 2 genes maximally up-regulated at day 5 and remained up-regulated; Cluster 3 genes continuously induced over 2 to 14 days; and Cluster 4 genes down-regulated after bleomycin administration.

How do these groupings correlate with the phenotypic changes encountered? Pulmonary fibrosis involves the accumulation of collagen, the growth of fibroblast cells, and a thickening of the lung septa. Cluster analysis revealed that genes involved in inflammatory response (e.g., complement C3, osteopontin) were induced in the early stages of the disease followed by the appearance of genes related to the fibrosis process activity (collagen, fibronectin) occurring at a later stage in the process. Therefore, the authors postulate that bleomycin-induced fibrosis may first involve an inflammatory response in which the damaged lung is then repaired by laying down additional extra-cellular matrix proteins such as collagen, thereby leading to the progression of fibrosis pathogenesis.

Cancer

One of the most extensive undertakings employing DNA microarrays for research on cancer models was that of Ross et al. (2000). In their studies, 60 tumor-derived cell lines (NCI 60) were profiled for alteration in gene expression when subjected to anti-cancer drugs. These cell lines maintained under the U.S. National Cancer Institute's Developmental Therapeutics Program have been assessed for drug sensitivity against over 70,000 compounds. The cDNA slide microarray used in this study included over 8000 different cDNAs of which 3700 were well identified to previously characterized human protein (gene) products. The other elements on the array were ESTs (2400) or homologs (1900) from other organisms. Essentially, about 80% of the genes are correctly identified in the cDNA clones.

The experimental design is rather straightforward. The mRNA from each test cell line was converted to Cy5-labeled cDNA by reverse transcriptase. These were compared against a reference of pooled mRNA from 12 separate cell lines that had been reverse transcribed to produce a Cy3-labeled cDNA reference. This pool represented the maximum diversity across the 60 cell lines and was used as the reference throughout the study. Thus the hybridization signal intensity ratio Cy5/Cy3 served to normalize for each cell line, permitting comparison across all 60 cell lines. To minimize differences due to culturing conditions and cell density, cells were grown out to 80% confluency and mRNA isolated 24 hours after transfer of the cell line into fresh media. Results were clustered in hierarchical fashion to allow groupings of similarly expressed genes in relation to their tissue of origin. However, approximately 1200 genes exhibited wide variation in expression across the 60 cell lines, and these were of most interest (Figure 5.12).

What did this study accomplish? First, the clustering approach was found to be generally valid in that cell lines derived from the same tissue were found to group together while those from different tissues occurred in separate branches. For example, colon and ovarian derived cell lines were observed in separate branches. On the

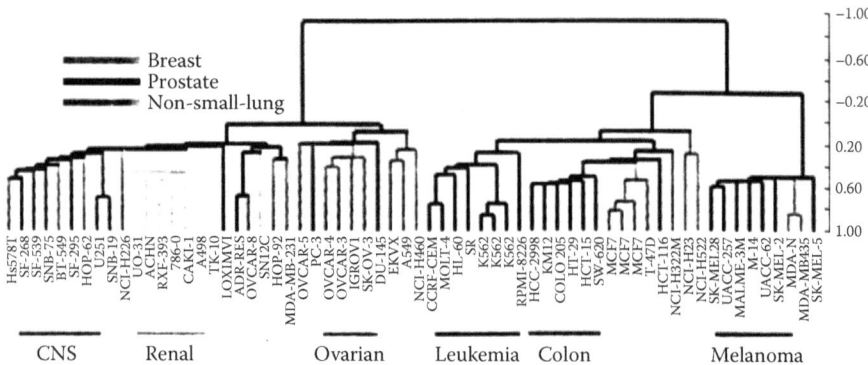

FIGURE 5.12 Hierarchical clustering of expressed genes from the NCI-60 tumor cell lines. (From Ross DT, Scherf U, Eisen MB, et al. Systematic variation in gene expression patterns in human cancer cell lines. *Nat. Genet.* 24: 227–235, 2000. With permission.)

other hand, breast tumor cell lines were found to distribute across several branches suggesting a higher degree of heterogeneity in gene expression. Other cell lines such as those from melanoma tissue exhibited clusters containing as many as 90 genes with high levels of expression, many of which are involved in melanin metabolism. When cell line expression patterns were found among different tissues of origin, it was also observed that these carried with them genes from these different tissues. One cell line may show a relationship with epithelial derived cells while another cell line may be more stroma cell like in its gene expression pattern. A gene cluster including cells derived from colon carcinoma and ovarian and breast cancer shared genes involved in formation of the basolateral membrane of epithelial cells. In another branch were found all glioblastoma-derived cell lines in which many clustered genes were associated with stroma cellular function. Yet, no one gene was distributed across all of the clustered cell lines. However, each cell line held a characteristic expression pattern related to the regulation of the extracellular matrix proteins. This is a good case supporting the rationale to perform hierarchical clustering. A single gene expression pattern is generally not sufficient to track down relationships.

Unger et al. (2001) examined the gene expression profiles of adjacent breast tumors using microarray analysis. The occurrence of multiple tumors in tissue is problematic for diagnosis and appropriate treatment because of the potential for metastasis. The tumors may have originated from different primary tumors from different tissues. In that case treatments of the individual tumors may have to be different or compromised. Current methodology based upon X-chromosome polymorphisms that are used to check for tumor clonality apparently are limited in their scope to access the genomic character of individual tumors. For this reason the value of examining the gene expression profiles was assessed using clinical samples.

The Affymetrix U95A GeneChip containing 12,500 known human genes was used for this purpose. Two adjacent breast tumors from an 87-year-old woman were removed and used in the study along with an additional five breast tumors obtained from different individuals. Thus, seemingly related tumors from one individual could be compared with tumors from different sources to validate the microarray-based prognosis. In addition to validation of the microarray for this purpose, the researchers examined a number of important aspects related to microarray performance. We will first discuss the issue of system performance that ultimately impacts the clinical interpretation presented by this study.

First, the total cRNA representation on the chip was found to be approximately 50% of the available gene content. This of course means that the chip could not read half of the sample's content. While arguably assessment of 50% of the gene population is significant, it is also difficult to accept the fact that we are from an analytical position starting out at a disadvantage in not being able to see the complete profile. Simply put, significant information that could support or invalidate the author's conclusions may as likely lie within the missing genes. The chip-to-chip variation between duplicate hybridizations of the same sample was low, resulting in a pair-wise correlation of 0.995 for both adjacent tumors. The small difference (0.5%) between duplicates represented 50 genes from tumor 5A and 36 genes from tumor 5B. None of the 36 genes that were found to be at a twofold expression for 5B replicates were found among the 50 genes expressed by 5A replicates. Examining the two adjacent tumor samples (5A and 5B), 149 genes were differentially expressed in common ($r = 0.987$). However, based upon the level of variation in duplicates, some of the observed expression could be attributed to experimental noise. Couple this with a reported sample-to-sample variance ($r = 0.915$) being larger than the observed chip variance (for tumor 5A: 50 genes from hybridization replicates and 100 genes differentially expressed between sample replicates), it would be difficult to believe one way or another that the tumor expression profiles were the same or different. The authors suggest that because the number of differentially expressed genes between the two tumors was less than that of genes expressed from replicates of the same sample, that these tumors are virtually indistinguishable. I would argue that the results are inconclusive, and as the authors later point out, "a simple pair-wise correlation comparison may not fully represent the relationship between gene expression profiles" (Unger et al., 2001, p. 339). In fact, when the researchers focused on those genes implicated in breast cancer, significant differences among various tumors were revealed based upon microarray analyses that were not evident based upon pathology (Table 5.1).

Wang et al. (2000) wished to identify new cancer markers for potential use as antigens in tumor-specific immunotherapy, especially for treatment of lung squamous cell carcinoma (LSCC). Non-small cell lung carcinoma makes up about 80% of all lung cancers with a 5-year survival rate of less than 10%. LSCC is a member of this disease pathology that currently lacks sufficient markers for early diagnosis and treatment. Thus, the object of this study was to use microarrays to identify overexpressed genes in lung tumors relative to normal lung tissue in the hope of finding candidate markers. However, Wang et al. recognized that the presence of high abundance genes on the microarray would limit the representation of lower copy number genes. As a result, the researchers sought to combine the process of subtractive hybridization that would eliminate highly abundant transcripts while enriching lower abundant genes of interest for use in the generation of microarray probes.

TABLE 5.1
Genes of Known Interest in Breast Cancer

	Tumor		
Gene	5A/5B	5A/7	5A/9
ER	NC	D(12.7)	NC
PR	NC	NC	I(2.7)
Androgen receptor	NC	NC	D(2.5)
Epidermal growth factor receptor	NC	NC	NC
ERBB2	NC	NC	NC
VEGF	NC	I(2.0)	D(2.2)
TP53	NC	NC	NC
Ataxia telangiectasia (ATM)	NC	NC	NC
FHIT	NC	NC	NC
BRCA2	NC	NC	NC
RAD50	NC	NC	NC
BARD-1	NC	D(2.9)	D(3.5)
Retinoblastoma-1 (RB1)	NC	NC	NC
Amplified in breast cancer (AIB1)	NC	NC	NC
Breast cancer transcription factor (ZaBCI)	NC	NC	NC
Thymidylate synthetase	NC	D(3.0)	NC
Multidrug resistance gene (MDR-1)	NC	NC	NC
Thrombospondin-1	NC	NC	D(2.2)
KI-67 antigen	NC	NC	NC
Breast epithelial antigen (BA46)	NC	I(4.3)	NC
Human mammaglobin	NC	I(3.6)	I(36.9)
Human mammaglobin β precursor	I(40.1)	NC	I(29.1)

Source: From Unger MA, Rishi M, Clemmer VB, et al. Characterization of adjacent breast tumors using oligonucleotide arrays. *Breast Cancer Res.* 3: 336–341, 2001. With permission.

The subtractive hybridization process is outlined in Figure 5.13 and will be briefly described here. First, total RNA pools are isolated from normal and diseased tissues and the corresponding poly A+ RNA (mRNA) purified. The mRNA is converted by reverse transcriptase to cDNA to create a tester cDNA library from LSCC tissue and a driver library from normal lung cells. The driver library is used to remove common and highly abundant transcripts from the tester population. This is accomplished by first biotinylating the driver cDNA that will be bound later to streptavidin and precipitated out. Tester and driver cDNAs are mixed with the driver being in excess of the tester cDNA in order to efficiently capture the more abundant and common transcripts. Following mixing to hybridized tester + driver, streptavidin is added and the mixture precipitated. cDNA not binding to the streptavidin-biotin-driver is isolated and thus enriched after several rounds. The enriched tester cDNA (Lung Squamous Tumor-Specific Subtracted cDNA Libraries, LST-S1...) are then ligated

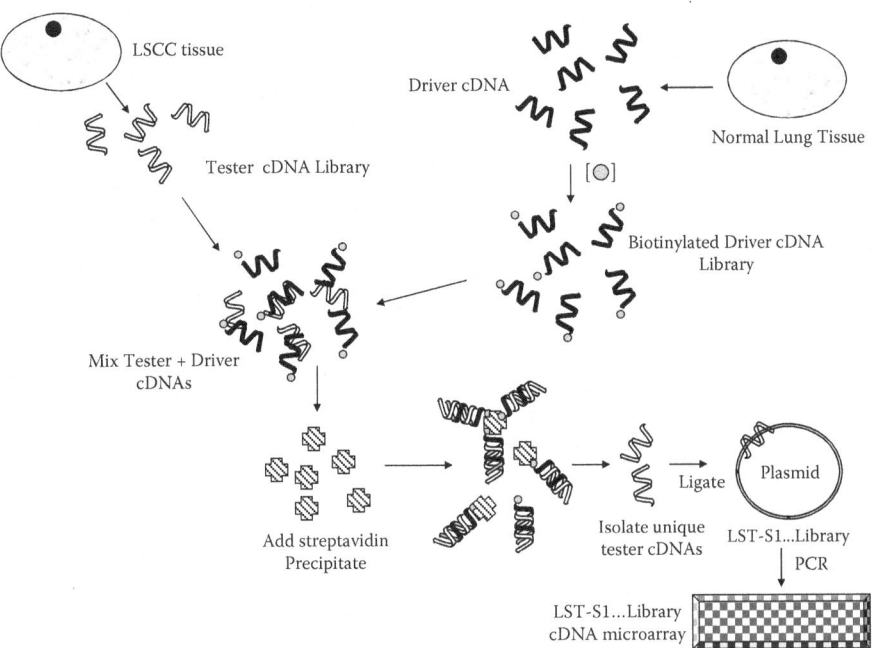

FIGURE 5.13 Subtractive hybridization. (From Wang T, Hopkins D, Schmidt C, et al. Identification of genes differentially over-expressed in lung squamous cell carcinoma using combination of cDNA subtraction and microarray analysis. *Oncogene* 19: 1519–1528, 2000. With permission.)

into plasmid and amplified by PCR. The LST-S1, S2, S3 libraries donated the various levels of stepwise subtractive hybridization and enrichment of tumor-specific clones. The resulting cDNA amplicon subtracted library (LST-S1...) can then be spotted down on a glass slide microarray to create tumor-specific gene probes. The relative gene expression levels of genes present in tumors can then be assessed and overly expressed genes identified as potential cancer markers.

Wang et al. were able to isolate approximately 2000 cDNAs as LST-S1, S2, S3 libraries and prepare probes. Using the competitive hybridization-dual labeling (Cy3/Cy5) method, 17 genes were found to be differentially overexpressed in LSCC. This included 13 known genes and 4 unknown genes (Figure 5.14). These expression results were confirmed by Northern analysis and real-time RT-PCR. The genes could also be classified as tissue or tumor-specific markers. Tumor-specific genes included cell signaling, enzymes, and antigens, while tissue-specific genes included cytoskeletal and squamous cell specific markers (Table 5.2).

Chen et al. (2001) validated the use of cDNA microarrays for gene expression studies involving an established cell-line model for lung metastasis. Recognized as the leading cause of mortality in cancer patients, metastasis is a complex process in which cancer cells from the primary tumor move (invade) into other tissues and organs. Cell-to-cell interactions and influences from the surrounding microenvironment are important factors governing the tumor's invasiveness. Certain molecules

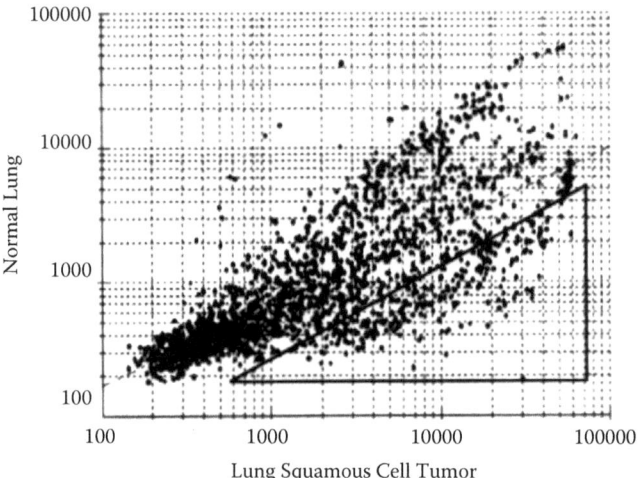

FIGURE 5.14 Lung squamous tumor-specific subtracted cDNA libraries. (From Wang T, Hopkins D, Schmidt C, et al. Identification of genes differentially over-expressed in lung squamous cell carcinoma using combination of cDNA subtraction and microarray analysis. *Oncogene* 19: 1519–1528, 2000. With permission.)

are known to promote metastasis such as laminin receptor and CD44, while others are inhibitory such as cadherin. However, genetic instability of cancer cells has proven to be a challenge in the further identification of genes involved in metastasis. Chen and co-workers used human lung adenocarcinoma cell lines (CL1-0 and sublines CL 1-1 and CL 1-5) varying in their "metastatic potential" and assessed their gene activity under defined growth conditions using microarrays.

The microarray included 9600 nonredundant ESTs from the IMAGE collection of human cDNA clones arrayed onto nylon membrane. Of these, 1875 clones (19.5%) were at the time verified by resequencing. In this study, 110 of the 589 genes expressed (18.7%) that were correlated with the metastatic potential were among those verified by sequencing.

In the study, invasiveness of the CL lines was measured by in vitro and in vivo methods. The in vitro monitoring process included the movement of cells across a membrane of defined pore size within a specially designed growth chamber or MICS (membrane invasion culture system). The 10 micron diameter nucleopore membrane was coated with a mixture of laminin (promotes invasion), collagen, and gelatin. Cells were added to the topside of the chamber in media and the extent of cell movement into the bottom of the chamber (invasion) through the membrane was determined. The in vitro invasive assay consisted of grafting rat tracheas by implanting into severe combined immunodeficiency (SCID) mice. The tracheas were first injected with the adenocarcinoma cells to promote tumor growth. Invasiveness was determined by histochemical staining of the isolated tracheas to reveal tumor growth and invasion of epithelial cells into the basement membrane. Satisfied with the ability to characterize invasiveness under defined conditions, the gene expression profiles of these cells were examined using the microarray.

TABLE 5.2
Lung Squamous Cell Carcinoma Differential Gene Expression

cDNA Clone	Gene	Functionality	Hits	Northern	RT-PCR	Specificity
520	SPRC	Cell marker	1	++		Tissue
513	PVA	Cell marker	2	+++	+++	Tissue
521	SPR1	Cell marker	2			Tissue
525	Plakophilin	Cytoskeleton	2		±	Tissue
527	Cytokeratin	Cytoskeleton	2			Tissue
529	Connexin	Cytoskeleton	2			Tissue
516	ARH	Enzyme	1			Tumor
523	KOC	Antigen	1	+++	+++	Tumor
524	PTHrP	Cell signaling	1	0	+++	Tumor
526	ATM	Cell signaling	1			Tumor
515	IGF-β2	Cell signaling	2			Tumor
522	ADH7	Enzyme	5	0	++	Tumor
528	NMB	Antigen	14		+++	Tumor
514	Novel	Unknown	1	+++	+++	Unknown
531	Novel	Unknown	1	0	++	Unknown
530	Novel	Unknown	2	+++	++	Unknown
519	Novel	Unknown	10	++	++	Unknown

Source: From Wang T, Hopkins D, Schmidt C, et al. Identification of genes differentially over-expressed in lung squamous cell carcinoma using combination of cDNA subtraction and microarray analysis. *Oncogene* 19: 1519–1528, 2000. With permission.)

Isolated mRNA from cell lines was biotin labeled, hybridized, and signal generated using a colorimetric reagent. Significant levels of expression were seen in 8525 of 9600 "genes." These were arranged in 100 SOM (self-organizing map) clusters, of which four clusters correlated with the promotion cell invasiveness (277 genes) and an additional four clusters (312 genes) correlated with inhibitory effects (Figure 5.15). These gene clusters were then rearranged for hierarchical clustering across the four cell lines used in the study and grouped in terms of cellular function (e.g., adhesion molecules, motility proteins, cell cycle regulators, signal transduction, angiogenesis related). Many of the genes identified in this study were found to be consistent with other reports that recognized these genes as participating in various aspects of metastasis including the role of angiogenesis in blood vessel formation. One important outcome of this study was the finding that the tumor-associated surface antigen L6 was highly associated with tumor invasion in this lung metastasis model. L6 appears to be highly expressed in human lung, breast, and colorectal carcinomas and may serve as a useful diagnostic marker.

The results of the microarray gene expression studies from the CL cell lines were also confirmed by Northern analysis with the application of flow cytometry providing an additional level of confidence to the data. For Northern analysis, five sequence-verified genes and 5 ESTs found to have a positive correlation for metastasis were

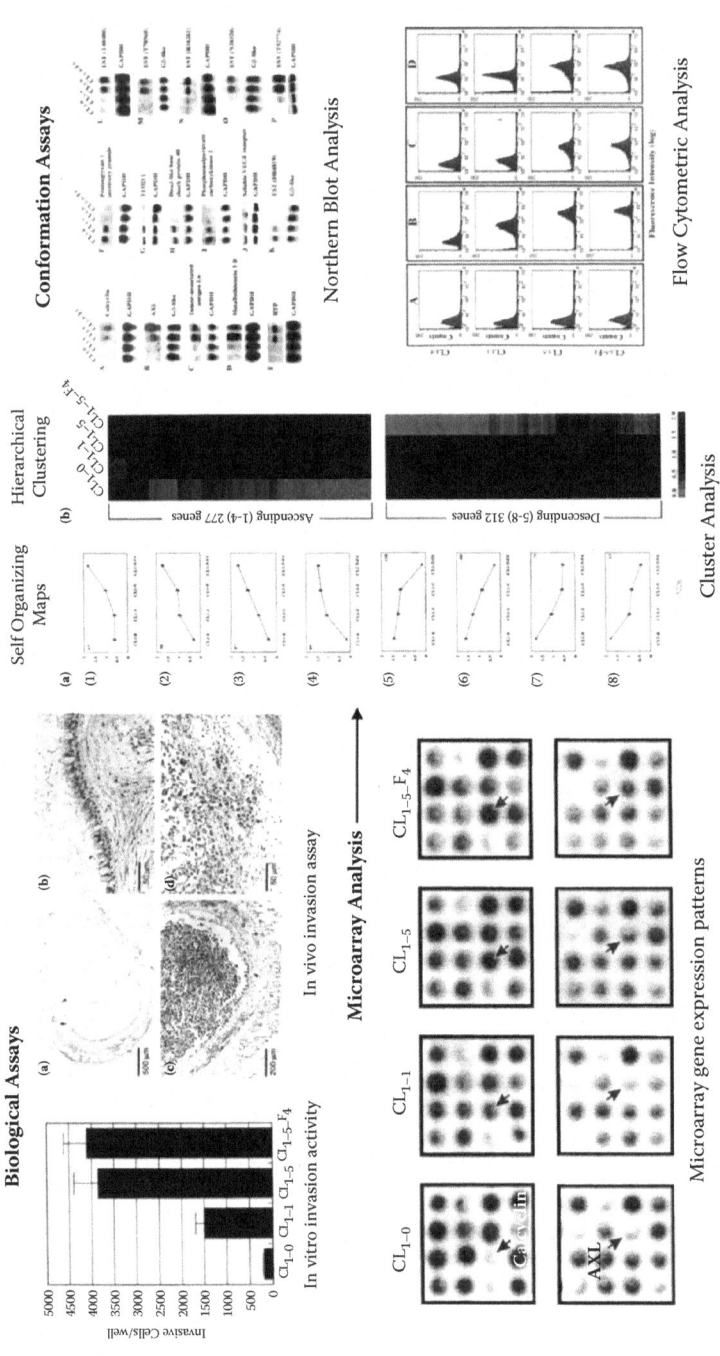

FIGURE 5.15 Validation of microarray-based gene expression analysis. (From Chen JJW, Peck K, Hong T-M, et al. Global analysis of gene expression in invasion by a lung cancer model. *Cancer Res.* 61: 5223–5230, 2001. With permission.)

selected for comparison across the four cell lines. In addition, five known genes and one EST having a negative correlation were examined. The results of the 16 genes were found to be in agreement with the relative gene expression levels determined by the microarray experiments.

Finally, fluorescently labeled monoclonal antibodies directed toward L6 antigen, integrins α-3 and α-6 were used to interrogate the four cell lines. Antibody labeled cell populations were analyzed in a flow cytometer. In conformation of both the microarray and Northern blot analysis, the IL-6 antigen and integrins were most prominent in the cell line demonstrated to have the greatest degree of tumor invasion.

In summary, this study represents one of the most thorough attempts at validating the utility of microarrays for clinical research: characterizing tumor cell line invasiveness by two independent methods with histochemical verification, performing replicate microarray experiments and cluster analysis, and then confirming the microarray results on gene expression at both the transcriptional and translational levels.

While the work of Chen et al. (2001) broadly defined the expression of genes associated with metastatic potential in the lung cancer model, others have more narrowly focused upon relationships for specific genes. For example, Pinheiro et al. (2001) examined gene expression profiles from patients with colorectal adenocarcinomas with paired normal tissue. Using microarray filters including 18,376 cDNAs (Incyte-Genomics), the study quickly focused upon the overexpression of a single gene, oligophrenin-1. Data from three independent microarray experiments showed fold changes from ~10- to 100-fold depending upon the tumor pool used. The results were confirmed by RT-PCR that oligophrenin-1 was consistently overexpressed in colorectal tumor but not significantly detected in normal tissue. The sequencing of 10 PCR products verified a 100% identity with the gene. A subsequent inquiry of the SAGE database (National Cancer Institute, Cancer Genetic Anatomy Project) revealed overexpression in a prostrate cancer cell line as well. Surprisingly, the authors found no reports on the overexpression of this gene associated with colorectal tumors. Oligophrenin-1 is known to be involved in X-linked mental retardation and encodes a protein having a Rho-GTPase-activating protein (rhoGAP) domain (Billuart et al., 1998). While at first glance mental retardation and colorectal adenocarcinoma would be seemingly unrelated, rho-GTPase is involved in the regulation of Rho and Ras proteins (Chen et al., 2003). The activation of the K-ras oncogene is well known for its involvement in colorectal cancer (Matson, 2000).

Mullan et al. (2001) used an oligonucleotide G110 array (Affymetrix) including 6800 genes and ESTs to examine expression profiling in BRCA1-induced cell lines in an effort to identify downstream targets. The G110 array contained approximately 1700 cancer-associated genes.

The BRCA1 mutation is associated with the occurrence of 10% of all human breast cancer and thereby is implicated in the predisposition of breast and ovarian cancers. What is not well understood is the mechanism(s) by which the tumor suppressor gene acts upon other genes. In order to study these effects, two cell lines (one derived from an osteosarcoma cell line and the other from a breast cancer cell line) were established to exhibit inducible, tetracycline-regulated expression of BRCA1. Thus, an exogenous gene could be switched off with the addition of (+ tet)

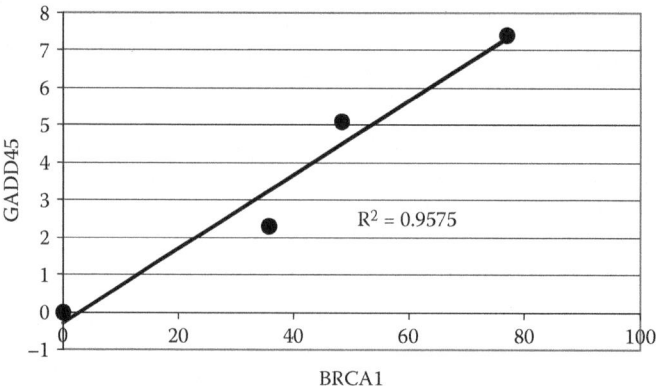

FIGURE 5.16   Association of the DNA damage-inducible gene (GADD45) expression during tetracycline induction of the BRCA1 gene. (From Mullan PB, McWilliams S, Quinn J, et al. Uncovering BRCA1-regulated signaling pathways by microarray-based expression profiling. *Biochem. Soc. Trans.* 29(6): 678–683, 2001. With permission.)

tetracycline or turned on by removal (– tet) of the antibiotic without significantly altering the genomic background expression levels.

In reality 23/6800 genes (0.3%) were induced during the tet switch. However, the DNA damage-inducible gene (GADD45 or Growth Arrest after DNA Damage) was found to be expressed 10-fold during BRCA1 induction (77-fold). Northern blots confirmed the increase in GADD45 expression 6 hours following BRCA1 induction in both cell lines. Fold-changes in expression with GADD45 and BRAC1 showed a linear correlation ($r^2 = 0.96$) over 24 hours following the tet switch (Figure 5.16). This was confirmed over the same time course by Northern blot analysis. These results strongly suggest that GADD45 is a transcriptional target of the BRCA1 gene.

Bouras et al. (2002) undertook a comprehensive study of differential gene expression focused upon genes associated with estrogen (ER-positive) responsiveness in a human breast cancer cell line correlated with clinical tumor samples. First, ER-positive human breast cells (MCF-7 cell line, ATCC) were grown in the presence of 17β-estradiol or the antiestrogen (ICI 182 780) and gene expression measured from the mRNA populations. In addition, mRNA from 25 primary tumors (13 ER+; 12 ER–) were analyzed. An Affymetrix 43K GeneChip set, including 10,000 known genes and 25,000 ESTs, was used to measure differential gene expression. The outcome of the cell culture experiments was that 299 genes (<1%) were significantly ($P \leq 0.0005$) regulated by estrogen or antiestrogen. The expression profiles for these 299 genes were then assessed in the 25 primary tumors and the 10 most highly differentially expressed were subjected to hierarchical cluster analysis. Among these 10 genes, Stanniocalcin 2 (STC2) was singled out as a potential diagnostic candidate for determining the estrogen responsiveness in breast tumors. This was based upon the following observations:

1. STC2 differential expression between ER+ and ER– breast tumors
2. STC2 mRNA levels correlate with ER mRNA and protein levels

3. A threefold expression of STC2 in estrogen stimulated MCF-7 cells within 3 hours, remaining elevated up to 48 hours
4. A threefold reduction in expression of STC2 following antiestrogen treatment of cells with 6 hours of administration

Of particular interest was the observed strong correlation obtained between STC2 mRNA levels and its cognate protein levels. Using tissue microarrays from 216 breast tumor samples, in situ hybridization with a probe to STC2 mRNA was performed with 75 tumors showing a positive staining. 83% of the mRNA positive tumors also were identified for STC2 protein by immunohistochemical staining with STC2 antibody. Thus, microarray-based expression profiling of STC2 was corroborated by both in situ hybridization and immunostaining for ER+ tumors. A related gene, STC1, showed a similar correlation.

The study by Bouras et al. (2002) clearly demonstrated a rather strong correlation between transcription and translation of a single gene, Stanniocalcin 2. As a result, there is a high likelihood that STC2 may serve as a diagnostic/prognostic marker for breast cancer because one would be able to monitor both mRNA levels and the protein product during the various stages of tumor growth.

In summary, we find gene expression profiling with microarrays to be an exceptionally powerful and profound analytical tool. Not only are they most useful for a global analysis (e.g., metabolic pathways and their inter-relationships) but also for their ability to focus upon (albiet, assisted by clustering) and track important singular events that would otherwise remain hidden under a genomic backdrop.

However, in general how well does gene expression (as measured by the microarray) mirror biological outcome? After all, microarrays offer only a transient (global) view of biology. Cellular function on the other hand is mostly the work of proteins. And, it is well known that post-translational modifications are important regulators. Yet, at this point, we have the equivalent protein expression microarray tools available. Can we therefore rely upon the gene expression microarray as a surrogate tool to adequately track cellular endpoints from gene activity? Many advocate that gene expression does not correlate with protein expression, but in reality this is too simple of an answer. Life is more complicated.

Take for example the work of Chen et al. (2002) in which there was an effort to correlate mRNA levels with protein expression levels in lung adenocarcinomas. In this study, 57 stage I and 19 stage III lung adenocarcinoma tissues together with 9 non-neoplastic lung tissues were compared. Gene expression was measured using the Affymetrix GeneChip HuGene FL oligonucleotide arrays including 6800 known human genes. Protein was estimated by spot densitometry after two-dimensional polyacrylamide gel electrophoresis (2D-PAGE) separation and silver staining. Proteins were identified from MALDI-MS of peptides obtained by tryptic digest of protein spots from preparative 2D gel separations of extracts of a well-characterized lung adenocarcinoma cell line, A549. Certain proteins were confirmed by Western blot analysis. The 2D PAGE analysis can resolve up to 2000 proteins. In this study, 820 spots were mapped to proteins, but only 165 spots were used in the protein to gene analysis. Presumably these represent the

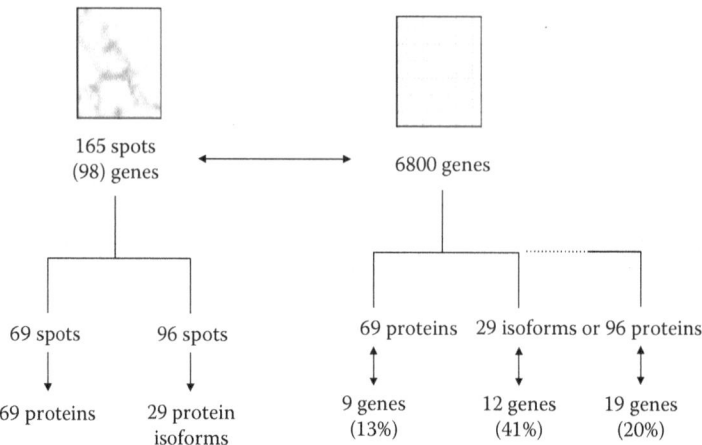

FIGURE 5.17   Correlations between mRNA and protein relative abundance. (From Chen G, Gharib TG, Huang C-C, et al. Discordant protein and mRNA expression in lung adenocarcinomas. *Mol. Cell. Proteomics* 1(4): 304–313, 2002. With permission.)

highest fold expressed genes from the lung tumors that were also visible on the gels by silver staining.

Several approaches to the analysis were undertaken to achieve correlation: pairwise individual protein to gene; protein isoforms to gene; average protein to gene expression values; and tumor stage related changes, protein to gene. As outlined in Figure 5.17, of the 165 protein spots, 98 represented genes. Of these, 69 proteins were mapped to single genes and 96 proteins appearing as 2-5 isoforms were mapped to 29 genes. However, only 9/69 (13%) of the single gene-protein pairs were significantly correlated ($p < 0.05$) in terms of relative abundance. For those proteins existing as isoforms, 19/96 (20%) showed correlation in relative abundance to the gene expression level. Thus, ~29% (28/98) of the genes showed a good correlation between mRNA abundance and protein abundance.

Proteins appearing as isoforms varied in their relationship to mRNA abundance. This is most likely the result of the effect of post-translational modification of the nascent protein and possibly degradation of the message. However, it is very difficult to generalize because subtypes had different levels of correspondence to mRNA abundance. For instance, 3/4 isoforms of OP18 were significantly correlated, while one isoform did not correlate. In the case of cytokeratin 8 only 1/5 isoforms correlated. What was very clear from this study is that using a generalized approach such as comparing averages of protein levels and mRNA abundance levels does not work. And, while many of the genes did not vary in their transcriptional to translational correlation relative to tumor stage, specific genes did show differentiation on the basis of tumor stage. From our discussion of these particular papers, it would seem that one cannot generalize about the relationships among transcriptional, translational, and post-translational events.

Infectious Disease

Even though the monitoring of infectious biological agents (bacteria, viruses) using DNA probe array technology is well known, much of that application involves identification of allele-specific targets. On the other hand, examining differential gene expression in the host is useful in elucidating mechanism(s) leading to virulence.

For example, Lane et al. (2001) examined gene expression in *Candida albicans*, a common yeast leading to human commensal infections. Pathogenicity is related to morphological changes in the organism characterized by a switch from the yeast form to the filamentous hyphal form. These changes in phenotype are obviously under control of various regulatory genetic elements and should be recognizable by comparing the gene expression patterns between the two morphological states. For example, signaling pathways are involved in the regulation of the filamentation process. In particular, the Cph1 transcription factor is known to be involved in hyphal formation. In turn, Cph1 is regulated by the mitogen-activated protein (MAP) kinase pathway. Other regulatory elements include an Egf1-mediated cAMP-protein kinase A as well as a Cph2 pathway. However, it is uncertain how these multiple cascade pathways act to shift growth into the hyphal state. Do they act independently or sequentially or are they convergent? What other genes may be important?

To more fully understand the mechanism(s) leading to the morphological switch, an array of PCR products representing ~1000 *C. albicans* genes (~10% of the genome represented by 7000 ORFs) was constructed on nylon membranes in triplicate. Gridding was accomplished using a Biomek 2000 equipped with a 384-pin HDRT (high-density replicating tool) system. Each spot was overprinted three to five times to assure full surface saturation and uniformity at each probe location. Membranes were then UV cross-linked prior to use. Each could be stripped in boiling 0.5% SDS and re-probed up to seven times without effecting a signal. As a result, the same filter sets could be used for multiple experiments with appropriate controls, thereby eliminating effects due to inter-filter array variations.

A series of yeast mutants grown under conditions selectively inhibiting or promoting growth of hyphal were compared with the wild type. For example, Lee's medium at 25°C promotes the yeast form while at 37°C induces hyphal growth. In addition, SS (synthetic succinate) medium permitted the transformation of the yeast form into the hyphal form for the wild type but not for the single cph1/cph1 mutant. It is known from other work that double mutants (cph1/cph1, efg1/efg1) are not virulent in a mouse model, while the single mutants cph1/cph1 and efg1/efg1 are virulent (hypha form). These mutants were grown under the above conditions, and their gene expression profiles were compared with the wild-type yeast form using the grid arrays. From this detailed study of comparing mutant and wild-type strains under a variety of growth conditions it was found that in fact Cph1, Cph2, and Efg1 regulation is convergent (Figure 5.18). On the basis of cluster analysis (Figure 5.19) and conformation by Northern blot analysis (Figure 5.20), these genes were found to regulate the expression of a set of hypha-specific genes (e.g., HYR1, ECE1, HWP1, SAP5,6 (Figure 5.21). In addition, two new genes (DDR48, YPL184) were discovered to be differentially expressed and under regulation by the convergent pathways, while the key regulator TEC1 gene was found to be under the influence of Cph1 and Egf1.

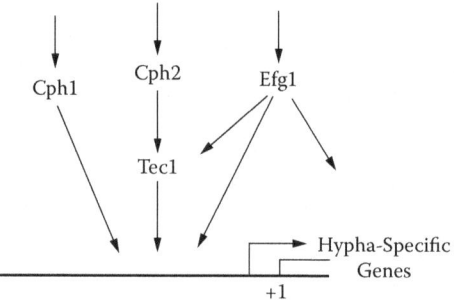

FIGURE 5.18 Convergent regulation of *Candida albicans* genes during the switch from wild type to the virulent form. (From Lane S, Birse C, Zhou S, Matson R, Liu H. DNA array studies demonstrate convergent regulation of virulence factors cph1, cph2 and efg1 in *Candida albicans. J. Biol. Chem.* 276(52): 48988–48996, 2001. With permission.)

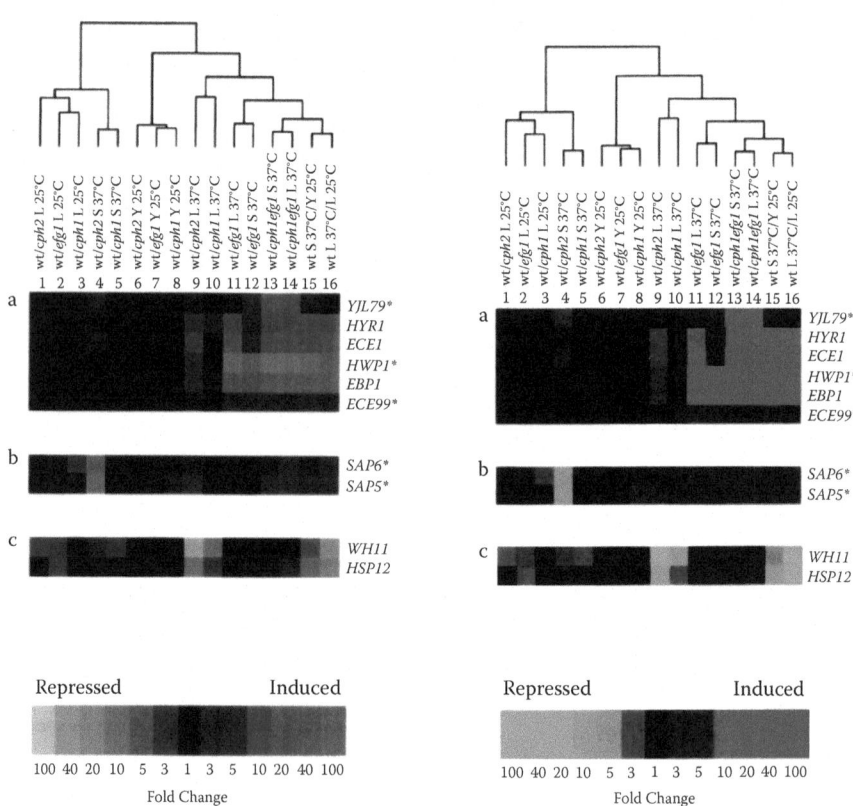

FIGURE 5.19 Gene expression cluster analysis. (From Lane S, Birse C, Zhou S, Matson R, Liu H. DNA array studies demonstrate convergent regulation of virulence factors cph1, cph2 and efg1 in *Candida albicans. J. Biol. Chem.* 276(52): 48988–48996, 2001. With permission.)

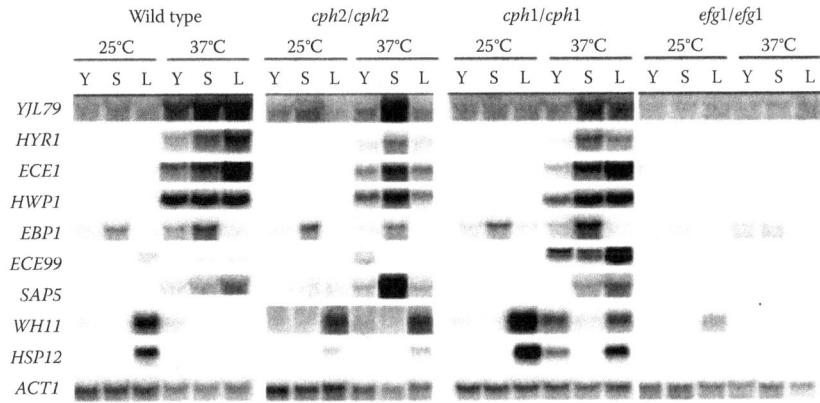

FIGURE 5.20 Northern blot analysis. (From Lane S, Birse C, Zhou S, Matson R, Liu H. DNA array studies demonstrate convergent regulation of virulence factors cph1, cph2 and efg1 in *Candida albicans. J. Biol. Chem.* 276(52): 48988–48996, 2001. With permission.)

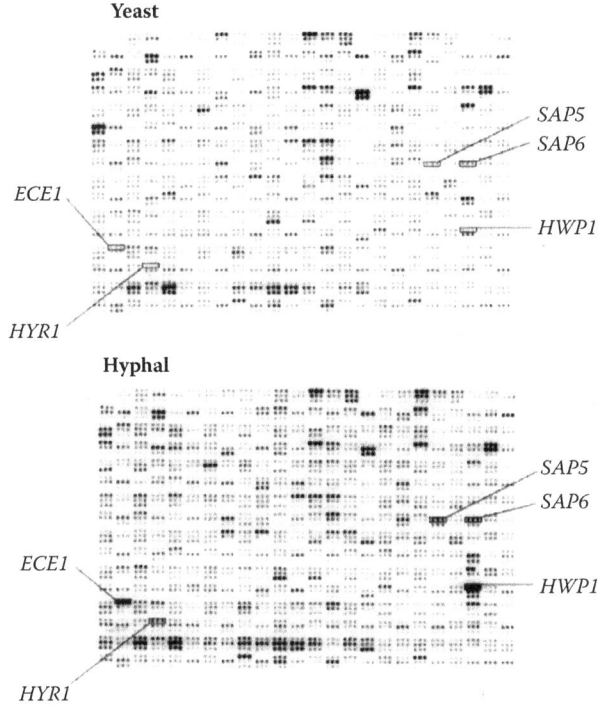

FIGURE 5.21 Comparison of wild type versus hyphal (virulent) gene expression patterns. (From Lane S, Birse C, Zhou S, Matson R, Liu H. DNA array studies demonstrate convergent regulation of virulence factors cph1, cph2 and efg1 in *Candida albicans. J. Biol. Chem.* 276(52): 48988–48996, 2001. With permission.)

Thus, even with the limited representation of the *C. albican*'s genome on the array, it was possible to probe into the regulation of virulence factors using a combination of well-designed biological approaches (mutations, growth media, temperature shifts) and gene expression tools (DNA arrays, Northern blots, clustering alogrithms).

In a similar manner, Maeda et al. (2001) examined *Helicobacter pylori* alteration of gene expression in gastric cancer cells. *H. pylori*, a Gram-negative bacterium, is well known for its infection of the human gastric mucosa. However, the pathogenesis of the associated gastroduodenal diseases (e.g., peptic ulcers) in the host is not well understood. The activation of various transcription factors (e.g., NF-κB) and induction of such inflammatory cytokines as IL-8 have been implicated. The activation of NF-κB in turn is believed to be under regulation of the cag PAI (pathogenicity island) genes that are found in a high percentage of *H. pylori* strain isolates.

cDNA microarray gene expression analysis was utilized to further investigate *H. pylori*–mediated induction of signaling pathways leading to an inflammatory response in the host. The glass slide microarray of 2304 cDNAs included a human cDNA library of 2280 sequence-validated cDNAs (Research Genetics) as well as a number of housekeeping genes as internal controls and luciferase genes as negative controls. RT-PCR and Northern blots were used for conformation of the microarray results.

A human gastric cancer cell line, MKN 45, derived from a gastric adenocarcinoma was co-cultured with either an *H. pylori* strain (cag PAI-positive) or a cagE-knockout strain. Of the 2300 genes only 8 genes (0.3%) were considered up-regulated at greater than a twofold expression change. These genes were not differentially expressed in the knockout strain suggesting the importance of the cag PAI involvement. Il-8 showed the greatest level in fold-expression (11.8 infected/control) followed by IκBα (fivefold). IκBα protein binds to NF-κB. In order to activate NF-κB, the phosphorylation of IκBα is required. Phosphorylated IκBα is subsequently degraded resulting in the release of active NF-κB from the complex. NF-κB is regarded as one of the major transcriptional factors for IL-8 induction which leads to the inflammatory response in the gastric mucosa. Of the remaining six genes showing at least twofold expression, the A20 gene (2.2-fold) was viewed as an important discovery in understanding the pathogensis of *H. pylori*. The A20 protein inhibits NF-κB activation.

In summary, the gene expression analysis of the gastric cancer cell host revealed that a small but significant set of genes within the host associated with inflammatory response were induced by the bacterium. As in the case of Lane et al. (2001), the analysis of differential gene expression between normal and diseased states uncovered key regulatory genes in specific pathways involved in the pathogenesis.

Other Disease States

While it would be difficult to provide a comprehensive overview, the following survey should provide ample proof of the continued expansion of microarrays into diverse fields.

Hearing Loss

Lomax et al. (2000) explored the potential for microarrays to investigate the involvement of genes in the recovery of hearing loss following noise trauma. The chick basilar papilla model was used. Noise exposure is known to cause the loss of hair cells in the basilar papilla. However, birds have the ability to regenerate these hair cells on the auditory epithelium and thus serve as a useful model for studying hearing loss and recovery. A low-density microarray containing 588 genes arranged in subgroups according to tissue (Clontech, Rat Atlas cDNA Array nylon membranes) was used to first examine which genes represented on the array were present in the cochlea and auditory regions of the brain. While preliminary, the investigation revealed three genes differentially expressed between two neuronal regions of the auditory system—the inferior cochlea (IC) of the brain and the cochlea modiolus (MOD). Two of the three proteins identified are known to be present in high abundance (mRNAs, moderate abundance class). Myelin proteolipid protein (PLP) is abundant in the brain, and peripheral myelin protein (PMP22) is localized in the peripheral nerves. PLP was observed to be differentially expressed 2.5 times higher in the IC (brain) region than the MOD, while PMP22 was 5.8 fold higher in MOD. The third gene, the plasma membrane calcium transporting ATPase (PMCA2), was twofold higher expressed in IC. Mutations to this rare gene are associated with deafness and imbalance in mice. The fact that the microarray could detect significant levels of PMCA2 was an unanticipated find. As the authors relate, "This exciting and gratifying result suggests that gene arrays may have a profound impact on the analysis of differential gene expression in the mammalian auditory system" (Lomax et al., 2000, p. 300).

Bone Pathology

Apert (Ap) syndrome is a craniofacial malformation thought to occur by mutation of the fibroblast growth factor 2 (FGFR-2) gene. This mutation has been reported to increase osteoblast differentiation that leads to premature calvaria ossification. Lomri et al. (2001) applied the use of cDNA microarrays in order to elucidate signaling pathways involved in osteoblast differentiation. Calvaria osteoblast cells were isolated from the bones of normal and Ap human fetuses and transformed using SV40 large T-antigen. The immortalized cell lines were maintained in culture and polyA+ mRNA was isolated from confluent cells. The resulting cDNA was radiolabeled by incorporation of 32P-dATP during the reverse transcriptase reaction and hybridized to a cDNA nylon membrane array (Atlas Human Expression Array, Clontech). The array contained 588 PCR cDNAs arranged in various human gene families. Greater than 40% of the 588 genes produced signal under high-stringency hybridization conditions. Of these, 27 genes were differentially expressed in Ap versus control cells: 22 up-regulated and 5 down-regulated.

In particular, significant differential expression was observed for three genes in Ap cells versus control cells: GTPase RhoA (3.6-fold), protein kinase Cα (2.9-fold), and the cytokine IL-1α (3-fold). All three putative proteins were confirmed to be overexpressed in the Ap mutant cell line relative to control cells by immunohistochemical staining (IL-1α) or Western blots (PKC, RhoA). These proteins are now implicated as serving an effector role on osteoblast differentiation resulting from the initial FGFR-2 mutation.

Glaucoma

Glaucoma is one of the most common diseases of the eye and leads to destruction of the optic nerve if left unchecked. Astrocytes, the major cell type in the

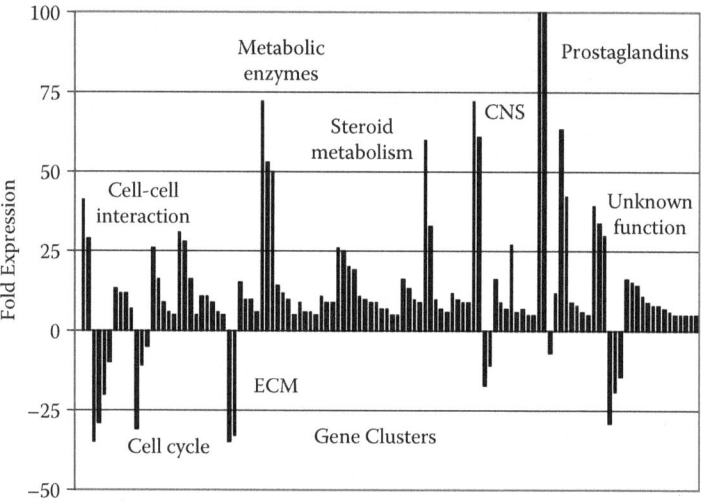

FIGURE 5.22 Investigating the pathogenesis of glaucoma based upon microarray gene expression clustering. (From Hernandez MR, Agapova OA, Yang P, Salvador-Silva M, Ricard CS, Aoi S. Differential gene expression in astrocytes from human normal and glaucomatous optic nerve head analyzed by cDNA microarray. *GLIA* 38: 45–64, 2002. With permission.)

optic nerve head, are believed to undergo phenotypic changes during the onset of glaucoma. Hernandez et al. (2002) used microarrays to monitor changes in gene expression patterns during phenotypic shifts in cultured astrocytes in an effort to better understand the pathogenesis of the disease. Optic nerve head astrocytes obtained from either normal human eyes or glaucomatous eyes were maintained in primary cell culture. Total RNA was extracted from the cultured cells, purified, and converted to labeled cRNA for hybridization to the U95A Human Genome GeneChip (Affymetrix). Differential gene expression between normal and glaucomatous astrocyte populations was compared. In many studies of this kind we have reported the differential expression of relatively few genes at between 2-fold and 10-fold expression. However, in this case at least 99 genes were overexpressed by at least fivefold in astrocytes from diseased eyes and 53 genes either absent or down-regulated by the same degree. In fact, many genes were found to be at least 25-fold differentially over- or underexpressed in reactive astrocytes (Figure 5.22). In particular, prostaglandin D2 synthase was differentially expressed by greater than 100-fold. While the mechanism of action of prostaglandin during glaucoma is uncertain, it is known that PGD2 is abundant in the central nervous system (CNS), especially in spinal fluid and ocular tissues. Astrocytes are known to proliferate at the site of neural damage and may be the source of the synthase enzyme. Another important finding from this study was that steroid metabolism is significantly up-regulated in reactive astrocytes, suggesting a possible relationship between glucocorticoid metabolism and glaucoma.

Multiple Sclerosis

Multiple sclerosis (MS) is a devastating disease of the CNS resulting in demyelination and inflammation. It is believed that the pathogenesis of the disease involves an immune reaction against various components of the myelin sheath. In the following study, the gene expression patterns in brain lesions obtained during autopsy of MS patients were examined by microarray cluster analysis (Lock et al., 2002). A total of 1080 genes (from 7026 represented on the GeneChip) were twofold or greater differentially expressed. The following observations were made concerning differential gene expression in MS lesions (relative to normal tissue) from the cluster analysis:

1. Migration of lymphocytic cells—Along with the presence of T cells, IL-17 a T cell transcript found to be elevated
2. Macrophage invasion—The up-regulation of macrophage mannose receptor, cathespin S, macrophage capping protein
3. Up-regulation of immune response genes—Overexpression of MHC and IgG
4. Inflammatory cytokine activity—IL-1 receptor and TNF receptor up-regulated
5. Down-regulation of myelin synthesis pathway genes

In addition, clustering revealed that not only were several genes significantly up-regulated but also they were differentially expressed for cases of acute and chronic MS.

In the next phase of the study, transcripts of up-regulated genes G-CSF and IgG were introduced into a mouse model commonly used to test out potential therapies—experimental autoimmune encephalomyelitis (EAE). From microarray analysis, granulocyte colony-stimulating factor (G-CSF) was found to be up-regulated in acute MS but not the chronic state of the disease. Subcutaneous injection of G-CSF prior to induction of EAE prevented onset of the disease in mice. The reversal of EAE by C-GSF has also been described (see Lock et al., 2002; Zavala et al., 2002).

Now in the case of immunoglobulin, microarray data indicated the opposite effect, that the Fc-receptor is elevated in chronic MS but not in acute lesions. Using Fcγ-receptor knockout mice, the disease was found to be absent. Intervenous immunoglobulin therapy in the EAE mouse model has been reported (see Lock et al., 2002, ref. 29).

In summary, Lock et al. were able to apply the results of microarray-based gene expression clustering of a human disease pathological state (acute verses chronic MS) to successfully identify therapeutic targets for an animal model (EAE) potentially applicable to the human condition.

MICRO-RNA

Simply put, micro-RNAs (miRNAs) are small, non-coding RNA species that are involved in the regulation of gene expression. They are single strands between 18 and 24 nucleotides in length. They bind to mRNA to regulate protein expression. They are in some manner involved in cellular differentiation, proliferation, and death. miRNAs are not to be confused with other noncoding RNA species such as small interfering RNA (siRNA) or short hairpin RNA (shRNA). While siRNA and shRNA may also be involved in regulation, they have different targets and different mechanisms of action.

The "global" profiling of miRNA populations is underway to elucidate their mechanism of action or functionality in various disease states, most notably cancer (Fridman et al., 2010; Zhao et al., 2010; Poláková et al., 2011). There are 1527 human miRNA sequences currently listed on *miRBase* (the officially recognized annotated micro-RNA database). The miRBase (www.mirbase.org) is currently maintained at the University of Manchester, United Kingdom.

The Utility of Gene Expression Microarrays in micro-RNA Analysis

A group from Agilent (Ach et al., 2008) undertook a study in order to compare results from qPCR and of the Agilent miRNA microarray. While the expression profiles across nine different tissues (normal human tissue, Ambion, Austin, TX) for 470 miRNAs were examined, the study really focused on a subset of 61 miRNAs that varied widely in expression level, GC content, and a few that were known to be difficult to detect. They also examined the impact of using different RNA preparation methods: TRIzol (Invitrogen), miRNeasy (Quigen), and mirVana (Applied Biosystems).

First, the TaqMan qPCR correlated well with the miRNA microarray for the miRNA subset. The cycle threshold (C_t) for qPCR and the $\log_2$ expression values from the microarray could be correlated directly because both measures exhibit a 1 unit change per twofold difference in miRNA concentration. The plot, C_t versus $\log_2$ signal, yields a linear relationship of $R = -1$; slope $= -1$ at equivalency. In these experiments, 56/60 miRNA expressions gave R values ranging from -0.8 to -1 upon analysis across nine tissues, separately for each miRNA species. A single miRNA (miR-637) was not detected and therefore not correlated; thereby the subset was reduced 60 miRNAs. While no explanation was provided about the sampling size (approximately 6% representation of the microarray), it is presumed that the primer sets were not available or the associated cost in running 470 PCRs was prohibitive for the study.

Analysis of the 60 miRNA expression population by the two methods across nine tissues is problematic because the methods vary greatly in sensitivity levels. Thus, the relative expression ratio across tissue pairs for qPCR and microarray was used for correlation scoring purposes. Pair-wise, linear plots: microarray [$\log_2$ tissue1-$\log_2$ tissue 2] versus qPCR [C_t tissue 1-C_t tissue 2] for each of 60 miRNAs and resulting 36 pairs were constructed. Again, good correlation was indicated from R values ranging from -0.82 to -0.98. The investigators determined a level of variability of 17% between the methods based upon the intercepts that reflect the consistency of the fold-changes measured.

However, three miRNAs did not correlate well. Spiked-in synthetic miRNAs of these were detected with good sensitivity by both qPCR and microarray. This suggests an endogenous interference that is most likely tissue specific. For example, correlation of at least one of these discrepant miRNAs was greatly improved by removing the placenta tissue data set.

The study is an impressive validation of the microarray approach, at least for the 60 miRNA determinations. Unfortunately, it would be difficult to extrapolate these findings for the complete microarray based upon the rather low representation of this subset.

The second part of this study aimed to address variation due to differences in sample preparation in the isolation of total RNA from tissue. Cell pellets from Hela (cervical carcinoma; 5×10^6 cells) and ZR-75-1 (breast carcinoma; 1×10^7 cells) cell lines were extracted for total RNA using the three commercial kits. The yield and total RNA quality were measured by spectroscopy (A260; 220 to 350 nm spectral scans; NanoDrop Technologies, Wilmington, DE) and size distribution (2100 Bioanalyzer, Agilent). RNA was of high quality (RIN values, 9.1 to 9.9) by all three methods and of similar yield: Hela total RNA, 34 ± 5 μg to 46 ± 7 μg; and ZR total RNA, 17 ± 6 μg to 24 ± 1 μg. TRIzol yields were consistently higher than those of mirVana and miRNeasy preparations. Spectral analysis indicated potential contaminates adsorbing in the A260 nm range which would lead to an overestimate of the RNA content, especially in the case of TRIzol.

Replicates of the prepared total RNAs (100 ng, Cy3 end-labeled, Agilent) obtained by the three extraction methods were hybridized to the Agilent miRNA microarrays. These were performed in triplicate for six different conditions, a total of 42 hybridizations. The hybridizations were done at 55°C for 20 hours under the following conditional sets: by prep method (same versus different) × prep replicate × hybridization replicate × hybridization day. The root mean squares (RMS) of hybridization signals were then compared as a measure of variability. Of little surprise was that greatest variability was associated with hybridizations of different preps with different replicate preps on different days, and the least variability was found among hybridizations of the same prep, the same replicate on the same day (Figure 5.23). Also, it was found that the TRIzol method provided the most consistent preparation

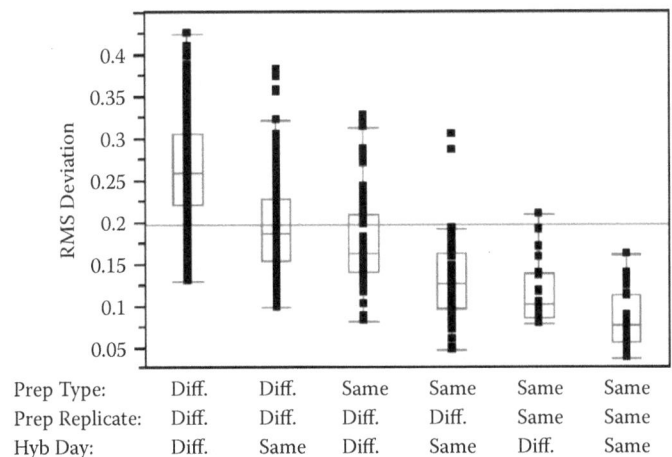

FIGURE 5.23 Variability in miRNA preparation, hybridization, and replicates. (From Ach RA, Wang H, Curry B. Measuring microRNAs: Comparisons of microarray and quantitative PCR measurements, and of different total RNA prep methods. *BMC Biotechnol.* 8:69, 2008. With permission.)

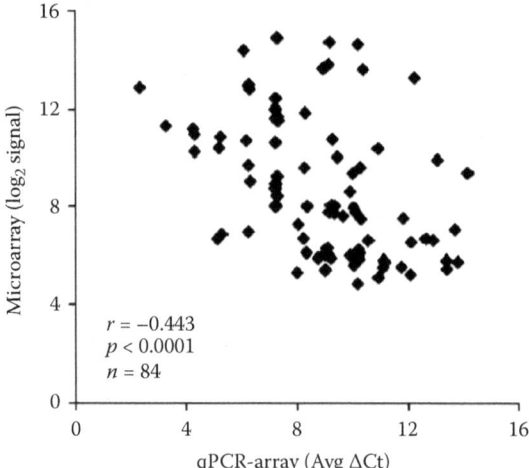

FIGURE 5.24   Concordance between microarray and qPCR in miRNA expression. (From Chen Y, Gelfond JAL, McManus LM, Shireman PK. Reproducibility of quantitative RT-PCR array in miRNA expression profiling and comparison with microarray analysis. *BMC Genomics* 10:407, 2009. With permission.)

of the three methods. The prudent strategy is to use the same extraction and labeling method throughout the evaluation.

Chen et al. (2009) found very little correlation ($r = -0.44$) between an miRNA expression microarray (LC Sciences, Houston, TX) and the TaqMan qPCR array (Applied Biosystems) for 84 miRNAs (Figure 5.24). Both platforms demonstrated excellent correlation in inter-assay evaluations. The average correlation for the qPCR was $r = 0.978$, while that for duplicate microarrays was $r = 0.974$.

The nature of the variation leading to the poor correlation between the two platforms was explored. GC-content and C_t values of miRNAs were examined and found not to impact the outcome. The largest variation on both platforms was observed in the measure of the lower abundant miRNAs in the sample population. Such variation may be attributed to differences in sensitivity between the platform technologies and could explain the increased incidence of false positives associated with microarrays. However, further analysis that examined the expression levels of individual miRNAs on both platforms led the researchers to conclude "that a low expression level was not always responsible for the false positive differential expression of miRNAs in the microarray analysis" (Chen et al., 2009, p. 6).

THE NATURE OF PLATFORM-TO-PLATFORM DISPARITY

Sato et al. (2009) examined both intra-platform and inter-platform performance of miRNA microarray systems from five commercial sources: Agilent, Ambion, Exiqon, Invitrogen, and Toray Industries. Two tissue types were selected representing low miRNA content (human liver total RNA; Ambion) and high miRNA content (human prostate total RNA; Ambion) for the analyses. For the cross-platform study,

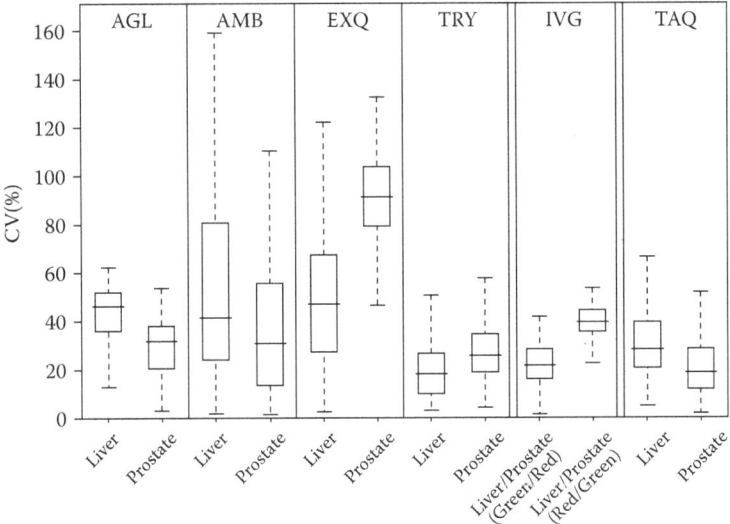

FIGURE 5.25 Variability in measuring miRNA expression among commercial platforms. (From Sato F, Tsuchiya S, Terasawa K, Tsujimoto G. Intra-platform repeatability and inter-platform comparability of microRNA microarray technology. *PLoS ONE* 4(5): e5540, 2009. With permission.)

309 miRNAs common to all platforms were measured. In addition, each platform was compared against the TaqMan qPCR system (Applied Biosystems) for a subset of 142 miRNAs.

For within a single platform (intra-platform) $n = 3$ replicates for liver and prostate miRNA expression were measured, and the results for positive signal calls were evaluated. As predicted, the miRNA expression level in prostate was greater than that of liver. Broader CV distribution (greater imprecision) of replicates was observed on those platforms that had lower thresholds for detection of a weakly positive signal. In particular, the Spearman's correlation (R_s) for prostate was significantly better than for liver. So, reproducibility on platforms with low call stringency would be more dependent upon the concentration range of miRNAs in a sample. In this study, the Agilent, Torey, and Invitrogen platforms exhibited good repeatability for both liver and prostate (comparable to TaqMan results), while Ambion and Exiqon were highly variable (Figure 5.25).

Inter-platform concordances for the five platforms were compared by rank correlation of the log-ratio of paired sample types (i.e., the measure of relative expression of liver versus prostate miRNAs on a single platform compared to the same measure on another platform) (see also Ach et al., 2008). This transformation is necessary in order to normalize differences in signals on platforms due to labeling, probe design, and hybridization. While results from Agilent and Torey systems correlated well (R_s = 0.87), the median rank across all platforms was $R_s = 0.55$, indicating poor agreement between the five systems. In fact, the call rate on positives ranged from 56.3 to 97.9% (liver) and 58.2 to 95.9% (prostate) across the platforms.

But what factors contribute to the disparity among these microarray platforms? The authors discuss the following issues—normalization, detection criteria, and hybridization:

1. Normalization—For gene expression, a general assumption has been that the total RNA (DNA) content is essentially constant among different samples. However, as pointed out in this study, for the equivalent amount of total RNA in liver and prostate tissue, there was a significant difference in total miRNA content between tissues.
2. Positive Call Stringency—This varies among platforms and was found to lead to a lack of concordance among platforms.
3. Hybridization—Dynamic (mixing) hybridization led to greater reproducibility than static (no mixing) hybridization.

Yauk et al. (2010) also performed a cross-platform performance evaluation of vendor products used in determination of micro-RNA differential expression. In this case, four commercial microarray-based platforms (Agilent, Exiqon, Invitrogen, LC Sciences) were compared against TaqMan PCR arrays (Applied Biosystems). Reference RNA was prepared from mouse total RNA (Ambion) and processed by the manufacturer's protocols to label the miRNA population in the sample for analysis. The following amounts of total mouse reference RNA required were for: Agilent (100 ng), Exiqon (1 µg), Invitrogen (0.5 µg), LC Sciences (1 µg), and TaqMan qPCR (1 µg).

Both one-color (Agilent, LC Sciences) and two-color (Invitrogen, Exiqon) label formats were represented. It was necessary therefore to validate two-color platforms for one-color analysis as well as undertake dye swapping to ensure optimal performance. Four sample replicates, split into two different reference pools, were analyzed by all platforms. Because the miRNA capture microarrays varied in probe content, it was necessary to evaluate a subset of 189 probes that were in common. These probes were also selected based upon *matched sequences* because different miRBase databases had been used among the vendors in probe design. For this particular reference RNA only 54/189 (<30%) probes were *scored positive* or "signal present" across all platforms including TaqMan. Although the authors report analyses based upon both subsets, for the purpose of our discussion concerning platform concordance we will refer only to the "54" probe set results.

The authors were rigorous in their statistical treatment examining both inter- and intra-platform performance by Spearman ranking and CCC (concordance correlation coefficient). Their findings are summarized in Table 5.3. Essentially, this study finds that the various platforms performed at a similar level of reproducibility, exhibited high intra-precision, and all correlated well with the qPCR results.

Pradervand et al. (2010) from the Center for Integrative Genomics at the University of Lausanne, Switzerland, undertook a comparative study of micro-RNA expression as measured by microarray, qPCR, and DNA sequencing. Digital gene expression (DGE) profiling of micro-RNA (miRNA) using the Illumina-Solexa sequencing technology (Genome Analyzer IIx, Illumina, Inc.) was compared with qPCR (7900HT Sequence Detection System, Applied BioSystems) and three microarray

TABLE 5.3

Correlation between miRNA Arrays and TaqMan Arrays

	One or Two-Color Analysis	Number of miRNAs Considered	CCC	95% Confidence Intervals
Agilent	One	122	0.416	0.337–0.488
		54	0.405	0.268–0.527
LC Science	One	131	0.545	0.454–0.624
		54	0.610	0.451–0.732
Exiqon	Two	131	0.461	0.371–0.542
		54	0.532	0.377–0.657
	One	125	0.409	0.329–0.483
		54	0.459	0.320–0.579
Invitrogen NCode	Two	118	0.314	0.241–0.383
		54	0.435	0.315–0.540
	One	124	0.225	0.225–0.281
		54	0.344	0.243–0.437

Source: From Yauk CL, Rowan-Carroll A, Stead JDH, Williams A. Cross-platform analysis of global microRNA expression technologies. *BMC Genomics* 11: 330, 2010. With permission.

Note: CCC, concordance correlation coefficient.

gene expression analysis platforms from Agilent (miRNA v2 microarray), Illumina (Expression BeadChips), and Affymetrix (GeneChip).

Replicate RNA samples (maximum value and purity (MVP) human normal adult tissue total RNA, Stratagene, Inc.) from brain and heart were analyzed across all platforms. There were significant differences in the amounts of total RNA required for labeling by these platforms: Agilent microarray (100 ng), Illumina microarray (500 ng), Affymetrix microarray (1 µg), ABI qPCR (700 ng), and Illumina sequencer (10 µg).

Concordance in $\log_2$ fold differences of miRNA expression with DGE and qPCR was strongest (correlation coefficient, $r = 0.9$; regression slope, $a = 0.87$). The Agilent microarray results also correlated well with both qPCR ($r = 0.9$; $a = 0.79$) and DGE ($r = 0.83$; $a = 0.7$). However, the Affymetrix and Illumina platforms did not correlate as well ($r = 0.66$ to 0.88) and showed significant regression slope ($a = 0.26$ to 0.46) suggesting bias (Figure 5.26). The investigators determined that the Illumina platform exhibited a fold-change compression that was likely introduced during PCR amplification of the target, while the Affymetrix GeneChip simply lacked the required sensitivity leading to the scoring of false negatives.

All platforms provided highly reproducible data sets from replicate analyses. However, the microarray platforms showed significant variation. This inter-platform variance among commercial microarrays has been well noted (see discussion in Chapter 2). While problematic, it is understandable based upon fundamental differences in probe design and software analysis programs utilized by various microarray

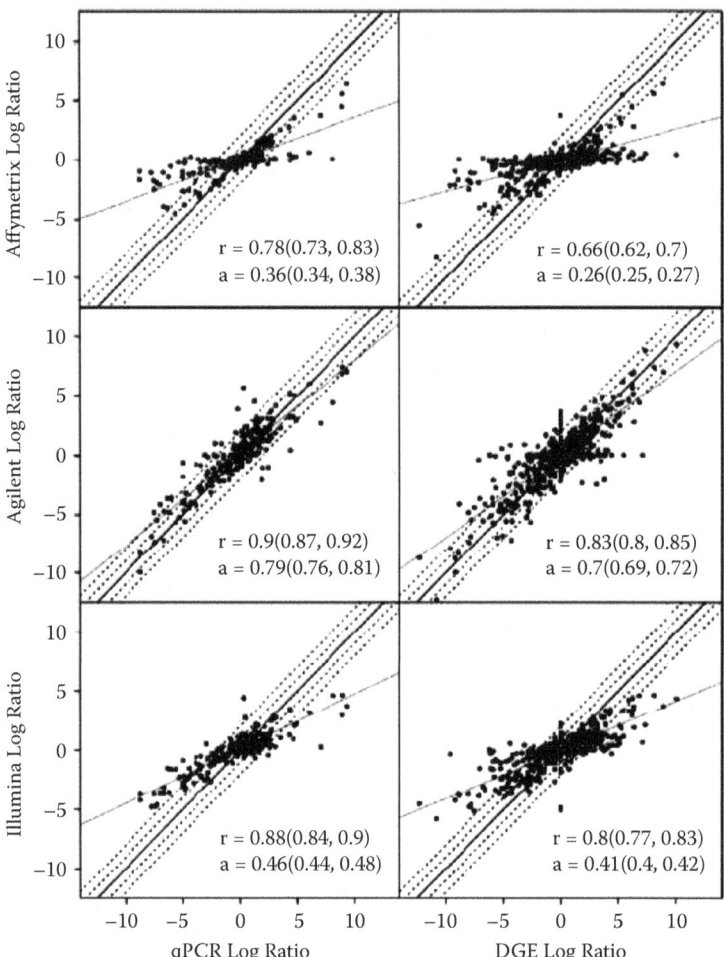

FIGURE 5.26 Concordance between miRNA microarrays, qPCR, and digital gene expression (DGE). (From Pradervand S, Weber J, Lemoine F, et al. Concordance among digital gene expression, microarrays, and qPCR when measuring differential expression of microRNAs. *BioTechniques* 48: 219–222, 2010. With permission.)

vendors. In this study, we learn how subtle differences in the probe sequence GC content and resulting T_m may influence the outcome. For example, in the case of the Affymetrix probe array, a mean $T_m = 51.9°C$ was ascribed to "false negative" probes (low signal intensity cutoff; mean GC content, 42.4%), while "true positives" were largely associated with a mean $T_m = 55.1°C$ and 50.5% GC content ($P < 0.001$).

ARRAY-BASED COMPARATIVE GENOMIC HYBRIDIZATION

Down Syndrome (Trisomy 21) is the most common single cause of human birth defects. Humans have two copies each of 23 chromosomes. In the case of Trisomy

21, an extra (third) copy of chromosome 21 is found. The presence of the extra copy results in abnormalities in brain development that often lead to mental retardation. Like Downs, for which there is no cure for the afflicted individual, there are a number of chromosomal-level genetic defects caused by either an increase or decrease in chromosomal copy number (or portions of a chromosome) with similar catastrophic effect.

Comparative genomic hybridization (CGH) was developed to map all such aberrations in a single test. Here, genomic DNA (gDNA) from the patient or "test" cell sample is fluorescently labeled with a dye. Likewise, a different dye is used to label gDNA from a well-characterized "reference" cell population. The two labeled populations are then mixed along with cot-1 DNA (in order to reduce interference from highly repetitive sequences in the genome) and applied under hybridization conditions to normal metaphase chromosomes. The karyotype therefore becomes differentially labeled creating a map of fluorescent intensities distributed across the chromosomal population. The resulting fluorescent dye ratio serves as an indicator of relative copy number of gene sequences (Kallioniemi et al., 1992).

Solinas-Toldo et al. (1997) and Pinkel et al. (1998) first introduced array-based comparative genomic hybridization (aCGH) using first bacterial artificial chromosome (BAC) clone arrays, followed by oligonucleotide microarrays. The aCGH approach greatly improved the chromosomal sequence resolution. CGH can detect loss or duplication in copy number at the chromosomal level at a resolution of about 5 Mb in segment length. However, higher resolution, less than 1 Mb sequence, is required to see microscopic aberrations. BAC arrays are capable of resolution to 100 kb, while oligonucleotides can determine a chromosomal imbalance in sequence regions under 100 bp.

The limitation of comparative genomic hybridization approaches is that there has to be an imbalance. That is, one is pairing a normal genomic population to a test population and then looking for a difference or imbalance between the two populations in terms of copy number. The overlying assumption is that the populations being compared are diploid. Thus, a gain of one copy number would indicate trisomy in the test karyotype, while a loss of one copy number would represent a monoploid. However, tumor cells in particular may also be polyploidal. For example, a tetrasomy might be balanced or imbalanced making it difficult to distinguish copy number differences. In this case, other approaches such as the use of the allelic ratio from single nucleotide polymorphism (SNP) mapping arrays should be considered (Gardina et al., 2008).

Karyotype, Fluorescence in situ Hybridization (FISH), Comparative Genomic Hybridization (CGH), or Array-CGH?

For the detection of submicroscopic chromosomal aberrations, cytogenetic examination by karyotype or fluorescence in situ hybridization (FISH) are often not adequate. An increase in chromosomal segment copy number or in the occurrence of micro-deletions is often undetected. Shaw-Smith et al. (2004) discuss this issue in patient populations with learning disabilities associated with such chromosomal imbalance. Routine karyotyping did not reveal chromosomal abnormality, while

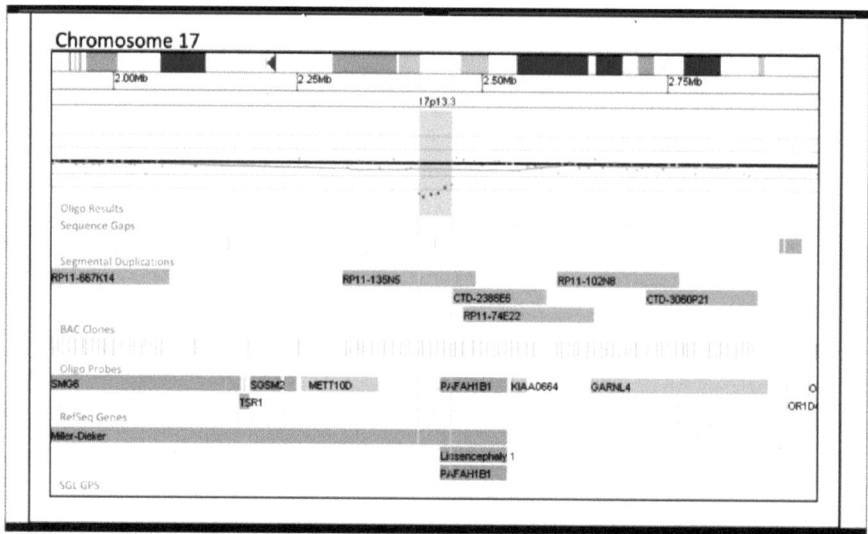

FIGURE 5.27   aCGH and BAC array detection of a deletion in chromosome 17. (From Neill NJ, Torchia BS, Bejjani BA, Shaffer LG, Ballif BC. Comparative analysis of copy number detection by whole-genome BAC and oligonucleotide array CGH. *Mol. Cytogenet.* 3: 11, 2010. With permission.)

FISH diagnosis revealed defects in 6% of the patients, and comparative genomic hybridization (CGH) with a 3 Mb resolution identified 10% of the population having micro-deletions or duplications. On the other hand, microarray-based comparative genomic hybridization (a-CGH) offering higher-resolution mapping of chromosomal imbalance at 1 Mb of chromosomal segment length achieved a detection rate of 24% for chromosomal defect.

Neill et al. (2010), a group from Signature Genomic Laboratories (Spokane, WA), compared two different CGH platforms, BAC oligo-array and array-CGH (aCGH) platform performance in the detection rate for clinically significant chromosomal copy number abnormalities (Figure 5.27). The BAC array included more than 4600 clones, arranged in overlaps (contigs with a mean gap of 1.6 Mb) to cover 1543 loci. This included regions of known micro-deletion and micro-duplication as well as over 500 genes known to be involved in development. The aCGH was custom designed at one probe for every 10 kb region and contained 105,000 probes (manufactured by Aglient Technologies, Santa Clara, CA). Selected gene loci were covered by an average of 50 probes per loci, while the genomic backbone received on average one probe for every 35 kb segment.

Two studies were undertaken. One examined 466 consecutive clinical specimens submitted for aCGH analysis and was used for platform validations. The other was a larger prospective diagnostic comparison of over 3000 cases.

In the side-by-side comparison of the 466 clinical specimens, it was necessary to exclude 347/466 cases from the BAC-array analysis that were largely due to hits on non-gene regions or those that were identified as normal population genetic variations. Likewise, the aCGH analysis found about 1300 copy number variations within the 466 cases, of which about 1200/1300 were irrelevant hits for this study. FISH was performed on the remaining subsets to remove known copy number variants. This resulted in a final selection of "alterations of potential clinical significance" that presumably had not been previously identified. In that case, the BAC array found 67/466 (14.3%) new cases, while aCGH discovered 73/466 (15.7%). The aCGH was reported to provide a greater sensitivity in determination of abnormalities in individual cases, for example, identifying two deletions rather than the single deletion observed from the BAC-array data.

For the prospective analysis, BAC arrays identified 365/3443 (10.6%) clinically significant abnormalities, while the aCGH discovered 477/3096 (15.9%). These results were deemed to be statistically significant ($p < 0.0001$).

Thus, array-based comparative genomic hybridization that provides greater resolution in chromosomal mapping is able to detect more chromosomal abnormalities. Question Examined Determined.

REFERENCES

Ach RA, Wang H, Curry B. Measuring micro-RNAs: Comparisons of microarray and quantitative PCR measurements, and of different total RNA prep methods. *BMC Biotechnol.* 8: 69, 2008.

Achiron A, Margalit, IR, Hershkoviz R, et al. Intravenous immunoglobulin treatment of experimental T Cell-mediated autoimmune disease upregulation of T Cell proliferation and downregulation of tumor necrosis factor a secretion. *J. Clin. Invest.* 93: 600–605, 1994.

Billuart P, Bienvenu T, Ronce N, et al. Oligophrenin-1 encodes a rhoGAP protein involved in X-linked mental retardation. *Nature* 392 (6679): 923–926, 1998.

Bouras T, Southey MC, Chang AC, et al. Stanniocalcin 2 is an estrogen-responsive gene coexpressed with the estrogen receptor in human breast cancer. *Cancer Res.* 62: 1289–1295, 2002.

Chen G, Gharib TG, Huang C-C, et al. Discordant protein and mRNA expression in lung adenocarcinomas. *Mol. Cell. Proteomics* 1(4): 304–313, 2002.

Chen JC, Zhuang S, Nguyen TH, Boss GR, Pilz RB. Oncogenic ras to rho activation by activating the mitogen-activated protein kinase pathway and decreasing rho-GTPase-activating protein activity. *J Biol. Chem.* 278 (5): 2807–2818, 2003.

Chen JJW, Peck K, Hong T-M, et al. Global analysis of gene expression in invasion by a lung cancer model. *Cancer Res.* 61: 5223–5230, 2001.

Chen Y, Gelfond JAL, McManus LM, Shireman PK. Reproducibility of quantitative RT-PCR array in miRNA expression profiling and comparison with microarray analysis. *BMC Genomics* 10: 407, 2009.

Cleary JD, Rogers PD, Chapman SW. Differential transcription factor expression in human mononuclear cells in response to amphotericin B: Identification with complementary DNA microarray technology. *Pharmacotherapy* 21: 1046–1054, 2001.

DeRisi J, Iyer VR, Brown PO. Exploring the metabolic and genetic control of gene expression on a genomic scale. *Science* 278: 680–686, 1997.

DeRisi J, Penland L, Brown PO, et al. Use of a cDNA microarray to analyze gene expression patterns in human cancer. *Nat. Genet.* 14: 457–460, 1996.

de Saizieu A, Certa U, Warrington J, Gray C, Keck W, Mous J. Bacterial transcript imaging by hybridization of total RNA to oligonucleotide arrays. *Nat. Biotechnol.* 16: 45–48, 1998.

Drmanac R, Strezoska Z, Labat I, Drmanac S, Crkvenjakov R. Reliable hybridization of oligonucleotides as short as six nucleotides. *DNA Cell Biol.* 9(7): 527–534, 1990.

Eisen MB, Spellman PT, Brown PO, Botstein D. Cluster analysis and display of genome-wide expression patterns. *PNAS* 95(25): 14863–14868, 1998.

Ferea TL, Botstein D, Brown PO, Rosenzwig RF. Systematic changes in gene expression patterns following adaptive evolution in yeast. *PNAS* 96: 9721–9726, 1999.

Fountoulakis M, Berndt P, Boelsterli UA, et al. Two-dimensional database of mouse liver proteins: Changes in hepatic protein levels following treatment with acetaminophen or its nontoxic regioisomer 3-acetamidophenol. *Electrophoresis* 21(11): 2148–2161, 2000.

Fridman E, Dotan Z, Barshack I, et al. Accurate molecular classification of renal tumors using micro-RNA expression. *J. Mol. Diagn.* 12(5): 688–696, 2010.

Gardina PJ, Lo KC, Lee W, et al. Ploidy status and copy number aberrations in primary glioblastomas defined by integrated analysis of allelic ratios, signal ratios and loss of heterozygosity using 500K SNP mapping arrays. *BMC Genomics* 9: 489, 2008.

Hernandez MR, Agapova OA, Yang P, Salvador-Silva M, Ricard CS, Aoi S. Differential gene expression in astrocytes from human normal and glaucomatous optic nerve head analyzed by cDNA microarray. *GLIA* 38: 45–64, 2002.

Ivanov I, Schaab C, Planitzer S, et al. DNA microarray technology and antimicrobial drug discovery. *Pharmacogenomics* 1(2): 169–178, 2000.

Iyer VR, Eisen MB, Ross DT, et al. The transcriptional program in the response of human fibroblasts to serum. *Science* 283: 83–87, 1999.

Jain KK. Applications of biochip and microarray systems in pharmacogenomics. *Pharmacogenomics* 1(3): 289–307, 2000.

Katsuma S, Nishi K, Tanigawara K, et al. Molecular monitoring of bleomycin-induced pulmonary fibrosis by cDNA microarray-based gene expression profiling. *Biochem. Biophys. Commun.* 288: 747–751, 2001.

Kallioniemi A, Kallioniemi OP, Sudar D, et al. Comparative genomic hybridization for molecular cytogenetic analysis of solid tumors. *Science* 258(5083): 818–821, 1992.

Lane S, Birse C, Zhou S, Matson R, Liu H. DNA array studies demonstrate convergent regulation of virulence factors cph1, cph2 and efg1 in *Candida albicans. J. Biol. Chem.* 276(52): 48988–48996, 2001.

Lapteva N, Nieda M, Ando Y, et al. Expression of renin-angiotensin system genes in immature and mature dendritic cells identified using human cDNA microarray. *Biochem. Biophys. Res. Commun.* 285, 1059–1065, 2001.

Lashkari DA, DeRisi JL, McCusker JH, et al. Yeast microarrays for genome wide parallel genetic and gene expression analysis. *PNAS* 94: 13057–13062, 1997.

Lock C, Hermans G, Pedotti R, et al. Gene-microarray analysis of multiple sclerosis lesions yields new targets validated in autoimmune encephalomyelitis. *Nature Med.* 8(5): 500–508, 2002.

Lomax MI, Huang L, Cho Y, Gong T-WL, Altschuler RA. Differential display and gene arrays to examine auditory plasticity. *Hearing Res.* 147: 293–302, 2000.

Lomri A, Lemonnier J, Delannoy P, Marie PJ. Increased expression of protein kinase Ca, interleukin-1a, and RhoA guanosine 5′-triphosphatase in osteoblasts expressing the ser-252trp fibroblast growth factor 2 apert mutation: Identification by analysis of complementary DNA microarray. *J. Bone Miner. Res.* 16(4): 705–712, 2001.

Maeda S, Otsuka M, Hirata Y, et al. cDNA microarray analysis of *Helicobacter pylori*-mediated alteration of gene expression in gastric cancer cells. *Biochem. Biophys. Res. Commun.* 284: 443–449, 2001.

Mahadevappa M, Warrington JA. A high-density probe array sample preparation method using 10- to 100-fold fewer cells. *Nat. Biotechnol.* 17: 1134–1136, 1999.

Marton MJ, DeRisi JL, Bennett HA, et al. Drug target validation and identification of secondary drug effects using DNA microarrays. *Nature Med.* 4: 1293–1301, 1998.

Matson RS. Chapter 64: Oligonucleotide arrays for the detection of ras mutations in *Nonradioactive Analysis of Bio-Molecules*, Kessler C, ed., Heidelberg, Springer-Verlag, 2000.

Al Moustafa A-E, Alaoui-Jamali MA, Batist G, et al. Identification of genes associated with head and neck carcinogenesis by cDNA microarray comparison between matched primary normal epithelial and squamous carcinoma cells. *Oncogene* 21: 2634–2640, 2002.

Mullan PB, McWilliams S, Quinn J, et al. Uncovering BRCA1-regulated signaling pathways by microarray-based expression profiling. *Biochem. Soc. Trans.* 29(6): 678–683, 2001.

Neill NJ, Torchia BS, Bejjani BA, Shaffer LG, Ballif BC. Comparative analysis of copy number detection by whole-genome BAC and oligonucleotide array CGH. *Mol. Cytogenet.* 3: 11, 2010.

Nishizuka S, Winokur ST, Simon M, Martin J, Tsujimoto H, Stanbridge EJ. Oligonucleotide microarray expression analysis of genes whose expression is correlated with tumorigenic and non-tumorigenic phenotype of HeLa X human fibroblast cells. *Cancer Lett.* 165: 201–209, 2001.

Pinheiro NA, Caballero OL, Soares F, Reis LFL, Simpson AJG. Significant overexpression of oligophrenin-1 in colorectal tumors detected by cDNA microarray analysis. *Cancer Lett.* 172: 67–73, 2001.

Pinkel D, Segraves R, Sudar D, et al. High resolution analysis of DNA copy number variation using comparative genomic hybridization to microarrays. *Nat. Genet.* 20: 207–211, 1998.

Poláková KM, Lopotová T, Klamová H, et al. Expression patterns of micro-RNAs associated with CML phases and their disease related targets. *Mol. Cancer* 10: 41, 2011.

Pradervand S, Weber J, Lemoine F, et al. Concordance among digital gene expression, microarrays, and qPCR when measuring differential expression of micro-RNAs. *BioTechniques* 48: 219–222, 2010.

Reilly TP, Bourdi M, Brady JN, et al. Expression profiling of acetaminophen liver toxicity in mice using microarray technology. *Biochem. Biophys. Res. Commun.* 282: 321–328, 2001.

Ross DT, Scherf U, Eisen MB, et al. Systematic variation in gene expression patterns in human cancer cell lines. *Nat. Genet.* 24: 227–235, 2000.

Saiki RK, Walsh PS, Levenson CH, Erlich HA. Genetic analysis of amplified DNA with immobilized sequence-specific oligonucleotide probes. *PNAS* 86: 6230–6234, 1989.

Sato F, Tsuchiya S, Terasawa K, Tsujimoto G. Intra-platform repeatability and inter-platform comparability of micro-RNA microarray technology. *PLoS ONE* 4(5): e5540, 2009.

Schena M, Shalon D, Davis RW, Brown PO. Quantitative monitoring of gene expression patterns with a complementary DNA microarray. *Science* 270: 467–470, 1995.

Schena M, Shalon D, Heller R, Chai A, Brown PO, Davis RW. Parallel human genome analysis: Microarray-based expression monitoring of 1000 genes. *PNAS* 93: 10614–10619, 1996.

Shalon D, Smith SJ, Brown PO. A DNA microarray system for analyzing complex DNA samples using two-color fluorescent probe hybridization. *Genome Res.* 6: 639–645, 1996.

Shaw-Smith C, Redon R, Rickman L, et al. Microarray based comparative genomic hybridisation (array-CGH) detects submicroscopic chromosomal deletions and duplications in patients with learning disability/mental retardation and dysmorphic features. *J. Med. Genet.* 41: 241–248, 2004.

Solinas-Toldo S, Dürst M, Lichter P. Specific chromosomal imbalances in human papillomavirus-transfected cells during progression toward immortality. *PNAS USA* 94, 3854–3859, 1997.

Unger MA, Rishi M, Clemmer VB, et al. Characterization of adjacent breast tumors using oligonucleotide arrays. *Breast Cancer Res.* 3: 336–341, 2001.

Wang T, Hopkins D, Schmidt C, et al. Identification of genes differentially over-expressed in lung squamous cell carcinoma using combination of cDNA subtraction and microarray analysis. *Oncogene* 19: 1519–1528, 2000.

Winzeler EA, Richards DR, Conway AR, et al. Direct allelic variation scanning of the yeast genome. *Science* 281: 1194–1197, 1998.

Wodicka L, Dong H, Mittmann M, Ho M-H, Lockhart DJ. Genome-wide expression monitoring in *Saccharomyces cerevisiae*. *Nat. Biotechnol.* 15: 1359–1367, 1997.

Yauk CL, Rowan-Carroll A, Stead JDH, Williams A. Cross-platform analysis of global micro-RNA expression technologies. *BMC Genomics* 11: 330, 2010.

Zanders ED. Gene expression analysis as an aid to the identification of drug targets. *Pharmacogenomics* 1(4): 375–384, 2000.

Zavala F, Abad S, Ezine S, et al. G-CSF therapy of ongoing experimental allergic, encephalomyelitis, via chemokine- and cytokine-based immune deviation. *J. Immunol.* 165: 2011–2019, 2002.

Zhao H, Shen J, Medico L, Wang D, Ambrosone CB, Liu S. A pilot study of circulating miR-NAs as potential biomarkers of early stage breast cancer. *PLoS ONE* 5(10): e13735, 2010. doi:10.1371/journal.pone.0013735.

6 Protein Microarray Applications

INTRODUCTION

Protein microarrays are in widespread use for life science research, clinical research, environmental testing, and diagnostic applications. It is the purpose of this chapter to describe the development of protein microarray technology as well as to provide commentary surrounding key applications. We begin an historical accounting.

The use of protein microarrays stems from the earlier described works on gene expressions arrays. However, unlike its predecessor whose process format (mutation detection, polymorphism screening, gene expression analysis, etc.) is essentially based upon solid-phase hybridization of nucleic acid complementary strands, the protein array may play different roles and comprise a variety of formats. For example, analogous to a gene array we may print down antibodies onto a solid support and capture the complementary antigen. However, we may also wish to immobilize an array of protein kinases and simultaneously measure their respective substrate levels in a cell supernatant. Thus, protein microarrays can be classified into at least two distinct categories, as Kodadek (2001) has suggested: protein function arrays (measuring the activities of native proteins) or protein-detecting arrays (monitoring protein levels). Although several groups within the biomedical industry had initiated work on protein-detecting arrays during the late 1980s (microspot concept—Ekins et al., 1989) and 1990s (slide-based microarray immunoassay—Silzel et al., 1998; Mendoza et al., 1999), the first publication describing the use of protein microarrays on a scale comparable to the gene expression arrays described by Schena et al. (1995) must be credited to MacBeath and Schreiber (2000) at Harvard. The technology quickly became adopted for work on the proteome. The following examples will serve to highlight the application of protein microarrays in biomedical research.

SPOT THEORY

The analytical concepts and strategies developed by Ekins and co-workers (1990) for improvements in sensitivity for the immunoassay herald the beginnings of the antibody array. Ekins determined that only a small amount of capture (sensory) antibody need be used to measure antigen in solution. More precisely, only a few binding sites within the antibody spot need be occupied by antigen in order to determine antigen concentration. Ekin's ambient analyte immunoassay relies on this fact—that the fractional occupancy (of capture antibody with antigen) reflects the true "analyte" concentration of antigen in solution. This amount of antibody does not harvest a significant amount of antigen. Thus, the conventional practice of coating a microtiter

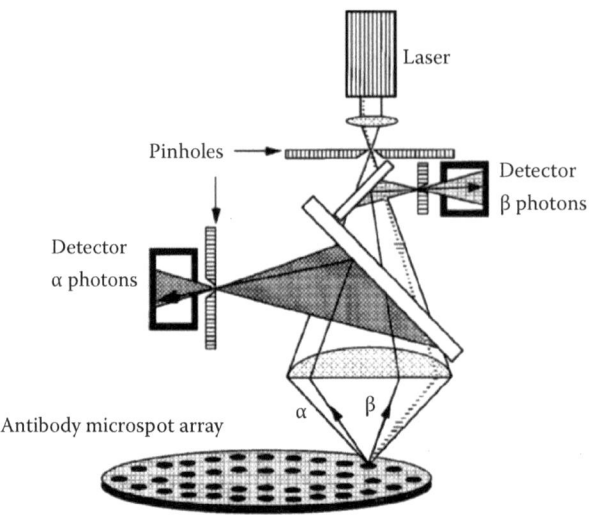

FIGURE 6.1   "Microspot" array detection. (From Ekins R, Chu F, Biggart E. Fluorescence spectroscopy and its application to a new generation of high sensitivity, multi-microspot, multianalyte, immunoassay. *Clin. Chim. Acta* 194: 91–114, 1990. With permission.)

plate well with excess antibody for use in an ELISA is not necessary. All that is needed is a small spot.

Because all that is required is a "microspot" of antibody for the assay, then multiple immunoassays may be run simultaneously using microspots of different analyte-specific antibodies in the same well. An early demonstration of the microspot immunoassay for determination of tumor necrosis factor (TNF) was achieved using a LaserSharp scanning confocal microscope (BioRad) (Figure 6.1). In this case, a ratiometric approach was used in which the capture antibody was labeled with Texas Red and the biotinylated secondary antibody was labeled with avidin-FITC. The FITC/Texas Red ratio for each microspot was determined and the specific binding activity was determined under ambient analyte conditions (Figure 6.2). While at the time it was only possible to create low-density microspot arrays, Ekins postulated that a microspot array of 50 micron diameter spots could yield over a million immunoassays in one square centimeter. However, implementation of the antibody microarray required additional considerations such as a means to deposit the microspots onto a substrate and the availability of high sensitivity detectors. As a result, in 1991 Ekins entered into a collaboration with Boehringer-Mannheim GmbH to pursue commercialization of the Microspot™ as a third-generation ultrasensitive immunoassay (Ekins, 1998). Boehringer-Mannheim constructed microarrays on small, disposable single-well polystyrene carriers using piezoelectric "ink-jet" technology to print down 100 to 200 spots on a 3 mm diameter well bottom (Figure 6.3). Each spot was approximately 80 microns in diameter. A prototype fully automated system built by the company in 1994 to 1995 produced 10,000 single-well carrier chips (200 spot array per chip) per hour. Lower limits of detection were reported to be approximately 0.01 detection molecules/μm^2, corresponding to an assay sensitivity

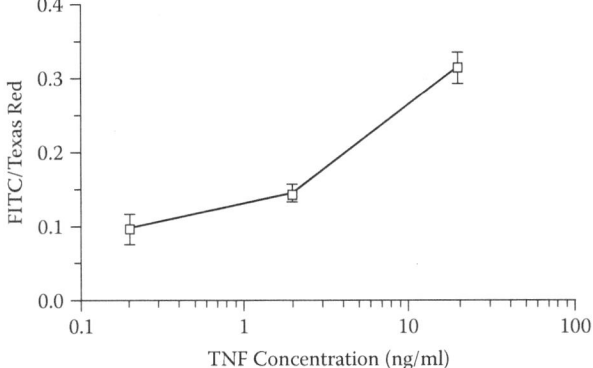

FIGURE 6.2 Tumor necrosis factor (TNF) microspot immunoassay based upon ratiometric signal detection. (From Ekins R, Chu F, Biggart E. Fluorescence spectroscopy and its application to a new generation of high sensitivity, multi-microspot, multianalyte, immunoassay. *Clin. Chim. Acta* 194: 91–114, 1990. With permission.)

approaching 10^{-17} M. For specific analytes high sensitivity was achieved (e.g., TSH, 0.01 µU/mL or total IgE, 0.01 U/mL) (Ekins and Chu, 2001).

An alternative approach to that of Ekins is that of the "mass-sensing" multianalyte microarray immunoassay first described by researchers at Beckman Coulter, Inc. (Silzel et al., 1998). As early as 1991 other groups within the company had adapted commercially available ink-jet printers (e.g., HP Deskjet) in order to deposit oligonucleotides or proteins such as streptavidin onto substrates to create arrays (Matson, unpublished). Later, piezoelectric jet printing was employed using a prototype drop on demand system (MicroFab, Inc.). Silzel and co-workers utilized these print technologies to deposit

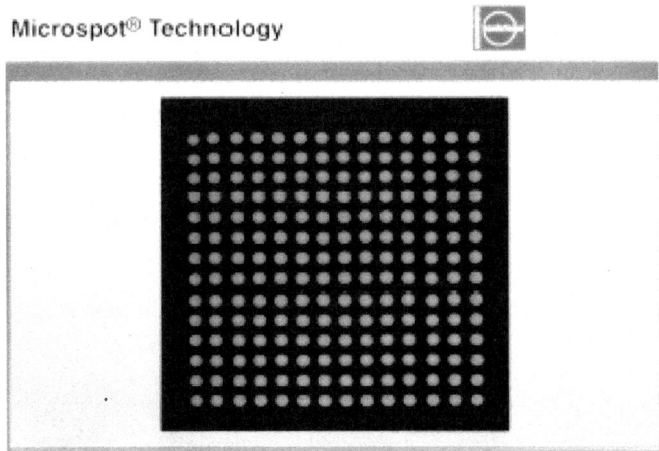

FIGURE 6.3 Boehringer-Mannheim's microspot array. (From Ekins RP. Ligand assays: From electrophoresis to minaturized microarrays. *Clin. Chem.* 44(9): 2015–2030, 1998. With permission.)

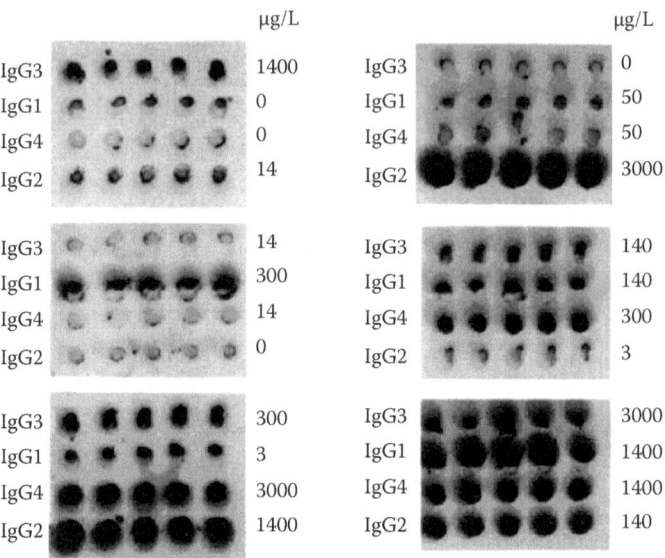

FIGURE 6.4   Human IgG subclass arrays. (From Silzel J W, Cerecek B, Dodson C, Tsong T, Obremski RJ. Mass-sensing, multianalyte microarray immunoassay with imaging detection. *Clin. Chem.* 44(9): 2036–2043, 1998. With permission.)

avidin for immobilization of biotinylated antibodies or for the direct printing of monoclonal antibodies. For example, an array of four human IgG subclasses (IgG1, IgG2, IgG3, IgG4) was immobilized onto a polystyrene slide as 200 micron diameter spots (Figure 6.4). In contrast to those of the ambient analyte immunoassay, their experiments demonstrated that capture antibody substantially depleted the sample of antigen analyte within a few hours. Thus, the antibody spot was thought to "harvest" the total analyte mass from solution, a process they termed "mass-sensing."

How different are the microspot ambient analyte and mass-sensing microarray immunoassays? Figure 6.5 compares these two formats. To be useful as a clinical assay, Ekins insists that microspot arrays obey "ambient analyte" conditions. That is, the solid-phase tethered capture (sensor) antibody must remove from the sample solution ≤1% of the analyte present in the sample in order for the assay to be valid. Provided the capture antibody's Ka is not altered during attachment, the ambient analyte condition would be satisfied in most cases by a surface antibody concentration <[0.01/Ka] M. So, if the antibody equilibrium association constant Ka ~ 10^{11}, then the antibody concentration required would be ~0.01/10^{11} M or 0.1 pM. That corresponds to roughly 6×10^7 antibody molecules or binding sites from a 1-mL sample volume. Adsorbed monolayers of antibodies (IgG) can be achieved in the range of 130 to 650 ng/cm^2 or from 10^9 to 10^{10} molecules/mm^2 depending upon the solid phase. Thus antibody microspots having spot diameters of about 50 to 120 microns would be sufficient. For a weaker binding antibody, Ka ~ 10^{10}, the ambient analyte condition would require larger microspot diameters (~170 to 380 microns). However, it is more likely that the sampling volume for an immunoassay would be 100 μL or less so that microspots on the order of 5 to 40 micron diameter would necessarily

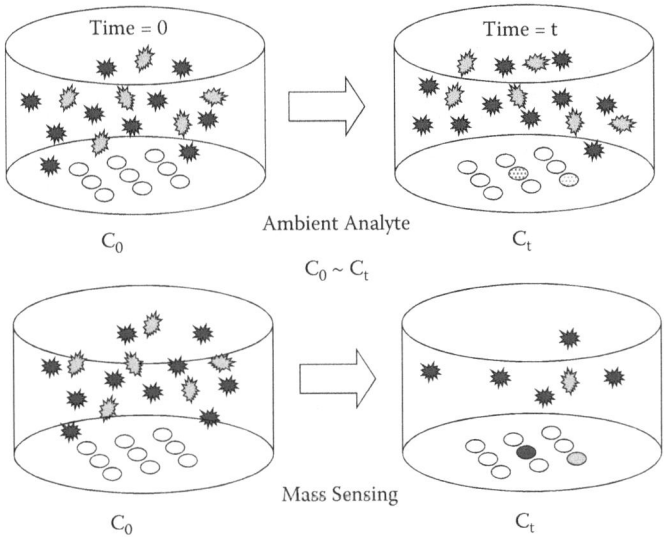

FIGURE 6.5   Ambient analyte versus mass-sensing immunoassay conditions.

be utilized depending upon antibody density or binding affinity. The smaller the diameter spot of antibody monolayer, the more likely it is that the ambient analyte condition would be maintained. From a practical standpoint the spot diameter cannot be vanishingly small because there would be a lower limit in detectability and acceptable capture rate.

In the work described by Silzel et al. (1998), a typical binding capacity for a 200 micron spot was reported to be on the order of 10^{10} analyte molecules based upon biotin DBCy5 dye-binding to avidin spots. Given a 100 μL sampling volume, this would correspond to an initial concentration of about 1.7×10^{-10} M or 170 pM assuming full depletion of the sample. In any case, this suggests a binding condition much higher than that of the ambient analyte condition of $0.01/Ka$, where $Ka \sim 10^{10}$. Silzel et al. then compared the mass assay relative to the ambient analyte assay based upon a hypothetical assay for TSH (Figure 6.6) and found a 60-fold improvement in absolute fluorescent signal when using the mass-sensing approach. For example, using TSH at a concentration of 10^{-15} M (60,000 TSH molecules) and an anti-TSH capture antibody with a $Ka \sim 10^{10}$, the ambient analyte condition would be satisfied with a total of 600 molecules (1%) bound to the microspot, while the mass-sensing assay would sequester 38,000 molecules (63%) for improved detection. However, under experimental conditions employing an antibody, array efforts at achieving such levels of sensitivity were compromised by nonspecific binding. The nonspecific signal was found to be associated with the capture antibody spot and estimated to be approximately 100-fold higher than the instrumental detection limit. As such, IgG_3 could be detected at approximately 15 μg/L (100 pM) providing an estimated capture of 3×10^8 IgG_3 molecules per 200 micron spot.

Sapsford et al. (2001) examined microarray-based antibody-antigen binding kinetics in real time to determine the effect of spot size. Capture antibodies were

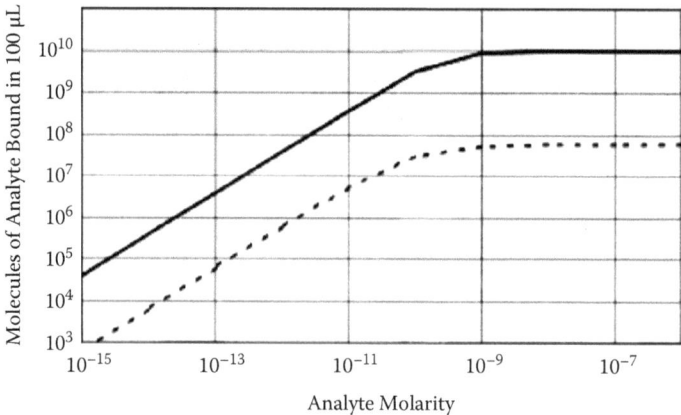

FIGURE 6.6 Hypothetical comparison of the ambient analyte and mass-sensing analysis of TSH. (From Silzel JW, Cerecek B, Dodson C, Tsong T, Obremski RJ. Mass-sensing, multianalyte microarray immunoassay with imaging detection. *Clin. Chem.* 44(9): 2036–2043, 1998. With permission.)

immobilized in an array pattern onto silver-clad microscope slides. An anti-mouse IgG was either directly attached to the surface or was obtained through use of Neutravidin capture of the biotinylated antibody. Cy5-labeled mouse IgG capture was monitored based upon signal generated from the excitation of the evanescent wavequide (slide) with a 635 nm laser source with detection by a charge-coupled device (CCD) camera system. Both static and flow-through conditions were employed. The binding kinetics were characterized in terms of the apparent time constant (K app = kf C + kr) where C = analyte concentration, kf = association rate constant, and kr = dissociation rate constant. In closed-loop experiments it was determined that a plateau value for K app of 0.0024 s⁻¹ was reached at a linear flow rate of 2.67 mL min⁻¹. K app was found to decrease with decreasing antigen [C] concentration with equilibrium achieved only at the highest level (1 μg/mL). The association rate constant Kf was calculated at 3.6×10^5 M⁻¹s⁻¹ for IgG binding. To determine the effect of spot size various photolithographic masks were used to create arrays of square patterns at spot widths from 80 microns to 1145 microns. The K app (± 1 standard deviation) did not vary over the spot size range of 80 to 1145 microns under constant flow conditions at fixed levels of antigen concentration. There was also little effect on the mean binding with variation in spot size.

This study does not support the Ekins ambient analyte model in that under flow conditions the rate of analyte capture is not increased as the spot size is decreased. However, as the authors point out, this study is not necessarily a contradiction because Ekins assumptions are based upon achieving equilibrium under static conditions. Under static conditions the role of diffusion is more likely an important limiting factor in influencing the binding rate. On a more operational note, these investigators also observed the problem in collecting sufficient signal from a very small spot relative to background noise. The other interesting observation from this study was that Kf for directly immobilized antibody was only

50% of the avidin-biotin bridged antibody. This suggests that random coupling reduces the number of effective binding sites relative to the oriented coupling using avidin-biotin.

In practical terms microarrays are operational by at least two means. One can create very small spots of capture ligands that bind only appreciable amounts of analyte and measure binding with ultra-sensitive signaling reagents and detectors, or one can create arrays of much larger spots of capture ligands that significantly deplete the sample of analyte and permit the use of less sensitive (and presumably less expensive) approaches to detection.

APPLICATIONS DEMONSTRATING PROTEIN MICROARRAY UTILITY

Researchers at Genometrix provided one of the first examples of a high-throughput microarray-based ELISA (Mendoza et al., 1999). They created arrays of antigens within wells of an optically flat glass substrate. The 96 wells were formed using a hydrophobic Teflon mask. The wells were chemically treated with NHS-active esters for covalent immobilization of proteins. Each well contained four replicates of a 6 × 6 element array (Figure 6.7). A uniquely designed capillary printer allowed for the simultaneous arraying of the 36 elements including various IgG antigens (Figure 6.8). Approximately 200 pL of each were delivered to the glass surface resulting in spot diameters of 275 microns at center-to-center spacings of 300 microns. The protein droplets were allowed to dry on the surface of the glass substrate prior to assay. Although this allowed for visualization of the print run, for quality control purposes this practice may not be advisable for other proteins. In this case, the immobilized IgG antigens most likely provided recognizable epitopes whether in the native or denatured state.

Assays included conventional ELISA processing except that the reagent volumes were greatly reduced ranging from 25 μL to 50 μL well additions. Following a 1 hour block in casein the monoclonal detection antibody was incubated an additional hour at room temperature. The assay sensitivity from the micro-ELISA was

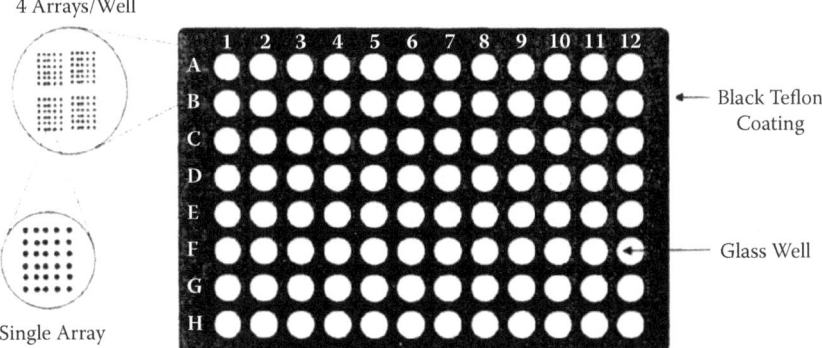

FIGURE 6.7 A 96-well microarray plate format. (From Mendoza LG, McQary P, Mongan A, Gangadharan R, Brignac S, Eggers M. High-throughput microarray-based enzyme-linked immunosorbent assay (ELISA). *BioTechiques* 24(4): 778–788, 1999. With permission.)

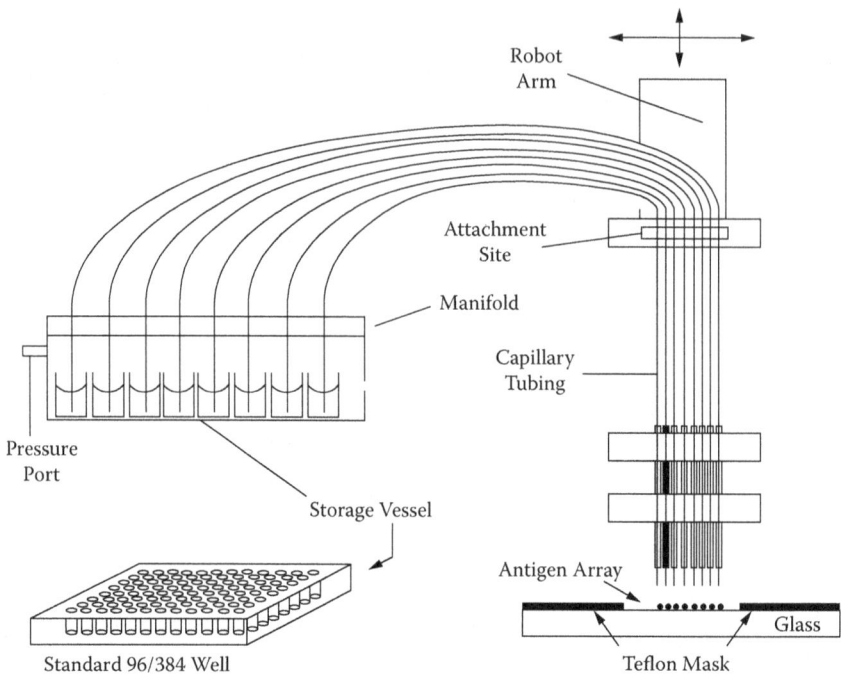

FIGURE 6.8 Capillary printer. (From Mendoza LG, McQary P, Mongan A, Gangadharan R, Brignac S, Eggers M. High-throughput microarray-based enzyme-linked immunosorbent assay (ELISA). *BioTechiques* 24(4): 778–788, 1999. With permission.)

approximately 13.4 ng/mL for rabbit IgG which is similar to that reported by Silzel et al. (1998). The benefit of the "array of arrays" approach is that 96 samples could be processed for multiple analytes (36 to 144 analytes) within the same time interval as a standard single analyte ELISA.

MICROTITER-BASED ANTIBODY ARRAYS

Conventional plastic microtiter plates have also been adopted for use with protein microarrays in the "array of arrays" format. Matson et al. (Oak Ridge Conference, 2001) printed anti-interleukin monoclonal antibodies in the bottom of a prototype shallow well, a vacuum-formed, 96-well polypropylene plate (Figure 6.9). The plastic was surface modified with acyl fluoride groups for rapid and efficient covalent attachment of the antibodies. Similar results in terms of sensitivity and dynamic range were obtained with printing of macroarrays (~500 micron dia spots) using a Biomek high-density replicating tool (HDRT) gridding tool or microarrays (200 micron features) prepared using an arrayer and quill pins. The micro-ELISA was performed with biotinylated secondary antibodies with streptavidin-alkaline phosphatase and enzyme-labeled fluorescence (ELF) (Molecular Probes, Inc.) signal amplification. In the case of IL-8, a dynamic range from 16 pg/mL (7.2% CV) to 1000 pg/mL (28.5% CV) was obtained with a minimal detectable dose estimated to be ≤1 pg/mL.

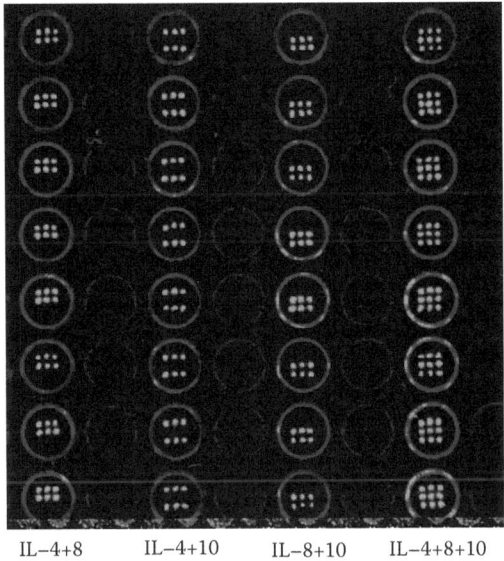

IL−4+8   IL−4+10   IL−8+10   IL−4+8+10

FIGURE 6.9 Monoclonal (anti-interleukin) antibody array in vacuum-formed 96-well microplate. (From Matson RS, Milton RC, Cress MC, Rampal JB. Microarray-based cytokine immunosorbent assay. *Oak Ridge Conference*, Poster No. 20, 2001. With permission.)

Moody et al. (2001) created a 3 × 3 "mini-array" of anti-cytokine monoclonal antibodies in wells of black Maxisorp™ 96-well plates (Nalge Nunc). The spots were approximately 0.4 mm diameter containing about 1 ng (7 fmol) of antibody. Assays were performed using biotinylated secondary antibodies with signal development using streptavidin-HRP and SuperSignal™ (Pierce Endogen) chemiluminescent reagent. Dynamic range was reported to be 0.8 to 200 pg/mL (IL-1α, IL-6, IL-10, INFα, INFγ); 0.4 to 100 pg/mL (IL-1β); and 1.6 to 400 pg/mL (TNFα). The average signal well-well intra-plate coefficient of variation (CV) ranged from 8 to 13.1%, while plate-to-plate CVs varied from 2.2 to 11.3%. The authors demonstrated the ability to measure cytokine levels in lectin-stimulated human peripheral blood mononuclear cells as well as lipopolysaccharide stimulated THP-1 cells for anti-inflammatory drug screening with dexamethasone. One drawback to the use of the chemiluminescent reagent is the blooming of light into adjacent spots. In this case, in order to expand dynamic range it was necessary to increase the spot center-to-center distances to at least 1.25 mm and limit the number of spots in the array to nine spots per well.

MEMBRANES

Perhaps a less sophisticated approach has been that of offset gridding onto membranes using robotic arms equipped with pin transfer tools (Figure 6.10). Lehrach's group at the Max Planck Institute (Lueking et al., 1999) created grids of expressed *His*-tagged protein from lysates (or purified by Ni-NTA metal affinity chromatography)

FIGURE 6.10 Early example of a robotic arm pin transfer tool for printing proteins. (From Lueking A, Horn M, Eickhoff H, Bussow K, Lehrach H, Walter G. Protein microarrays for gene expression and antibody screening. *Anal. Biochem.* 270: 103–111, 1999. With permission.)

on PVDF membrane. Note that *His*-tagging of proteins not only allows for purification from the lysates but also provides a useful quality control method for determining printing efficiency (e.g., anti-RGS-*His* antibody). At a pin-to-pin center distance of 4.5 mm and 250 micron tip pins, it was possible to array out 4800 samples onto 25 mm × 75 mm filter strips. Developed spots appeared to be sharp and uniform (Figure 6.11). The reported threshold sensitivity was calculated to be approximately 10 nM for purified G3PDH protein (10 pg/25 nL) spotted at several dilutions. The drawback to this approach has been the rather extensive washing required to remove nonspecific proteins and reagents from the membrane.

Joos and colleagues (2000) printed antigens onto either nylon filters or glass slides and compared the titers for the auto-antibodies in a micro-ELISA format. Antigens were delivered to the various substrates stabilized in 10% glycerol, 0.1% SDS, and 5 μg/mL bovine serum albumin (BSA). Protein arrays prepared in this manner were reported to be functional for up to 1 month when stored at room temperature in the dark (see exception below). According to the investigator's calculation, these arrays were most likely operating in the mass-sensing mode rather than by the ambient analyte immunoassay constraints. Nevertheless, the influences of spot size and antigen density on sensitivity were examined. As predicted from Ekins' microspot model (but under presumed mass-sensing conditions) the signal intensity decreased upon dilution of the antigen independent of spot diameter (area) or antigen density (Figure 6.12). The comparison of sensitivity (lowest detectable signal above

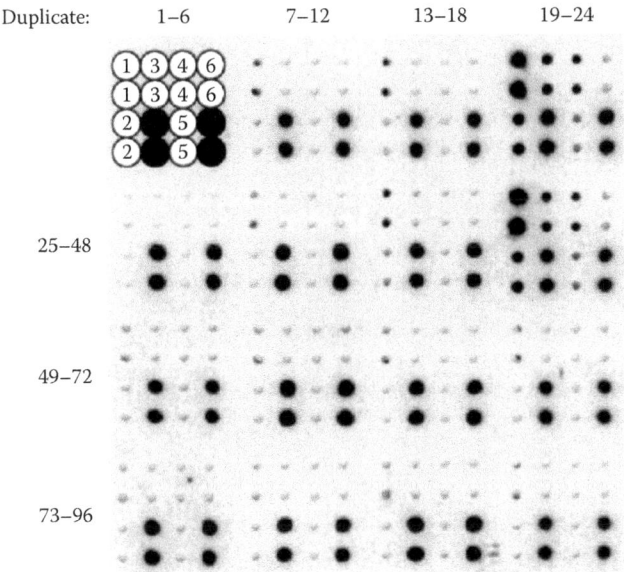

FIGURE 6.11 Membrane filter protein array. (From Lueking A, Horn M, Eickhoff H, Bussow K, Lehrach H, Walter G. Protein microarrays for gene expression and antibody screening. *Anal. Biochem.* 270: 103–111, 1999. With permission.)

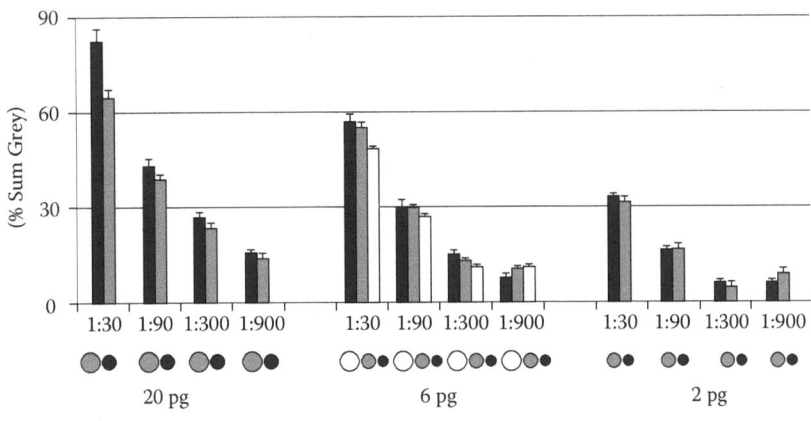

FIGURE 6.12 Antigen array: signal intensity versus spot size and loading density. (From Joos TO, Schrenk M, Hopfl P, et al. A microarray enzyme-linked immunosorbent assay for autoimmune diagnostics. *Electrophoresis* 21: 2641–2650, 2000. With permission.)

background) revealed that the nitrocellulose membrane provided a fivefold lower detection limit (8 fg huIgG/0.25nL spot volume ~0.2 nM) than an aldehyde-activated glass slide (40 fg huIgG). Understandably, most of the reported work was done using the nitrocellulose membrane with chemiluminescent signal detection. The membrane micro-ELISA gave comparable results to that of a conventional ELISA with some notable exceptions in which the microarray showed lower titer. This was found to be due to denaturation of the antigens on the microarrays during storage. Freshly prepared microarrays were found to perform equivalent to the ELISA.

Milagen's created antibody arrays for the purpose of target discovery by a process referred to as ANTIBIOMIX™ (antibody against biological mixture) in which polyclonal antibodies are used as tools to screen for both known and unknown gene products from clinical samples (Valle and Jendoubi, 2003). Milagen reported to have 61,000 polyclonal antibodies with 80,000 to 100,000 available by the end of 2002. The advantage of polyclonals over monoclonals in this application is that polyclonals have greater coverage for detection purposes (i.e., broader specificity and affinities provide a greater number of hits). This is particularly important in examining antigens with post-translational modifications. The antibody array panels generated by the ANTIBIOMIX process can be used for differential display (between diseased and normal states) to monitor the progression of a disease as well as its outcome (Figure 6.13).

GLASS SLIDES

Perhaps the first published demonstration of high-density applications for protein microarrays came from the work of MacBeath and Schreiber at Harvard (2000). Proteins were arrayed onto aldehyde-activated glass slides and analyzed in much the same manner as that first described in the creation of cDNA microarrays (Schena et al., 1995). This included the use of dual-color label detection of specific proteins.

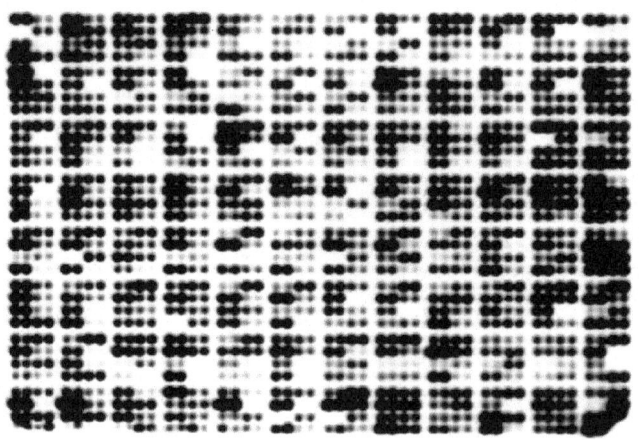

FIGURE 6.13   Milagen's antibody array. (From Valle RPC, Jendoubi M. Antibody-based technologies for target discovery. *Curr. Opin. Drug Discovery Dev.* 6(2): 197–203, 2003. With permission.)

Special care was taken to maintain proteins in their hydrated state so that there would be minimal surface denaturation. Proteins were printed in PBS containing 40% glycerol to prevent evaporation. Note that others such as Delehanty and Ligler (2002) have used a low salt buffer containing sucrose and BSA for this purpose.

The immobilization strategies are of particular interest. The author's reason that the use of aldehydes to tether proteins to the solid phase may be ideal for certain protein-protein interaction studies. Because many of the protein's lysine residues are available for coupling to the aldehyde via Schiff's base, a number of spatial orientations are possible for the protein. Such random oriented attachment would permit exposure of various surfaces of the protein to the solution. Potentially, new protein-protein interactions could take place. Another useful strategy is scaffolding. For example, immunoassays employ BSA not only as a blocking agent to reduce nonspecific adsorption of other protein but also as a scaffold. Essentially, BSA is first attached to the solid-support and then further derivatized for the coupling of additional capture ligands. MacBeath and Schreiber (2000) first formed a monolayer of BSA and then printed proteins on top of the monolayer. In this manner, small proteins were expressed on the surface and not buried by the BSA.

The specificity that can be achievable with the protein microarray was demonstrated by a number of powerful examples: First, a single spot of FRB protein printed down on the array was successfully detected among 10,799 spots of protein G (Figure 6.14). Second, enzyme-substrate reactions were possible using immobilized protein substrates of various kinases. The bound substrates were phosphorylated only by incubation of the slide with the specific kinase. Third, the microarray was used to screen for targets by immobilized protein–small molecule interaction studies. Here, immobilized protein receptors for steroid, biotin, or ketoamide ester were able to recognize their cognate partner. Finally, the ability to measure small molecule–induced conformational binding was shown by association of immunophilin with FRB only in the presence of the small molecule, rapamycin.

Brown's group at Stanford which earlier had introduced the slide microarray (Schena et al., 1995) turned to the large-scale immobilization of antibodies and antigens to study protein abundance (Haab et al., 2001). They examined the

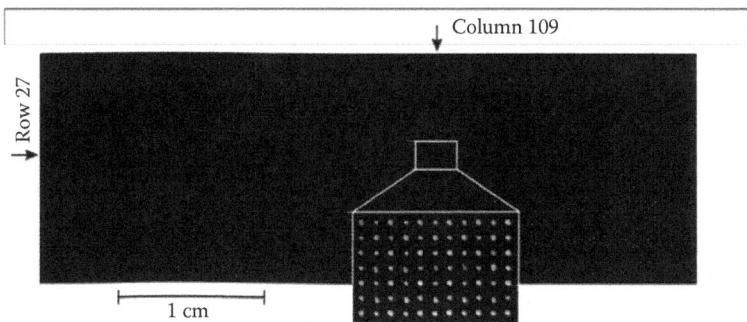

FIGURE 6.14 Protein microarray specificity. (From MacBeath G, Schreiber SL. Printing proteins as microarrays for high-throughput function determination. *Science* 289: 1760–1763, 2000. With permission.)

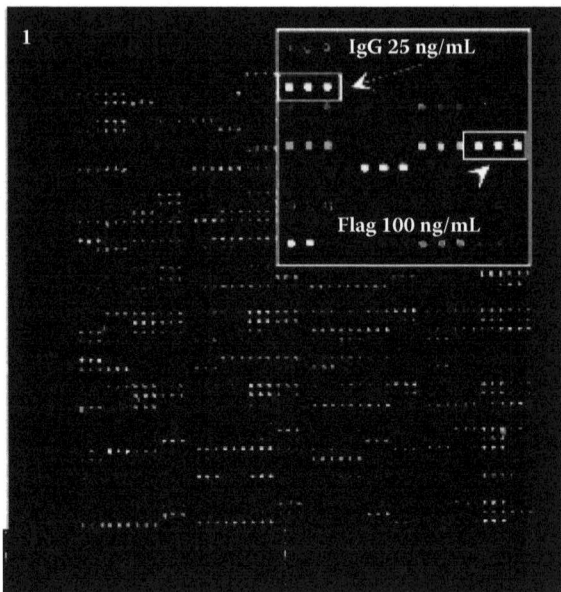

FIGURE 6.15   Applying the dual-color ratiometric "gene expression" labeling approach to protein expression analysis. (From Haab BB, Dunham MJ, Brown PO. Protein microarrays for highly parallel detection and quantitation of specific proteins and antibodies in complex solutions. *Genome Biol.* 2(2): research0004.1–0004.13, 2001. With permission.)

performance of 115 antibody-antigen pairs printing down arrays of either antibody or antigen onto poly-L-lysine coated glass microscope slides. Six to twelve replicates of each protein were placed on the slide. Comparative fluorescence labeling was used to measure performance. Cy3-labeled reference proteins (green fluor) were prepared at a constant concentration for all 115 proteins. The reference set was then mixed with the sample set having been labeled with Cy5 (red fluor) at specific abundance levels from 1 ng/mL to 1 µg/mL. The R/G ratio served as the calibrator for variation in binding between antibody-antigen on the array. The visual effect was observed in variation in color from red (higher concentration of protein in sample mix), yellow (equivalent concentration of sample and reference mixture), to green (higher concentration in reference set) (Figure 6.15). More quantitative information was obtained in plotting the log 10 (R/G) versus concentration to determine titer (Figure 6.16). In most cases the sample titer curves followed the predicted titer standard curve based upon known concentration ratios. Observed variations in linearity were traced back to inconsistencies in labeling rather than cross-reactivity or sample preparation issues such as pipetting or mixing errors. Overall, antigen arrays performed better than antibody arrays. While the exact reasons for the discrepancy between the two formats are not known, it most likely relates to protein stability and the noted inconsistencies in labeling. The authors suggest that while antibodies are structurally similar and relatively stable proteins that can be easily labeled at lysine residues in the Fc-region, antigens are much more variable both in structure and relative stability. An alternative explanation is

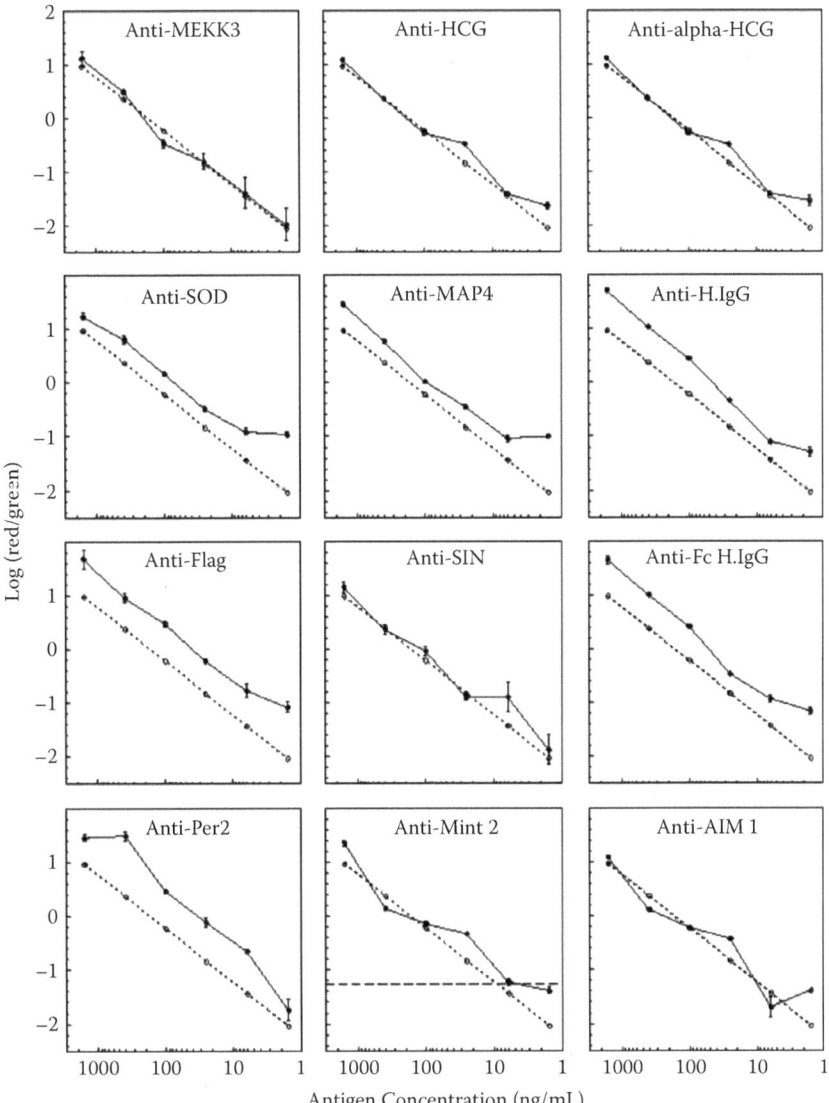

FIGURE 6.16 Microarray antibody-titer curves. (From Haab BB, Dunham MJ, Brown PO. Protein microarrays for highly parallel detection and quantitation of specific proteins and antibodies in complex solutions. *Genome Biol.* 2(2): research0004.1–0004.13, 2001. With permission.)

that antibodies likely recognize surface denatured antigens even better than native antigen. Thus, performance of antigen arrays may be more related to an increase in the number of epitopes exposed upon surface denaturation that are now recognized by the antibody. In terms of sensitivity, the detection limits on the microarray slide using the two-color system approached 0.1 ng/mL for antigen arrays and 1 ng/mL

TABLE 6.1

Properties of Commercial Membranes Useful as Protein Microarray Supports

| | | | Detection | | | |
| | | | IgGs | | Cytokines | |
Membrane	Manufacturer	Adsorption	Bkg	Sensitivity	Bkg	Sensitivity
Biotrans	ICN	Excellent	++	+++	+++++	?
Zeta-probe	Bio-Rad	Good	++	+++		
Colony/Plaque	NEN	Very Good	++	+++		
Hybond-N+	Amersham	Very Good	++	+++	+	+++
Magnacharge	MSI	Poor	+	+++	+++++	+
MagnaGraph	MSI	Excellent	−	++	+++++	+
Hybond ECL	Amersham	Excellent	++	+++	−	+++

Source: From Huang R-P. Detection of multiple proteins in an antibody-based protein microarray system. *J. Immunol. Methods* 255: 1–13, 2001. With permission.

for antibody arrays. Both were able to measure specific protein in a mixture of proteins at 1 part per million in total protein (partial concentration).

Studies by Huang (2001a) at Emory School of Medicine underscore an important issue regarding future work with protein microarrays—selection of substrates. In this case, membranes were selected based upon background and signal sensitivity for chemiluminescent detection (Table 6.1). For antigen arrays composed of various IgG species, the MagnaGraph membrane (MSI) provided excellent adsorption, negligible background, and moderate sensitivity. However, in a sandwich assay for cytokines this membrane could not be used because of its very high background. Instead, Hybond ECL (Amersham) was selected for cytokine assays based upon negligible background with good sensitivity.

In an effort to analyze cytokine expression at physiological levels directly from tissue culture or patient sera, Huang and co-workers modified their approach. To assay for specific cytokines in conditioned media at high sensitivity, membranes were pre-coated with the capture anti-cytokine antibody. Samples of conditioned media were printed down and the cytokine detected by ECL using a biotin-conjugated antibody recognizing a different epitope. For MCP-1 the detection limit was 4 pg/mL representing a 100-fold improvement in sensitivity over the standard ELISA using these "conditioned medium" arrays (Huang et al., 2001b).

Wiese et al. (2001) at Genometrix designed an antibody microarray based upon immobilization of monoclonal capture antibodies to activated, silanized glass plates (Mendoza et al., 1999). They compared the performance of this micro-ELISA to that of a conventional ELISA in the detection of prostate-specific antigen (PSA), PSA-ACT (α1-antichymotrypsin bound to PSA), and IL-6, all of which are indicators of prostate cancer. A good correlation ($r^2 = 0.88$) was obtained for 14 human serum PSA concentrations between the two assay formats (Figure 6.17). Note that Woodbury and colleagues (2002) also correlated the micro-ELISA with that of a standard ELISA and observed a similar correlation ($r^2 = 0.90$) during the analysis for

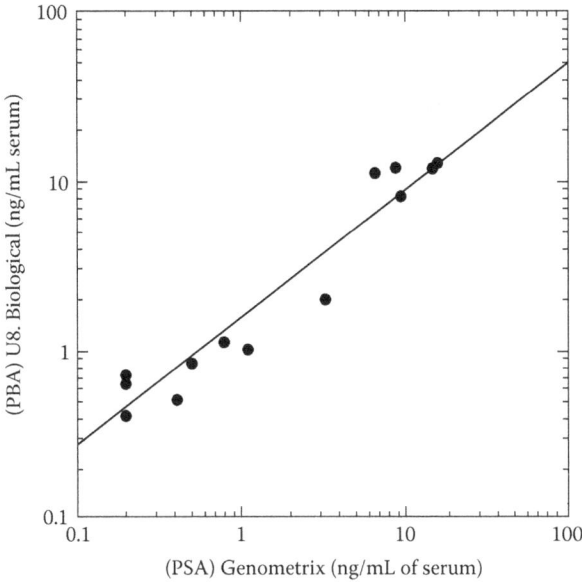

FIGURE 6.17 Concordance between standard ELISA and microarray-ELISA formats in PSA determinations. (From Wiese R, Belosludtsev Y, Powdrill T, Thompson P, Hogan M. Simultaneous multianalyte ELISA performed on a microarray platform. *Clin. Chem.* 47(8): 1451–1457, 2001. With permission.)

hepatocyte growth factor (HGF). In addition, there were certain design features in the Weise study that are noteworthy. First, capture antibodies were printed as a series of serial dilutions rather than at one concentration. This provides much needed flexibility in the dynamic working range for analyte analysis. It reduces the likelihood of having to repeat the assay due to signal plateauing with "out of range" samples. Both high and low concentration ranges for all analytes can be addressed within the same well. There is no need to adjust sample volumes in order to measure antigens at equivalent levels. The other important feature with the microarray design is that because all analytes are represented within each well, the calibration curves for each analyte can be run simultaneously. Finally, sufficient redundancy is built into the well array. This permits statistical treatment of the data and less reliance on weighting due to single data points and outliers.

Seong (2002) investigated the performance of various commercial slides for protein microarrays based upon different methods of protein immobilization. Two buffer systems were studied: PBS, pH 7.4 and carbonate buffer, pH 9.6 for immobilization of IgG antigen. All slide chemistries performed well in terms of binding capacity. However, each slide chemistry displayed slightly different loading isotherm profiles. Epoxy activated slides outperformed silyated (aldehyde) surfaces at the highest loading concentration of IgG. Carbonate buffer marginally improved IgG slide surface loading over slides in which the protein was printed in PBS as the medium. This was essentially the case for either silyated (aldehyde) or silanated (amine) slides. One explanation for the enhanced loading in carbonate buffer may be related to the ionization state of the protein. At pH above the isoelectric point, the protein would carry a slight negative charge favoring stronger interaction with the positive amine surface.

Angenendt and co-workers (2002) at the Max Planck Institute (Berlin, Germany) undertook similar studies comparing the binding characteristics of five different antibodies arrayed onto 11 different substrates. Performance was measured in terms of sensitivity as well as intra- and inter-microarray slide variation. Using a QArray System (Genetix, United Kingdom), antibodies were spotted down on the various substrates from 25 amole to 40,000 amoles in eight replicates. The antibodies included anti-human serum albumin (monoclonal, polyclonal), anti-fibrinogen (monoclonal, polyclonal), and anti-tubulin α (monoclonal). Antigens were labeled with either Cy5 or Cy3 for detection purposes, and microarrays were scanned using a laser confocal scanner (Model 428, Affymetrix). Substrates included plastic, glass, and hydrogel coatings. Of these, hydrogel coatings achieved the lowest limit of detection (LLD ~1300 to 1600 amole/spot) but exhibited significant assay variation (22% intra-slide to 37% inter-slide CV). Surface-modified polystyrene slides (Maxisorb black, Nunc) had an LLD = 1500 amole/spot at 15 to 32% CV, while reflective slides (mirror-like surface) coated with 3-aminopropyltriethoxysilane (Amersham) had the lowest variation with CVs at 11 to 14%. Polyacrylamide-coated slides showed wide variation in signal and lowered sensitivity. Pre-incubation in PBS buffer prior to spotting improved performance both in terms of LLD (~1875 amole/spot) and variance (14 to 15% CV). The use of immobilized ampholytes to alter the matrix pK_a were not beneficial. All antibody arrays could be stored for 8 weeks without loss in performance. Interestingly, all showed improved binding (maximal signal) after 2 weeks of storage. Two important observations were noted about the study. First, no single microarray substrate satisfied all performance criteria for all antibodies. Second, in other work certain antibodies worked well in conventional ELISA but not the microarray format. Thus, results from ELISA do not necessarily qualify an antibody for use on microarrays. It is best to validate each antibody for use on the intended substrate in a microarray-based assay.

Avseenko et al. (2001) immobilized antigens onto aluminum coated Mylar films by electrospray (ES) deposition. Various surface modifications of the metallized films were studied in their ability to enhance sensitivity. The plastic surfaces were first cleaned by plasma discharge treatment followed by coating with proteins (BSA, casein) or polymers such as poly (methyl methacrylate), oxidized dextran, or exposed to dichlorodimethylsilane to create a hydrophobic surface. Protein antigen was prepared in 10-fold excess sucrose and sprayed onto the surfaces to form arrays with spot diameters between 7 and 15 μm containing 1 to 4 pg protein. ES deposition is performed using proteins prepared as a dry powder. Sucrose protects the proteins during drying and deposition. Once deposited sucrose adsorbs moisture from the air creating droplets that re-solubilize the protein for attachment. Such microarrays could then be stored dry at –20°C for up to 8 months without loss as measured by ELISA. Of the various surface modifications studied, the greatest enhancement in sensitivity was achieved using dextran-coated Mylar. Dextran oxidation results in the formation of aldehyde groups that form a Schiff's base with primary amino groups on proteins. Oxidized dextran coating resulted in sensitivity to ~10 ng/mL, while reduction of the Schiff's base to form a stable covalent bond increased sensitivity to 1 ng/mL (~6.7 pM IgG). Contributing to this enhancement in sensitivity was also the low background achieved using dextran rather than protein blockers.

Mezzasoma et al. (2002) created an array of microbial antigens to determine the level of antibodies in human sera directed against *Toxoplasma gondii*, rubella virus, cytomegalovirus (CMV), and herpes simplex virus (HSV) ToRCH antigens. Microarray performance was measured against standard ELISA. Internal standards composed of IgG and IgM were printed on each slide in duplicate. The linear range was 2 to 50 pg for IgG and 0.4 to 8 pg for IgM. The detection limit (rabbit myosin negative control, mean intensity + 2 SD) was interpolated at 0.5 pg or an LLD corresponding to approximately 0.04 pM. Signal was generated using Alexa dyes (Molecular Probes): Alexa 546 labeled anti-human IgG monoclonal or Alexa 594 labeled goat anti-human IgM μ-chain. Detection was done using a laser confocal scanner system. Microarray performance metrics were studied. Inter-slide variance was from 1.7 to 8.6% CV, intra-slide variance ranged from 2.6 to 15% CV, while batch-to-batch variability was between 5.2 and 18% CV for all serum IgG reactive antigens. The antigen microarray was compared against a commercial ELISA (Radim S.p.A., Italy) using well-characterized human sera panels (54 to 56 samples) containing different levels of antigen-specific IgG. Agreement (80 to 90% concordance) between the microarray and the ELISA determinations was obtained except in the case for rubella virus. The microarray assay identified ~18% positive, while the ELISA scored 87.5% as positive for the virus. The discrepancy appeared to favor the results of the microarray assay when samples were evaluated by an independent method.

MEASURING MICROARRAY PERFORMANCE

We have described a number of early uses for protein microarrays. Others have suggested enhanced sensitivity when using vanishingly small spots to capture antigens (Ekins et al., 1990). How does the microarray perform relative to other well-established technologies such as the ELISA? Which microarray format or detection process is most accurate and sensitive? In order to compare performance criteria (accuracy, precision, sensitivity, dynamic range) it is important to define such in the same terms. Unfortunately, there is little consistency in reporting values. It is largely a matter of preference or accepted practice within a particular field of study. In this section we will attempt to review the results of those studies reporting on sensitivity using equivalent terms. Hopefully this will lead to a better understanding of what has been accomplished or not.

SENSITIVITY AND DYNAMIC RANGE

It is most important to agree on the definition of assay sensitivity. Operationally this is usually taken at the lower limit of detection (LLD or LOD) or minimally detectable dose (MDD) whose value must be greater than the precision (standard deviation or the coefficient of variation) in measuring the zero dose value (Ekins et al., 1990). For example, sensitivity in a noncompetitive immunoassay with $n = 3$ replicates would be defined as LLD = B0 + 3 SDB0, where B0 is the zero dose (minus analyte) sample value. For a competitive assay, LLD = B0 − 3 SDB0. However, fluctuation in the standard deviation due to such things as pipetting error or sample interferences can dramatically alter the LLD. It is therefore essential that the assay variability be known precisely (Ezan & Grassi, 2000). The assays' performance in terms of accuracy, reproducibility (CV, inter-assay variation), and repeatability (CV, intra-assay variation) should be determined. For an ELISA, accuracy in the range of 85 to 115% of the standard value and CVs in the range of 15 to 20% are common. The assay limit of quantification is then taken as the lowest concentration of analyte that provides CVs under, for example, ≤10% and accuracy within, for example, ±15% of the standard value. A discussion of factors leading to imprecision as they apply to microarrays will be addressed later.

First, however, we examine specific examples of measured sensitivity in microarray assays. These are predominately antibody or antigen microarrays in which a micro-ELISA has been evaluated. The antibody array format involves the immobilization of a library of different capture antibodies and is commonly employed in the standard sandwich immunoassay in which antigen is captured and detected by a labeled secondary antibody. This format requires that the capture antibody be capable of efficient binding of antigen (i.e., it has been immobilized largely in its native state). In case of the antigen array format, the antigen is attached directly to the surface and probed with a reporter (labeled) antibody. The assay requirements for this format are less stringent. Immobilized antigen may be surface denatured and still be recognized by the reporter. This is possible when surface denaturation leads to the linearization of previously unavailable epitopes permitting enhanced recognition, especially by polyclonals.

Table 6.2 summarizes many of the studies in this review. For comparison, the analyte LLD values (pg/mL) have been recalculated in terms of analyte concentration and reported in pM units. In doing so we can easily identify assay formats leading to higher sensitivity. Where possible we will also discuss dynamic range.

A standard sandwich ELISA for IL-4 is capable of detecting approximately 30 pg/mL at the MDD and may have a linear dynamic range to about 3000 pg/mL (R&D Systems). The working range for the assay is thus capable of detecting from nM to pM in IL-4 analyte concentration. Most reported work is based upon antibody microarrays operating well within this range. There are also a number of examples of sensitivity achieved at the fM level of detection. Certainly, second- and third-generation ultra-sensitive standard ELISA assays are able to achieve fM-aM sensitivity as well. The real advantage of the microarray-based assay over the standard assay is that of multiplexing and parallel processing. Miniaturization that reduces the amount of reagents and analyte are also attractive features.

It is apparent that signal amplification provides increased sensitivity over direct labeling. This is especially true for fluorescent-based assays. One of the most sensitive signal detection technologies is that of immunoRCA (Schweitzer et al., 2000). Here, rolling circle amplification (RCA) is combined with antibody detection. RCA involves the amplification of circularized oligonuceotide probes under isothermal conditions by DNA polymerase (Lizardi et al., 1998). In the case of immunoRCA the 5′ primer is attached to the reporter antibody. Initiation of the amplification starts when the circular DNA template binds to the attached primer. In the presence of

TABLE 6.2

Assay Sensitivity for Antibody/Antigen Microarrays and Standard ELISA

Analyte	Lowest Limit of Detection (LLD) (pg/mL)	LLD (pM)	Signal	Detection	Microarray Support	Reference
Assay Sensitivity for Antibody/Antigen Microarrays						
IgG	40 fg/spot	1000	ECL	CCD camera	Glass	Joos
	8 fg/spot	200			Membrane	
IgG	15,000	100	DBCy5	CCD microscope	Plastic	Silzel
IgG	13,400	89	ELF	CCD camera	Glass	Mendoza
IgG	5,000	10.7	Cy5/Cy3	Confocal scanner	Glass	Haab
FKBP12	150	12.5	Cy5	Confocal scanner	Glass	MacBeath
IL-2	4	~2	ECL	Film	Membrane	Huang
IL-4	10	~0.7	Cy5	PWG CCD camera	Glass	Pawlak
IL-8	1	0.15	ELF	CCD camera	Plastic	Matson
IL-6	4	0.15	ELF	CCD camera	Glass	Wiese
TSH	2	~0.1	Texas Red/FITC	Confocal microscope	Plastic	Ekins
IL-6	0.8	0.03	CL	CCD camera	Plastic	Moody
IL-8	0.25	0.03	RLS	CCD camera	Plastic	LaBrie
HGF	0.5	0.006	TSA-Cy3	Confocal scanner	Glass	Woodbury
PSA	0.1	0.0003	RCA	CCD camera	Glass	Schweitzer
Standard Sandwich ELISA						
IL-4	10	~0.7	HRP colorimetric	Microplate reader	96-well Plastic	R&D Systems
IL-6	0.04	~0.02				
IL-8	10	1.5				
IL-4	2	HRP colorimetric	Microplate reader	96-well Plastic	Pierce Endogen	
IL-6	1	~0.04				
IL-8	2	0.3				
TSH	0.2 μIU/mL	1.4	HRP colorimetric	Microplate reader	96-well Plastic	Biocheck
TSH	0.05 μIU/mL	0.36				Yes Biotech
TSH	0.01–0.02 μIU/mL	0.07–0.14	Third-generation tests			

Notes: CL, chemiluminescence; ECL, enhanced chemiluminescence; ELF, enzyme labeled fluorescence; RCA, rolling circle amplification; RLS, resonance light scattering; TSA, tyramide biotin amplification, sAV-Cy3.

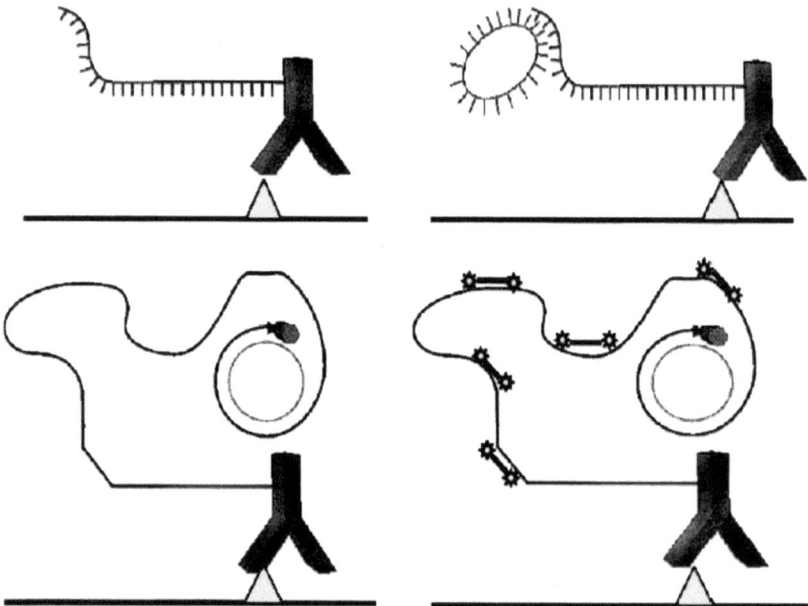

FIGURE 6.18 ImmunoRCA assay. (From Schweitzer B, Wiltshire S, Lamber J, et al. Immunoassays with rolling circle DNA amplification: A versatile platform for ultrasensitive antigen detection. *PNAS* 97(18): 10113–10119, 2000. With permission.)

DNA polymerase the 5′ primer is extended as the circular DNA template rolls along the extended primer. The result is the formation of replicate copies of the original template laid out in linear fashion and still attached to the antibody. Smaller labeled oligonuceotide probes can then hybridize to portions of the amplified primer at multiple sites. The signal is thus amplified by the number of labeled probes (Figure 6.18). For example, PSA was detected at 0.1 pg/mL (300 zeptomoles) on a microarray corresponding to a sensitivity of about 0.3 fM. This is three logs more sensitive than that of a PSA ELISA. In other experiments, IgE was detected at 1 pg/mL with a dynamic range of about five logs (Figure 6.19).

In other work, Schweitzer and co-workers (2002) created antibody arrays on thiolsilane-coated glass slides immobilizing a library of monoclonal antibodies directed toward various cytokines in quadruplicate. The glass slide was partitioned by Teflon barriers into 16 circular (0.5 cm diameter) subarrays with each subarray containing 256 features. Seventy-five cytokines could be determined in this manner at ~10% CV for single analyte replicates and ~25% CV between the various antibodies in terms of signal variation. Arrays stored dry at 4°C were stable for at least 1 month. Again, RCA was used for detection. The authors noted that signal amplification is necessary to achieve sensitivities below 1 ng/mL (fluorescence signal without amplification) for cytokines. The RCA immunoassay was able to achieve in specific examples 1000-fold sensitivity over direct detection (Figure 6.20). For the 75 cytokines tested, the following sensitivities were reported: 45 (≤10 pg/mL), 22 (≤100 pg/mL), and 8 (≤1000 pg/mL). Dynamic range was approximately three orders

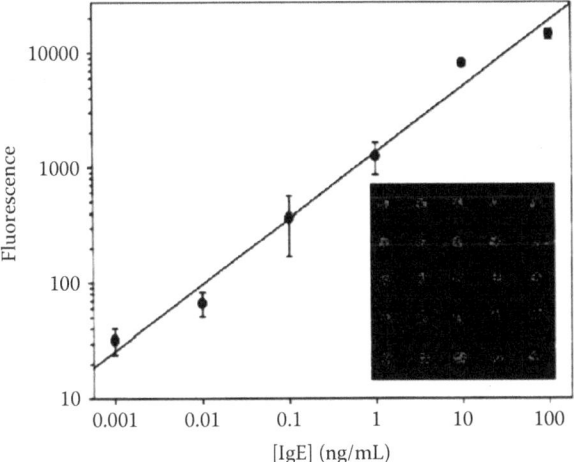

FIGURE 6.19 ImmunoRCA anti-IgE sensitivity. (From Schweitzer B, Wiltshire S, Lamber J, et al. Immunoassays with rolling circle DNA amplification: A versatile platform for ultra-sensitive antigen detection. *PNAS* 97(18): 10113–10119, 2000. With permission.)

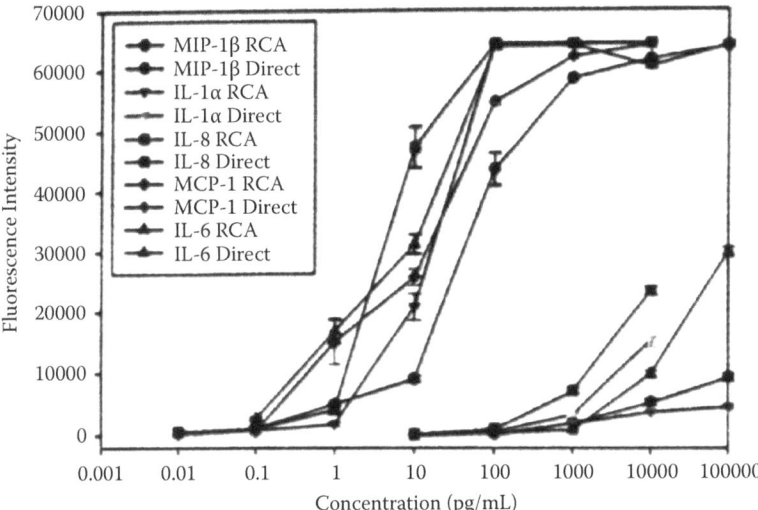

FIGURE 6.20 ImmunoRCA versus direct detection. (From Schweitzer B, Roberts S, Grimwade B, et al. Multiplexed protein profiling on microarrays by rolling-circle amplification. *Nat. Biotechnol.* 20: 359–365, 2002. With permission.)

of magnitude with a precision of ~5% CV reported for four assays. These results were from serial dilution of purified cytokine antigens. While in most instances the RCA-microarray immunoassay was shown to be comparable with published performance reports on commercial ELISA (Quantikine, R&D Systems), there were notable exceptions in which ELISA appeared to outperform the microarray by 10-fold

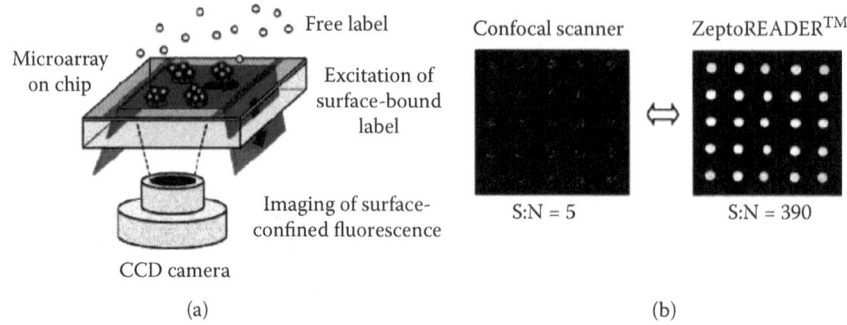

(a)   (b)

FIGURE 6.21   Planar waveguide (PWG) technology. (From Pawlak M, Schick E, Bopp MA, Schneider MJ, Oroszlan P, Ehrat M. Zeptosens' protein microarrays: A novel high performance microarray platform for low abundance protein analysis. *Proteomics* 2: 383–393, 2002. With permission.)

to 50-fold sensitivity. Such differences may be due to variations in binding affinities exhibited by the particular capture antibodies employed in the assay. In studies involving lipopolysaccaride (LPS)-induced secretion of cytokines from human dendritic cells, the RCA-microarray immunoassay was found to be concordant with published data from standard ELISA. Unfortunately, these investigators did not perform a side-by-side comparison with an ELISA using LPS-induced samples except in the monitoring of a single analyte. In that single example, however, the assays were concordant.

Another approach to enhanced sensitivity relies upon improved detector designs such as those employing planar waveguides (Pawlak et al., 2002). Planar waveguide (PWG) technology involves the excitation and detection of surface-confined fluorescence (Figure 6.21). The net result is to improve the signal-to-noise ratio (S/N), thereby lowering the detection limit. For example, Cy5-labeled IgG spotted down on a glass substrate was detected at a lower limit of detection (LOD) at 2 pM and over a dynamic range of 3.5 logs. The LOD for IL-4 was determined to be 10 pg/mL from a single Cy5 label. In this study the limit of quantification (LOQ) was also estimated from the dose precision profile (Figure 6.22). The inter-assay variation was <10%, while the intra-assay (chip to chip) variations were within 20% providing an LOQ for IL-4 of 10 pg/mL.

Detection by resonance light scattering (RLS) involves the use of small gold beads that are immobilized to reporters such as antibodies (Figure 6.23). Yguerabide and Yguerabide (1998, 2001) discovered RLS when they observed that colloidal gold particles when illuminated by a narrow beam of white light were able to scatter light giving rise to intense color. For example, they found that a 60 nm gold particle scattering power is equivalent to the fluorescence of 500,000 fluorescein molecules. Suspensions can be detected by eye in the fM range. The nanometer diameter beads possess an intrinsic vibrational resonance that transfers from one bead to another when they are in close proximity. The result is a change in color from red to blue when the beads are in close proximity to each other, such as in the case when the reporter (bead) antibodies are localized on a surface during binding to antigen. The

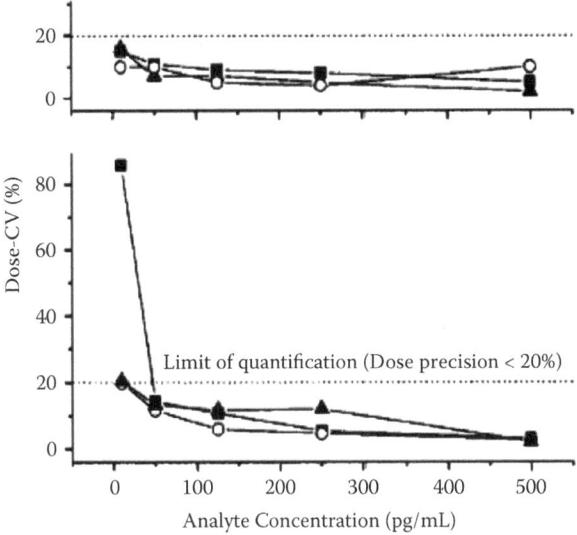

FIGURE 6.22 Planar waveguide (PWG) assay dose precision profiles. (From Pawlak M, Schick E, Bopp MA, Schneider MJ, Oroszlan P, Ehrat M. Zeptosens' protein microarrays: A novel high performance microarray platform for low abundance protein analysis. *Proteomics* 2: 383–393, 2002. With permission.)

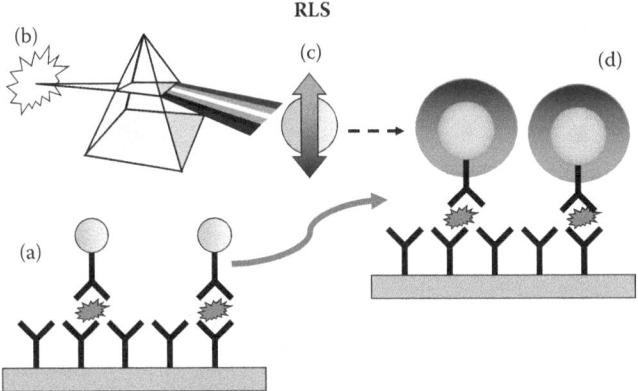

FIGURE 6.23 Resonance light scattering (RLS). (From Yguerabide J, Yguerabide EE. Light-scattering submicroscopic particles as highly fluorescent analogs and their use as tracer labels in clinical and biological applications. *Anal. Biochem.* 262(2): 137–156, 157–176, 1998. With permission.)

intensity (I) of the scattered light is particle size (diameter) dependent increasing by the sixth power of the radius, $I \propto r^6$. The resulting spectra obey Rayleigh theory for small particle ($\leq$40 nm diameter) light scattering. Silver and gold have the highest light scattering power. Silver scatters at 380 nm (purple), while gold scatters at 520 nm (green). In terms of a molar extinction coefficient (ε): for silver, $\varepsilon = 1.68 \times 10^{10}$;

gold, $\varepsilon = 5.88 \times 10^9$; copper, $\varepsilon = 2.59 \times 10^9$; and fluorescein, $\varepsilon = 6 \times 10^4$. Thus, RLS is capable of ultra-high sensitivity relative to fluorescein. In terms of particle size, for example, gold at 30 nm diameter, $\varepsilon = 5.9 \times 10^9$; 40 nm, $\varepsilon = 1.6 \times 10^{10}$; 60 nm, $\varepsilon = 5.3 \times 10^{10}$; 80 nm, $\varepsilon = 1.1 \times 10^{11}$. The dynamic range for the light scattering intensity is approximately three logs with fM sensitivity. In an ELISA for toxin A (causing dysentery) a sensitivity was reached at 10^{-14} M with 1 hour incubation at >95% specificity. Researchers at Incyte Genomics demonstrated sub-pM sensitivity in an antibody array employing RLS (LaBrie, 2001). Antibodies were arrayed in 96-well microtiter plates in a 10×10 pattern at a 290 micron center-to-center spacing. Assays were developed using anti-biotin RLS particles for detection of either the biotinylated sample proteins (cytokines) for direct detection or by biotinylation of a secondary (detection) antibody in a sandwich assay format. Similar results were obtained in terms of sensitivity and dynamic range. For IL-8 a limit of detection was achieved at 0.25 pg/mL with a linear dynamic range from 0.5 pg/mL to 10,000 pg/mL. This was compared to a standard ELISA that was found to have a sensitivity at 1 pg/mL with a linear dynamic range less than two logs.

TSA (tyramide biotin amplification, PerkinElmer Life Sciences, Inc.) and ELF (enzyme labeled fluorescence, Molecular Probes, Inc.) are related detection technologies. ELF is a soluble substrate (ELF-phosphate) for alkaline phosphatase which then cleaves the substrate into an insoluble and high fluorescent product (ELF-alcohol). For immunoassay purpose, the secondary antibody is first labeled with biotin which then allows for binding of streptavidin alkaline phosphatase conjugate. Addition of the ELF substrate results in the accumulation of the ELF-alcohol at the site of attachment, in this case, precipitating over the captured antigen (Figure 6.24). In the tyramide amplification process, a tyramide-biotin complex is produced by the action of horseradish peroxidase. The complex precipitates near the binding site and accumulates. The complex is detected by the use of streptavidin-Cy3/Cy5.

Various chemiluminescence (CL) signal amplification methods may also be used with microarrays. Most of these rely upon the formation of an unstable intermediate that decays to release light. The intermediate is a charged species that is sequestered as the result of surface charge. Thus the generated light becomes localized on the surface and can be detected using film, phosphorimagers, or CCD camera systems. Most CL detection is employed for work with membrane arrays that have intrinsic fluorescent backgrounds and light scattering issues.

OTHER MICROARRAY FORMATS USEFUL FOR PROTEOMIC APPLICATIONS

mRNA-Protein Fusions

The popularity of antibody/antigen microarray remains on the increase, but there are other approaches useful in proteomic studies. Phylos technology (Weng et al., 2002) makes use of mRNA-protein fusion products that self-assemble onto microarrays composed of complementary oligonucleotide capture probes (Figure 6.25). The PROfusion process involves in vitro translation of modified mRNAs using the rabbit reticulocyte system. The mRNA species are conjugated to puromycin via an oligonucleotide linker.

Enzyme Labeled Fluorescence Detection (ELF-97*, Molecular Probes)

ELF-97 is a soluble phosphorylated substrate that is cleaved by alkaline phosphatase into a highly fluorescent, insoluble product

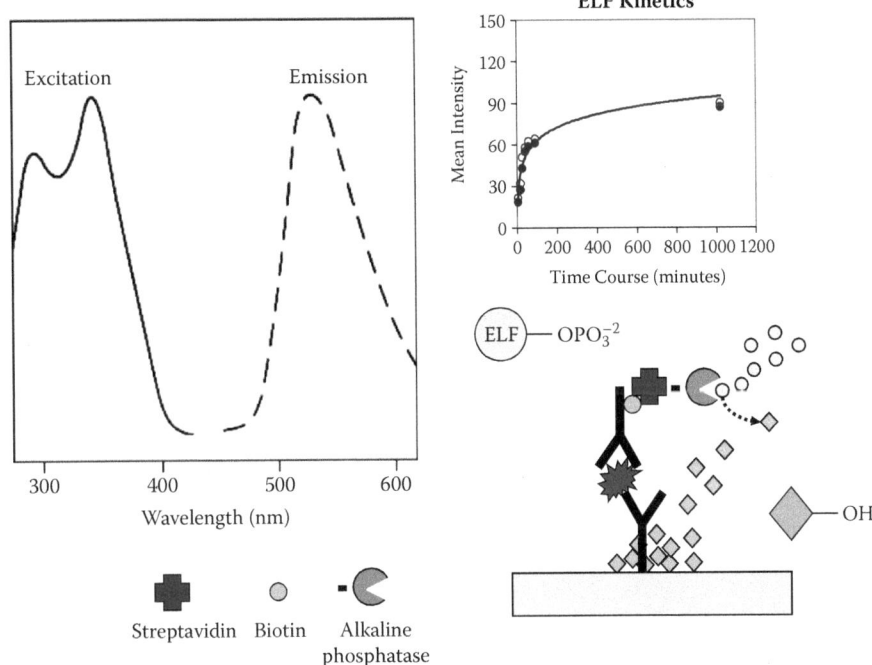

FIGURE 6.24 Enzyme labeled fluorescence (ELF). (From Molecular Probes, Inc., Eugene, OR. With permission.)

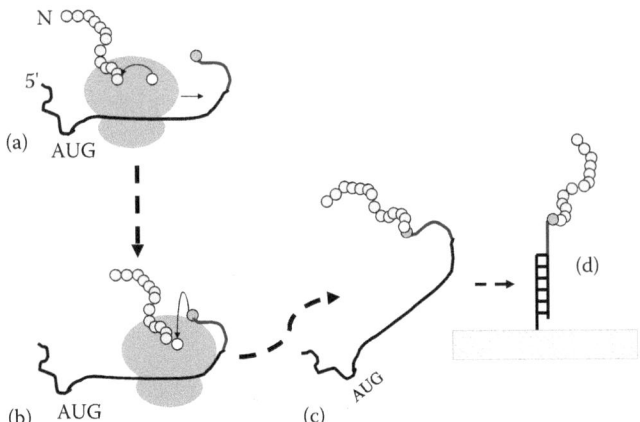

FIGURE 6.25 PROfusion. (From Phylos, Inc., Lexington, MA. With permission.)

The nascent polypeptide is assembled on the ribosome. Protein synthesis proceeds unchecked until reaching the RNA-oligonucleotide junction. At this point the puromycin and C-terminus of the peptide are linked halting further extension of the nascent peptide. The result is the formation of the mRNA-protein fusion product that is then released from the ribosome following the addition of metal chelators.

Capture oligonucleotide probes complementary to the mRNAs are arrayed onto a glass substrate with immobilization via a 3′ terminal amino group. Because the mRNA is 5′ to the polypeptide, capture results in orientation of the polypeptide away from the surface. The capture probes are tethered to the substrate indirectly through the use of a hydrophilic spacer arm including hexaethylene oxide-phosphodiester units that terminate with the 3′ amino group. The oligonucleotide array is converted into a protein array by the addition of the mRNA-protein fusion products to the support. The proteins are presented in proper orientation for efficient binding. The hydrophilic spacer arm further extends the protein away from the surface allowing greater access to binders and potentially lowering the level of nonspecific adsorption. Self-assembly can be accomplished either from the crude reticulocyte product or a partially purified form to further reduce the nonspecific background. The investigators were able to demonstrate specific protein capture in the range of 1 to 4000 attomoles per 200 micron diameter spot. The authors so aptly point out how "tremendous is the resolving power of the biochip format" to provide such efficient self-assembly to be accomplished even from a crude extract.

PISA

Another approach to creating arrays of proteins is PISA (protein in situ array) that was first described by researchers at the Babraham Institute, Cambridge, United Kingdom (He and Taussig, 2001). In their process the gene for a specified protein is first amplified by PCR and linked to, for example, *His*-tag domain sequence to create a tagged-gene PCR construct. An upstream primer (G/back) for gene-specific amplification containing a T7 promoter overlap is used with a downstream primer (G/for) containing a tag domain overlap sequence. The resulting amplicon therefore contains the gene sequence plus the T7 and *His*-tag overlaps. The tag sequence is amplified from plasmid pTA-*His* that provides a flexible linker and a double (*His*)6 tag. The two amplicons can then be assembled and amplified by using an upstream T7 primer containing the promoter + kozak sequence and start codon for translation. For downstream a T-term/for primer is used to amplify across the tag and linker. The resulting construct contains the gene, the promoter, and the tag (Figure 6.26). This is key to the technology, for the construct is then used for cell-free expression of the protein. Because the protein is now tagged with, for example, *His*, it can be directly immobilized to a Ni-NTA support such as the bottom of a microtiter plate well (Figure 6.27). The entire process from cDNA (or mRNA) PCR (RT-PCR) amplification coupled to in vitro transcription-translation (e.g., using a rabbit reticulocyte system) and immobilization of tagged protein requires 10 hours. In one example, a human single-chain antibody fragment (VH/K) anti-progesterone was obtained by this process. Approximately 120 ng from a 25 μL reactin volume was generated, of which approximately 50% of the tagged antibody was immobilized (He et al., 2001).

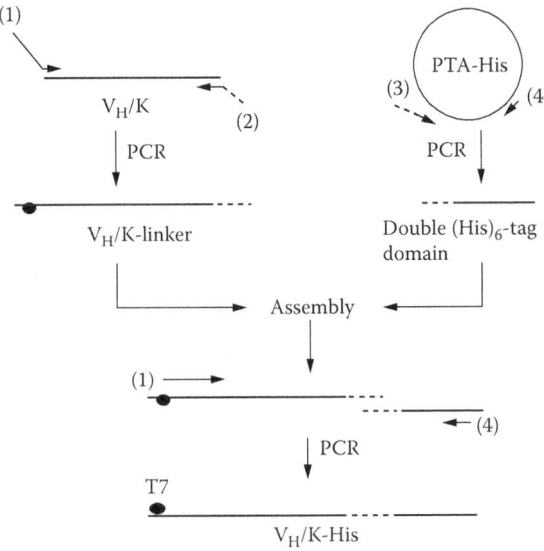

FIGURE 6.26 Protein in situ array (PISA) constructs. (From He M, Taussig MJ. DiscernArray technology: A cell-free method for the generation of protein arrays from PCR DNA. *J. Immunol. Methods* 274: 265–270, 2003. With permission.)

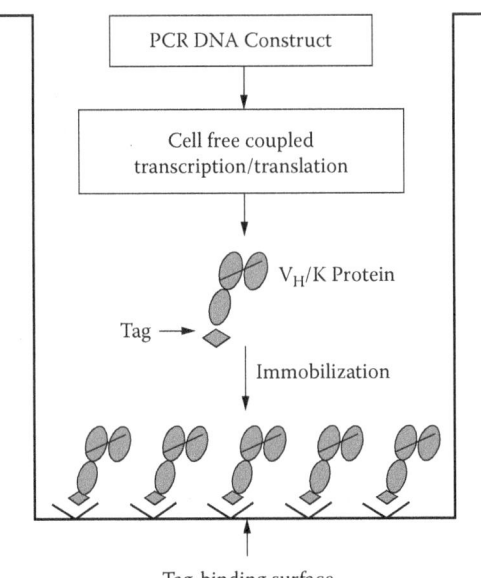

FIGURE 6.27 In-well protein in situ array (PISA) and *His*-tag expressed protein immobilization. (From He M, Taussig MJ. DiscernArray technology: A cell-free method for the generation of protein arrays from PCR DNA. *J. Immunol. Methods* 274: 265–270, 2003. With permission.)

A detailed protocol for the commercial product (DiscernArray™) has been provided (He et al., 2003).

APTAMERS

Another nucleic acid technology that has demonstrated utility in proteomics is that of aptamers and photoaptamers (Zichi et al., 2002). Aptamers (from the Latin word "aptus" translated "to fit") are oligonucleotides that selectively bind proteins. A comprehensive review of aptamer technology focusing on their diagnostic utility has been provided by Jayasena (1999). Pioneered by Larry Gold and his collaborators at the University of Colorado, Boulder (Tuerk & Gold, 1990), as well as Ellington and Szostak (1990) at the University of Texas, Austin, these ligand binding oligonucleotides are selected by a combinatorial process known as SELEX (systematic evolution of ligands by exponential enrichment). In this process a random library of oligonucleotides is prepared with additional fixed sequence for amplification. A purified antigen is mixed with the library and the complexes captured on nitrocellulose membrane. The bound oligonucleotides-proteins complexes are denatured, and the aptamer candidates are amplified for a second round with the putative protein. The selection rounds continue until the population of aptamers is reduced to high binding candidates that are then fully sequenced. Enrichment of high-affinity candidates is usually achieved in 8 to 15 rounds of SELEX. Each round takes approximately 2 days to perform. The process has been automated using robotic liquid handlers both for DNA aptamers (SomaLogic) and RNA aptamers (Cox et al., 2002). Next, the sequenced aptamer is prepared in bulk by conventional DNA synthesis chemistry, purified, and the aptamer arrayed onto a solid-support. Thus, within 2 to 3 months an aptamer is ready for application. Because the sequence is known, preparation of additional aptamer is easily accomplished using conventional oligonucleotide chemical synthesis.

How well aptamers perform relative to antibodies remains an open debate. In earlier work involving a survey of 100 aptamers, it was reported that more than 75% were characterized by $Kd \leq 1$ nM (Brody et al., 1999; Brody & Gold, 2000). Such affinities are well within the range exhibited by antibodies. So, it is quite plausible that libraries of aptamers could replace antibodies as a general tool for certain applications such as antigen screening, or specific aptamers could be selected with higher Kd for use in affinity chromatography purification processes. However, for diagnostic application both sensitivity and specificity need to be rigorous. In this case, Kd in the pM to aM range are required, and nonspecific binding to the oligonucleotide aptamer must be minimal.

A more recent version of this process known as photoSELEX addresses these concerns (Figure 6.28). During photoSELEX aptamers are produced with the photolabile Br-dUTP incorporated into specific regions on the oligonucleotide (Willis et al., 1994). When the protein becomes bound to the photoaptamer, the aptamer-protein are cross-linked by laser excitation of the Br-dUTP groups that are in close proximity to specific amino acid residues of the protein (Golden et al., 2000) (Figure 6.29). Because the cognitive protein is now covalently linked to the aptamer, nonspecific protein can be extensively washed from the array under harsh conditions. With the ability to substantially reduce the background, the array can be read with greater sensitivity. Because only the cognate protein remains following the rinse, it is now

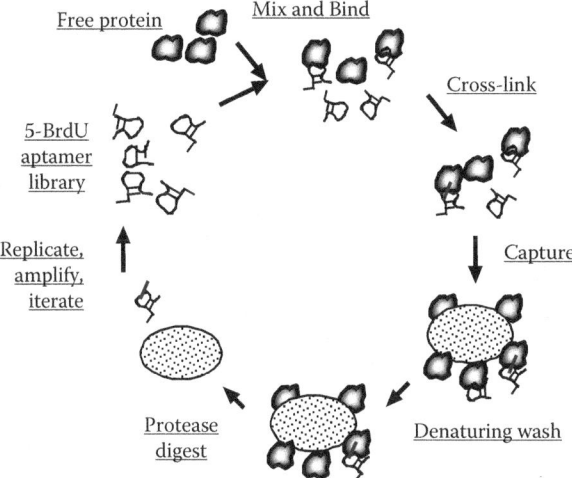

FIGURE 6.28   photoSELEX process. (Courtesy of SomaLogic, Inc., Boulder, CO.)

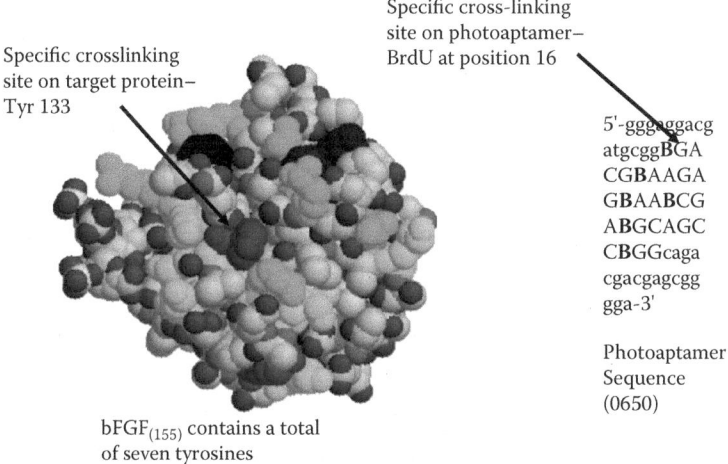

FIGURE 6.29   PhotoAptamer cross-link. (Courtesy of SomaLogic, Inc., Boulder, CO.)

possible to use general protein staining reagents for detection. For example, researchers at SomaLogic created aptamer arrays on glass slides and used NHS-Alexa 555
dye staining of the captured cognate proteins following a harsh rinse under denaturing conditions. In this manner, for example, thrombin could be detected at 100
pM in a multiplexed assay involving a total of eight different proteins. Total protein
concentration was 11.1 nM with the concentration of individual proteins ranging
from 0.01 nM to 10 nM (Figure 6.30). The combination of the photoSELEX process
with the denaturing rinse steps provides for increased specificity. Improvements in
general protein staining are need to further increase sensitivity into the low pM to

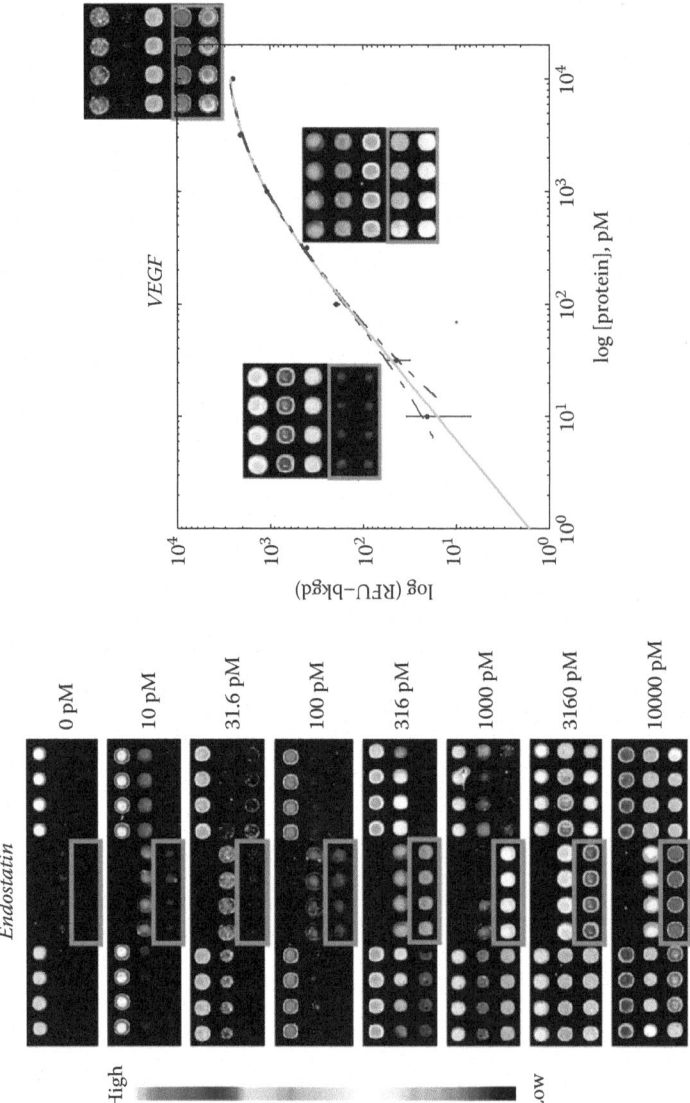

FIGURE 6.30 Aptamer dose response. (Courtesy of SomaLogic, Inc., Boulder, CO.)

sub-pM ranges in order to be competitive with current ELISA and antibody microarray assays (see Table 6.2). In recent studies, proteins have been detected in the pM range using an ALONA (antibody linked oligonucleotide assay) sandwich aptamer-antigen-reporter antibody. In this format, the aptamer provides capture specificity, while the secondary antibody coupled with signal amplification allows for increased sensitivity (Matson, unpublished).

UNIVERSAL PROTEIN ARRAY

A universal protein array (UPA) system has been carefully designed by Ge (2000) based upon the use of various transcription factors, their activators, and cofactors as probes. To create the UPA on nitrocellulose filters, 48 different, highly purified factors were used. Protein-protein interactions of various binding affinities could be assessed using different ionic strength buffers (e.g., 100 mM KCL versus 1000 mM KCL). For example, the relative binding of a radiolabeled (32P) GST-K-p52 protein to various transcription factors was studied. It was discovered that p52 specifically interacted with nucleolin but not topo I, thereby supporting the observation that p52 associates with nucleolin as a multiprotein complex in HeLa cells. Nucleolin is thought to be involved in pre-mRNA splicing and the unwinding of DNA-RNA duplexes, as well as mediating cell doubling time in human cancer cells. The p52 protein is a general transcriptional cofactor capable of potentiating activated transcription of class II genes. It also serves as a pre-mRNA splice regulator. The UPA system in the future may serve as a new tool in discovering new components involved in gene regulation and in the evaluation of therapeutic protein targets.

Zyomyx introduced a micro-fabricated microarray system based upon the use of pillars that serve as platforms for depositing capture antibodies. Relying on substrate features such as ultra-flatness and surface modification, they have created an environment for optimal protein-protein interaction. Precise delivery of reagents to each pillar enables robust assay development.

Peluso et al. (2003) investigated strategies for antibody immobilization using their chip format. Four immobilization strategies for placement of capture antibody were studied: random versus oriented coupling of IgG or random versus oriented coupling of Fab′ fragments (Figure 6.31). Oriented coupling of IgG antibody to solid-supports is well documented in the literature as having distinct advantage over the random tethering of antibodies. Foremost is the loss of antigen binding activity due to steric hindrance by being too near the surface or too close to adjacent antibodies, thereby blocking the binding region (Figure 6.32). There can also be surface denaturation of the antibody by excessive attachment at multiple sites on the surface. Oriented coupling of the antibody at an optimal surface density has been demonstrated to overcome many of these issues (Matson and Little, 1988). In this study, streptavidin (sAV) was used as a means to orient antibodies that had been site-specifically modified with biotin versus randomly biotinylated IgG and Fab′. In addition, two different surface modifications that yield streptavidin monolayers were studied: b-SAM (biotinylated self-assembled monolayer) and PLL-PEG-biotin (Figure 6.33). The b-SAM was formed on gold-coated glass slides treated with a SAM of oligo(ethyleneglycol) containing alkane disulfides terminating with NHS that then were reacted with

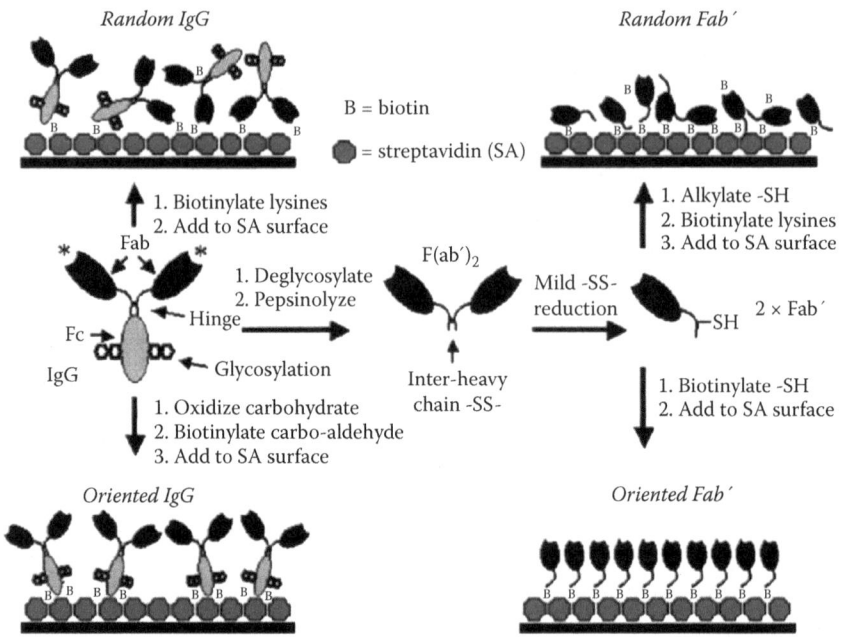

FIGURE 6.31   Strategies for immobilization of capture antibodies. (From Peluso P, Wilson DS, Do D, et al. Optimizing antibody immobilization strategies for the construction of protein microarrays. *Anal. Biochem.* 312: 113–124, 2003. With permission.)

tri(ethylene glycol) amino biotins. Incubation with streptavidin led to the creation of the streptavidin monolayer. This surface was used exclusively with the BIAcore system in SPR (surface plasmon resonance) studies. The PLL-PEG-biotin-sAV was used for microarray studies as described for the commercial product. A detailed description of the design and physical-chemical characterization of the poly-L-lysine-grafted poly(ethylene glycol) monolayer is provided by Ruiz-Taylor et al. (2001).

Streptavidin could be reproducibly tethered to the b-SAM surface at about 2×10^{12} molecules/cm^2 (~4 pmoles/cm^2). Not surprisingly, oriented Fab′ could be immobilized at the highest density with a 70% retention of the calculated specific binding activity. For the three antibodies studied, the oriented Fab′ was 1.8-fold to 5.6-fold more active than Fab′ biotinylated by a random process. In the case for immobilization of the full-length antibodies, while oriented antibodies did have higher binding activity, on average there was only ~1.3-fold increase over randomly biotinylated IgG. With the b-SAM surface, oriented Fab′ bound 49% more antigen than oriented IgG antibody.

For microarray work the PLL-PEG-biotin-sAV monolayer adsorbed onto an underlying titanium dioxide surface (pillars) replaced the b-SAM gold-coated glass surface. While oriented capture agents outperformed their random counterparts, there were distinct differences in Fab′ and antibody performance on this new surface relative to that of the b-SAM surface. Under certain conditions the oriented antibody

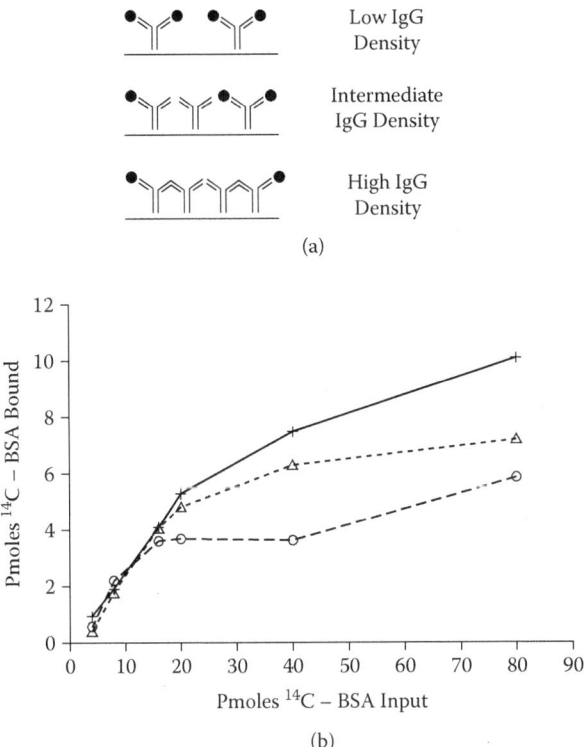

Low IgG
Density

Intermediate
IgG Density

High IgG
Density

(a)

(b)

FIGURE 6.32 Solid-phase antibody density and steric hindrance effects. (From Matson RS, Little MC. Strategy for the immobilization of monoclonal antibodies on solid-phase supports. *J. Chromatogr.* 458: 67–77, 1988. With permission.)

outperformed the oriented Fab′ system: At high antigen loading (100 nM) oriented antibody was threefold higher than oriented Fab′, while at 100 pM loading (Ka antibody ~100 pM) the Fab′ was threefold higher than antibody. One explanation offered by the authors is that the tethered antibody provided a higher number of antigen binding sites per surface area but that these were of lower antigen affinity (Ka′ >> Ka) than the Fab′ sites. This could be the result if the antibody were damaged during periodate oxidation while Fab′ biotinylation was less destructive. However, this does not explain why these differences were not observed on the b-SAM surface. It was observed that the PLL-PEG surface provided a higher fold change in random versus oriented tethering of the capture antibodies and Fab′ fragments. Perhaps it is the spatial distribution and relative density of antibody versus Fab′ that are more important here.

The Protein Profiling Biochip™ is composed of six chips assembled in a flow cell cassette device. Each chip provides 200 data points (200 pillars per chip) for a total of 1200 data points per cassette. Pillars are 50 microns in diameter. The mesa on each pillar is covered with a self-assembled monolayer (20 to 25 Å thickness) of biotin-derivatized poly-L-lysine-g-poly (ethylene glycol) groups. A constant

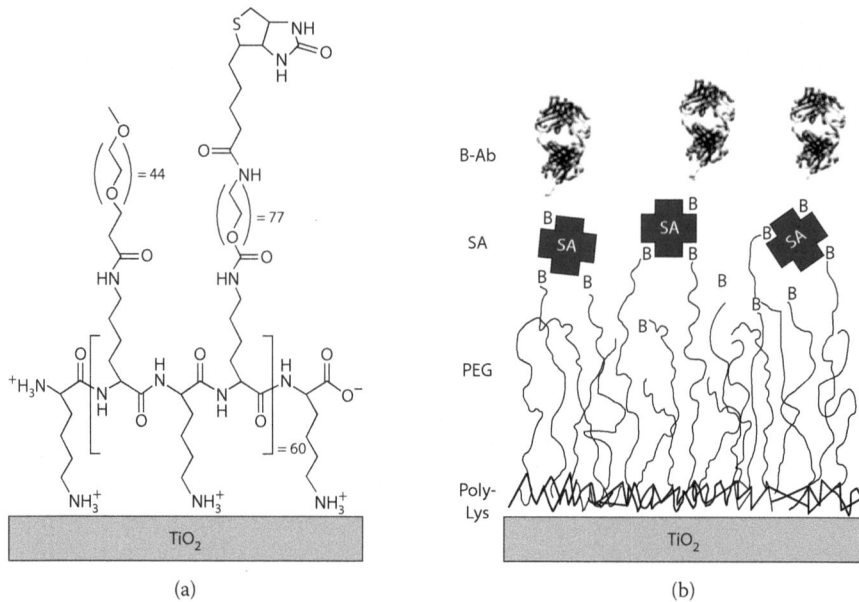

FIGURE 6.33 Zyomyx's b-SAM (biotinylated self-assembled monolayer) and poly-L-lysine-graft-(polyethylene glycol) (PLL-PEG)-biotin surface coatings. (From Nock S, Wilson DS. Recent developments in protein microarray technology. *Angew. Chem. Int. Ed.*, 42(5): 494–500, 2003. With permission.)

grafting ratio of 3.5 Lys : 1 PEG is maintained with variable biotin-PEG content. Streptavidin-antibody is immobilized at 0.5 to 2 pmole/cm² or Fab′ fragment at 4 pmole/cm². Each pillar is addressed by a separate capillary to apply sample and processing reagents. This is a noncontact process in which the capillary seats over the peg which has a hydrophobic coating, thereby allowing only the wetting of the mesa. Oriented coupling of antibody to the surface allowed for the lower limit of detection of IL-8 to 1 pg/mL (0.15 pM) at total signal above twofold background (Wagner, 2002). A multiplex assay for 30 human cytokines in each of the six channels has recently been introduced for analysis using as little as 40 µL sample volume.

Ciphergen (Fung et al., 2001) described the ProteinChip proteomics platform that combines the use of chromatographic support materials to capture proteins with the resolving power of mass spectroscopy. The ProteinChip in its current format more closely resembles a macro-array. Various surfaces borrowed from chromatographic use are prepared as large spots in a 1 × N array that can be read in a mass spectrophotometer (SELDI-TOF, surface enhanced laser desorption-time of flight). The Ciphergen system was especially useful for analysis of protein differential display (protein profiling) in which changes in the cellular content of proteins vary. For example, one can compare different growth conditions and then focus on the appearance or disappearance of cellular proteins that have been isolated onto the ProteinChip.

The elucidation of differentially expressed proteins is based upon well-known protein isolation and characterization strategies. For example, Thulasiraman et al. (2001) were able to identify virulence factors in *Yersinia pestis* (a bacterium causing human plague) by examining the m/z profile differences in bacteria grown at two different temperatures. The temperature shift from 26 to 37°C induces up-regulation in the virulence genes. Thus, cell extracts from the bacteria grown under the two temperatures were compared for protein content by first applying to different chip surfaces. Strong anion exchange (SAX) spots were used to adsorb negatively charged proteins, while metal affinity resin (IMAC) loaded with Cu^{+2} was used to capture *His*-proteins. SAX isolation identified the occurrence of 14.9 kDa and 78.8 kDa peaks that were present only in *Y. pestis* grown at the higher temperature. Because both proteins bound at pH 7.4, this suggests that both had a pI < 7.4. Furthermore, the 78.8 kDa protein bound to the IMAC spot, while the 14.9 kDa protein did not bind. This suggests that the 78.8 kDA protein had a free histidine imidazole group on its surface, while the 14.9 kDa protein did not have an available metal binding domain. Following a scale-up purification of bacterial lysate, the two proteins were excised from SDS-PAGE bands and subjected to tryptic digest. SELDI analysis of the peptide fragments with reference to the NCBI Protein Database resulted in a match for the 78.8 kDa protein as KatY protein. The KatY protein was also a good fit with the protein characterization obtained (i.e., induced at 37°C; 78.8 kDa mass; pI = 6.4, and active site motifs containing histadine imdiazoles). Similarly, the 14.9 kDa protein was identified by tryptic digest as the Antigen 4 protein that contains a fibrillar structure required for full virulence.

In other studies Boyle et al. (2001) used the Ciphergen SELDI protein chip to analyze the secretion and autoactivation of a cysteine protease, SpeB, from *Streptococcus pyogenes* that is implicated in the onset of group A streptococcal infections and may contribute to toxic shock symptoms. SpeB could be detected at ~0.75 ng protein in a 30 minute assay based upon SELDI-TOF analysis. This represents greater than a 6000-fold increase in sensitivity over that of the conventional Western immunoblot analysis. Zymogen and activated intermediate forms of the protease could be identified and easily monitored during the culture growth. This will enable future studies on the activation process to elucidate the virulence factor's mechanism of action.

The resolving power of mass spectroscopy is further exemplified in studies of post-translational modification studies of acyl carrier protein from *Mycobacterium tuberculosis*. Researchers were able to determine not only the nature of this modification but also the mechanism by which the protein was regulated (see Fung et al., 2001, ref. 7).

The protein profiling approach also provides the use of pattern recognition for discrimination of disease states. Biomarkers for prostate cancer were profiled and a panel assembled that could differentiate cancer patients from non-cancer populations (see Fung et al., 2001, ref. 11). Poon et al. (2003) utilized the ProteinChip to obtain tumor-specific proteomic signatures to detect hepatocellular carcinoma (HCC) in patients having chronic liver disease (CLD). The proteomic signature is based upon the separation, detection, and profiling of low molecular mass proteins in terms of their relative abundance using the SELDI-TOF (m/z) spectrum. Serum fractions from anion-exchange fractionation were placed on IMAC3 (immobilized metal affinity capture) ProteinChip Arrays loaded with copper Cu^{+2} ion as well as WCX2 (weak cation exchanger) arrays. Following processing of the arrays each spot was subjected to SELDI-TOF and various m/z peaks identified as being unique to either HCC or CLD cases. A total of 2384 mass (m/z) peak assignments were identified in samples with 1087 from IMAC3 and 1297 from WCX2 arrays. Of these, 250 markers were found to be significantly different in HCC and CLD. This was accomplished by applying two-way hierarchical clustering and artificial neural network (ANN) for differentiation. Because the number of peptides, proteins, and other polymers within a sample can lead to generation of 8000 to 10,000 m/z values in the SELDI-TOF spectrum, the ANN has proven to be very useful in predicting tumor grades (Ball et al., 2002). In the above study both methodologies returned a specificity and sensitivity of approximately 90% in detection of HCC. This is similar to the earlier work that identified ovarian cancer at 100% sensitivity, 95% specificity, and detected prostate cancer at 83% sensitivity. Wellmann et al. (2002) also examined prostate carcinoma with the aid of laser assisted microdissection and SELDI to analyze protein extracts from about 500 cells. A number of differentially expressed proteins in the 1.5 to 30 kDa range were found between normal prostate gland cells and those from prostate tumors. A prominent (threefold) up-regulated protein peak at mass 4299 daltons was observed in the case of prostate tumor. The average relative intensities were tumor (24.37), transitional zone (9.99), and normal prostate (7.26). Finally, Batorfi et al. (2003) used a similar approach combining laser capture microdissection and SELDI-TOF of ProteinChips to investigate the pathogenesis of gestational trophoblastic disease by differential protein expression. In their studies of normal and molar trophoblast (tumor) cells in placenta, the researchers identified three metal binding polypeptides (11.3, 13.8, 15.2 kDa) that were present at statistically significant lower levels in tumor.

Peptide Arrays for Antibody Detection

Melnyk et al. (2002) created peptide arrays for detection of antibodies in blood raised against various infectious agents (e.g., HCV core protein). Peptides were synthesized and terminated with diglyoxyl groups that were in turn oxidized to aldehydes using periodate. Glass slides were surface treated with 3-aminopropyltrimethoxysilane and converted to the semicarbazide form by a triphosgen mediated reaction with Fmoc-hydrazine. The glyoxylyl peptides were printed down onto the semicarbazide slide and the array held overnight at 37°C in a humid chamber. The peptide microarrays were stored at 37°C for 6 months without significant variation in signal or background. Slide-to-slide variation was 4.5% (SD). Oxidized antibodies were also immobilized without issue.

Diluted human serum (1 µL) was incubated with the peptide microarray and bound antibodies detected using a rhodamine-labeled anti-human IgG. Signal was detected using a slide scanner (Affymetrix model 418 scanner) with data collection in the Cy3 channel. A reported eightfold gain in sensitivity at 100% specificity over standard ELISA was achieved using the peptide microarray. Comparison of the peptide microarray and a standard ELISA was made using a reference collection of HCV infected and normal human sera. Concordance was observed in 117/130 sera. The remaining 13 samples were found to be reported as false positive or false negative by the ELISA. The length and position of spacer groups on the glyoxylyl peptides did not appreciably affect the binding of serum antibodies from HCV patient sera. However, the manner in which EBV peptides were immobilized did have significant impact on sensitivity. Thus, generalizations regarding tethering of peptides to the solid-support appear to be unwarranted in this case. This study clearly demonstrates the difficulty in selecting attachment chemistries for optimizing binding conditions.

PHAGE-DISPLAY ANTIBODY SELECTION

De Wildt et al. (2000) utilized phage-display techniques in order to screen for large antibody populations. The phage were selected to express scFv antibody fragments with binding regions for protein A and protein L. Thus, capture and detection of the antibody fragment were possible without interference with the antigen binding domains. Selected clones were gridded onto a large square tray and then coated with a NC filter and the phage grew up overnight. A second NC filter coated with protein L was overlayed with the colony filter to capture the expressed scFv phage. This filter could then be probed with protein A-HRP to determine the expression levels. Up to 18,342 different colonies could be arrayed in duplicate onto a 22 cm × 22 cm NC filter. Fifteen of these filters could be produced by robotic gridding. The scFv grids could be used as antibody arrays for large-scale protein (antigen) expression analysis. Selected scFv fragments could be released from the membrane and purified by protein A-Sepharose affinity chromatography. From the purified fragment the affinity binding constants could be determined by solution phase competition experiments using a BIAcore biosensor system.

PROTEIN KINASE MICROARRAY

We have discussed numerous applications involving primarily ligand binding (such as in the use of antibody to capture antigen), and there are other interesting and potentially valuable protein array formats. A case in point is that of the enzyme microarray in which the activity of the sequestered enzyme is used to access the sample for substrate content. In the array format a library of enzymes acting on a variety of substrate analytes would be immobilized. In this format, the metabolic activity of cells could be accessed globally simply by measuring the content of cell extracts or spent culture media.

Perhaps the most well-known example is the work of Zhu et al. (2000) from Yale on yeast protein kinase chips. Disposable microwell microarrays (later described as nanowell chips) were fabricated out of poly(dimethylsiloxane) elastomer (PDMS) onto glass slides. For example, an optimal design was to prepare wells of 1.4 mm diameter and 300 microns deep to hold approximately 300 nL. These were arranged in a 10 × 14 pattern on a 1.8 mm pitch between wells. Two arrays could be mounted onto a single slide such that 288 wells could be analyzed. Proteins were covalently attached to the wells via epoxide using 3-glycidoxypropyltrimethoxysilane (Figure 6.34). Conversely, substrates could be attached to the wells and the presence of enzyme in the sample determined.

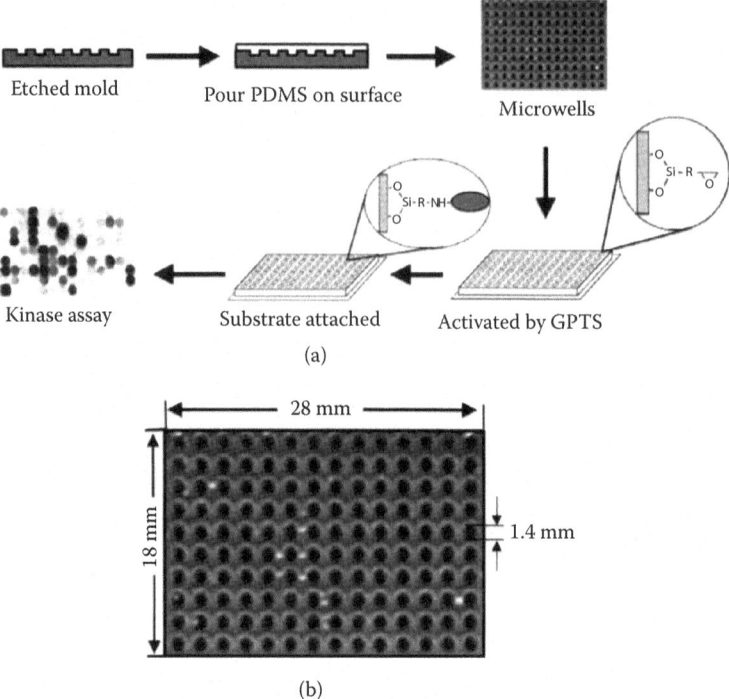

FIGURE 6.34 Elastomer-based microwell protein kinase assay array. (From Zhu H, Klemic JF, Chang S, et al. Analysis of yeast protein kinases using protein chips. *Nat. Genet.* 26: 283–289, 2000. With permission.)

For the studies involving the yeast kinases, entire coding regions for these were cloned into an expression vector generating GST-fusion proteins rescued in *Escherichia coli*. In this manner, 119 GST:kinase fusion proteins were successfully prepared, overexpressed in yeast and purified on glutathione Sepharose beads in a 96-well format. The purified forms were immobilized in nanowells for assay. In this case, the assay consisted of incubation of specific substrates with 33Pγ-ATP. Seventeen different assays were performed to simultaneously characterize the activity of 119 kinases. Following the reactions the microarrays were rinsed and phosphorylation signal of each kinase quantified using a phosphorimager. The investigators found that greater than 10-fold improvement in S/N was achieved with the nanowell array over that of the conventional microtiter plate assay while using reaction volumes at 1/20 to 1/40 those needed for 384-well assays.

PROTEIN MICROARRAYS USEFUL IN AUTO-ANTIBODY SCREENING

McBride et al. (2008) developed a new microarray format based upon a membrane filtration 96-well microplate. They were able to demonstrate utility in the development of an ANA (antinuclear antibodies) multiplex panel for the diagnosis of systemic rheumatic diseases. The **NALIA** methodology (Nanodot Array Luminometric

ImmunoAssay) involves printing of 10 auto-antigens commonly associated with rheumatoid diseases along with controls on the underside of the cellulosic filter plate (MultiScreen-HA, Millipore, Inc.) in a 5 × 5 array pattern. The advantage to this flow-through approach is that of greatly reducing sample diffusion while facilitating the washing steps. Thus, the assay can be completed within 1.5 hours making it very useful for high-throughput processing using standard robotic liquid handlers equipped with a vacuum filtration device. The plate was read from the bottom using a specially designed CCD camera that imaged the entire plate. A chemiluminescent substrate signal was generated using HRP conjugated to antihuman IgG or antihuman IgM. The read time was reported to be within 2 minutes.

The intra-assay imprecision was acceptable ranging from 10 to 14% CV based upon control spots of biotinylated human albumin and signal developed with streptavidin-HRP. However, the more important attribute, that of plate-to-plate comparisons (interassay), was not reported. Sensitivity measured from 3 SD variance of $n = 4$ negative serum samples was reported at approximately $1.8 × 10^4$ IU/L, dsDNA antigen (minimum detectable dose); with a positive cutoff value at $5 × 10^4$ IU/L. Levels of sensitivity for the other nine auto-antigens were not provided but reported as of "similar" result.

Concordance studies between the NALIA multiplex with that of commercial (EUROIMMUN) ELISA singlet assays ($n = 5$) were undertaken based upon patient sera. For the four auto-antigens tested, the agreement was 93 to 98% from analysis of +120 patients. In the case of dsDNA antigen evaluation of 182 patient sera, the agreement between the two formats was 83%. The authors attribute these differences in part due to differences in dsDNA antigens used, each assay therefore detecting a different subpopulation of the serum anti-dsDNA antibodies present. That assertion could easily have been answered by the investigators if they had prepared ELISAs using the same auto-antigens that were used for construction of the NALIA panel.

In conclusion, NALIA technology offers a promising format for high-throughput multiplex immunoassays. It has advantage over conventional well multiplex or singlet immunoassays by reducing the barrier to diffusion, while providing a rapid and efficient means for washing. Additional studies are needed to characterize its full utility for quantitative assay development.

According to the 2012 report by the Alzheimer's Association, one in eight persons aged 65 or older in the United States has Alzheimer's disease (AD). It is the sixth leading cause of death with related healthcare costs estimated to be $200 billion in 2012. There is at present no diagnostic test for an accurate, early detection of the disease. Nagele and co-workers (2011) sought to screen for serum auto-antibodies in those afflicted with the disease using a high-throughput protein microarray as the screening tool. Using a commercially available protein microarray (ProtoArray, v5.0, Human Protein Microarray, Invitrogen, Inc., Carlsbad, CA) the group examined serum samples from 30 AD patients and 40 nondemented (NDC) control subjects.

The protein microarray included 9,486 human proteins expressed as GST fusions in insect cells. A total of 149 serum samples were probed with the microarray. It was estimated that >1000 different auto-antibodies were present on average in each sample. A candidate panel of 10 diagnostic biomarkers was identified from microarray training sets leading to a 96% sensitivity and 92.5% specificity for AD in the full cohort. The biomarker panel was able to differentiate AD (50 samples) from a non-neurological disease (breast cancer, 30 samples) with a positive predictive value of 90.7% and a negative predictive value of 96.2%. To test differentiation of the panel for AD and a closely related neurological disease (Parkinson's disease), a panel of five biomarkers was selected. The panel was able to differentiate with an overall accuracy of 86% with 90% sensitivity and 79% specificity.

Stempfer et al. (2010) adopted the SEREX (**ser**ological identification of antigens by recombinant **ex**pression cloning) approach for selection of biomarkers for the screening of tumor associated auto-antibodies in serum. First, candidate antigen clones (cDNA expression library *E. coli* clones) from SEREX screenings of brain and lung cancer serums were isolated and then sub-cloned into an expression vector for histidine-tagging of fusion proteins. The recombinant proteins were expressed in *E. coli* following induction by IPTG (isopropyl β-D-1-thiogalactopyranoside). The resulting *His*-tagged protein isolates were subsequently purified using the Ni-NTA method (see Bornhorst & Falke, 2000).

At an expression rate of approximately 40% following IPTG induction, the purified proteins were found to be present at about 7 to 70 ng per mL of bacterial culture. This yielded 200 to 250 μg protein in 75 μL of elution buffer (50 mM sodium phosphate, pH 8 containing 500 mM imidazole, 0.01% SDS, 0.01% sodium azide). The antigens (96 different proteins) were then spotted down directly onto epoxy activated glass slides in triplicate along with positive (crude cell lysates) and negative controls (buffer) as three identical subarrays per slide.

For screenings, the recombinant antigen microarray slides were bathed in patient sera for 2 hours (optimized protocol). Following washing, Alexa 647 (Invitrogen) dye-labeled goat anti-human IgG was added for an additional hour. The labeled slides were then imaged on a microarray scanner (Axon Genepix 4000A, Molecular Devices). Optimization processes included assessment of low nonspecific binding of reporter antibody (low background signal obtained in absence of serum); specificity of antibodies to targeted proteins (confirmed by a lack of correlation of signal strength with mass of protein printed); and no false-positive signals (no signal on buffer spots indicating absence of any print-related carryover, i.e., lack of cross-contamination). Inter-assay precision was excellent. The pair-wise correlation plots of slide replicates were found to range, $r = 0.92$ to 0.97 at a median correlation of 0.957. However, the authors also offered some good advice in using microarrays constructed from expressed proteins: "while using (protein) microarrays for screening purposes defining biomarkers would be done best using the same batch of microarrays" (Stempfor et al., 2010, p. 6). This avoids any bias created by differences in protein purity between batch-to-batch preparations that could affect dose response. In this study, a clear differentiation of auto-antibodies present in brain and lung tumors from patient sera could be made using a multidimensional scaling model of selected highly significant antigens.

DUAL LABELING OF TARGETS FOR INCREASED SENSITIVITY AND SPECIFICITY

Schroder and co-workers (2010) took a different approach to large-scale microarray-based proteomic analysis. Selected on the basis of differential gene expression targets identified in pancreatic and colon cancer translational studies, 810 peptide-specific, polyclonal antibodies (rabbit host) were arrayed onto epoxy-silane slides (Schott Nexterion, Jena, Germany). From 10 μg of each antibody arrayed in duplicate at a print concentration of 1 mg/mL, 1080 arrays were produced. Proteins in the sample matrix (plasma, serum, urine) were labeled in situ using N-hydroxysuccinimide (NHS) esters of the fluorescent dyes, Dy-549 or Dy-649 (Dyomics, Jena, Germany) at a protein concentration of 4 mg/mL. Test samples were labeled with Dy-549 and a pool of control (reference) samples was labeled with Dy-649.

Dual-color and single-color labeling strategies were compared for assay robustness. Coefficients of variation (%CV) were determined for inter-array replicates as well as intra-array spot duplicates. The average inter-array CV as well as the CV distributions were found to be significantly reduced for dual-color labeling over that of one-color labeling. The mean CV for 20 slides that had been sampled across five slide batches (540 slides per batch) was 13%, and ranged from 9 to 14%. The majority of antibodies (96%) had CVs under 20%. In comparison, the authors cite other studies for inter-array CV ranges: single-label antibody microarray, 12 to 22% CV; bead-based assay, 18 to 44% CV; repeated mass spectrometry, 21% CV; and DNA microarray, 5 to 15% CV. Thus, in this study, assay precision based upon dual labeling of protein samples is similar to that obtained for dual labeling of nucleic acids in gene expression analysis and is more robust than single-label approaches for large multiplex assays.

THE DEPLETION OF HIGHLY ABUNDANT PROTEINS FROM SERUM DEEMED UNNECESSARY

Also, Schroder et al. (2010) examined the effect of serum depletion of high-abundance proteins ($n = 20$) to determine the impact on assay sensitivity and performance for enriched proteins. The depletion of high-abundance proteins from serum has been regarded as an important step in order to improve detection of the lower abundant and trace-level proteins found in the matrix. It is particularly useful in the characterization of low-abundance proteins by 2D-gel electrophoresis and by mass spectrometry. Yet, these researchers found that the signal-to-noise ratio (SNR) taken as a measure of sensitivity was not improved upon depletion. A significant impact on signal intensity was noted for about 25% of the targets, with approximately one-half of these detected at a higher intensity and the remainder at a lower intensity following depletion of the 20 most abundant proteins from the sample. Thus, in agreement with other reported studies such as that by Fountoulakis et al. (2004), the co-depletion of trace proteins with highly abundant proteins is most likely, thereby introducing an unwelcome bias into the results. Schroder et al. demonstrate that there is no need to undergo protein depletion from serum with microarrays.

COMPETITIVE ELISA BY PROTEIN MICROARRAY

Zhong et al. (2010) examined the performance of a protein microarray in the monitoring of illicit drugs found in animal food products. The presence of two such veterinary drugs, clenbuterol (CL) and sulfamethazine (SM2), fortified in chicken muscle tissue were examined by a competitive indirect ELISA (ci-ELISA) method. Results were compared against the protein microarray format. Even though the microarray is a 2-plex, albeit with 49 replicates, this study represents one of the few examples where the format is employed as a competitive ELISA.

These small molecule drugs (<1000 Daltons) first required their conjugation to a carrier protein (ovalbumin). This was found to be necessary in order to elicit a strong immunogenic response from these haptens that would provide suitable polyclonal antibodies (rabbit host) for the ELISA. Thus, artificial antigens (haptens) were immobilized by adsorption onto a conventional 96-well microplate at 5 μg per well for the ci-ELISA, or spotted onto microarray slides at 20 ng per 7 × 7 array. In either case, pooled antiserum and competing haptens were applied to the microplate and slide. Goat anti-rabbit Cy3 conjugate was then applied to the slide and the resulting signal read in a microarray slide scanner (Axon GenePix 4000B, Molecular Devices). For development of the microtiter plate ELISA, goat anti-rabbit HRP conjugate was applied with colorimetric signal generated with TMB and was then read using a standard microplate reader.

The IC_{50} for each drug was determined in both formats by nonlinear regression analysis of dose-response. The microarray-based competitive ELISA was found to be three- to fivefold more sensitive than the ci-ELISA. For CL, the ci-ELISA measured IC50 = 190.7 ng/mL, while the microarray ELISA measured 39.6 ng/mL; and for SM2, ci-ELISA (156.7 ng/mL) versus microarray ELISA (48.8 ng/mL). More importantly, the analytical recovery of drug assayed from muscle samples of chicken that had been exposed to CL tainted feed remained higher in microarray ELISAs than those tissues analyzed by ci-ELISA. The authors conclude that "the protein microarray displayed a much more consistent recovery and higher sensitivity"; while they "consumed far less samples that ci-ELISA" (Zhong et al., 2010, p. 6).

THE ISSUE OF CROSS-REACTIVITY IN A PROTEIN MICROARRAY SANDWICH ELISA

Pla-Roca et al. (2012), concerned with the vulnerability of multiplex immunoassays toward cross-reactivity, explore the virtues of their antibody colocalization microarray (ACM) approach. Here, capture antibodies are first printed down on a microarray substrate in an array format as with any conventional antibody microarray. The microarray is then removed from the arrayer for assay development. A well-forming gasket is temporarily applied to the microarray slide. Samples are added to wells containing the subarray and the analyte captured by the immobilized, cognate (analyte-specific) capture antibody. At this point, following washing of the residual sample from wells, the gasket is removed and the slide is returned to the arrayer for site-specific delivery of detection antibody over the respective capture antibody spot. The authors describe this as a "colocalization" step (i.e., physically locating the detection antibody with the capture antibody by on-spot delivery). They refer to the microarray as the antibody colocalization microarray (ACM) format. The new format was then compared to conventional ELISA and multiplex sandwich assays (MSAs).

A similar approach was first described by Matson et al. (2007) for high-throughput, multiplexed immunoassay development. In that case, all components of the sandwich immunoassay were site-specifically dispensed to a planar surface in a parallel fashion. Their "overprint immunoassay" utilized protein A to sequester analyte-specific capture antibodies onto plastic surfaces. Because of the hydrophobic character of the surface-modified plastics, it was possible to develop a complete immunoassay without the need for wells. Sensitivity was comparable to that of a conventional ELISA but at a 1000-fold reduction in reagent consumption and sample.

In the Pla-Roca et al. study, efforts are first directed at demonstrating cross-reactivity associated with the MSA format, and then at validation of the ACM approach as a viable alternative when cross-reactivity is an issue. A vulnerability index is introduced to define the number of "liability pairs" that *might* result in a false-positive signal due to various cross-reactivites. Five scenarios were described for cross-reactivity: detection antibody-antigen (dAb-Ag), detection antibody-capture antibody (dAb-cAb), detection antibody-detection antibody (dAb-dAb), capture antibody-antigen (cAb-Ag), and antigen-antigen (Ag-Ag) associations. Thus, the number of liability pairs in a sandwich immunoassay multiplex having N targets can be defined by a pair-wise vulnerability index of $N(N - 1)/2$ pairs. The authors contend that liability pairs do not exist in the case of the single-plex ELISA format ($N = 1$), while the greatest number of pairs is associated with the MSA format having $4N(N - 1)$ pairs with which to contend. There is an $N - 1$ factor to account for the reactivity of the target analyte with its cognate capture antibody.

A 14-plex MSA was evaluated for cross-reactivity with antigens spiked into dilution buffer (PBST) at a challenge of 32 ng/mL (32,000 pg/mL), except for IL-6,-8 which were spiked at the more physiological level of 2 ng/mL (2000 pg/mL). Detection antibodies were delivered at 1 μg/mL; signal reporter, streptavidin-Cy5 was delivered at 5 μg/mL. Of the 728 (4 × 14 × 13) theoretical cross-reactivities from liability pairings, this study demonstrated a rather limited number of measurable levels of cross-reactivity across the sandwich assay 14-plex matrix (Figure 6.35). This is not surprising given the antigen loadings used. In particular, the detection antibodies for EFG and CEA cross-reacted in a significant manner with the majority of targets and would consequently have been replaced with more suitable antibodies during an optimization process. Note that this study should have been performed in serum diluent rather than the PBST buffer to avoid other potential artifacts (e.g., nonspecific streptavidin-Cy5 binding). We have observed this in a number of other studies when streptavidin-dye conjugates are delivered in a simple buffer without a protein blocker present. Both streptavidin and dyes can contribute to nonspecific interactions with antibodies. These authors have also studied the microarray optimization process for sandwich immunoassays (Luo et al., 2011) by the Toguchi method in efforts to improve sensitivity. Surprisingly, while they reported in the earlier paper that "the SA-Cy5 concentration was the most important factor and contributed to more than 50% of signal variation" (Luo et al., 2011, p. 5771), in the present study the dye-conjugate was applied at fivefold higher concentration. It is also important to closely match the assay buffer composition to that of the sample matrix. Of course, diluents need to be screened as well for the

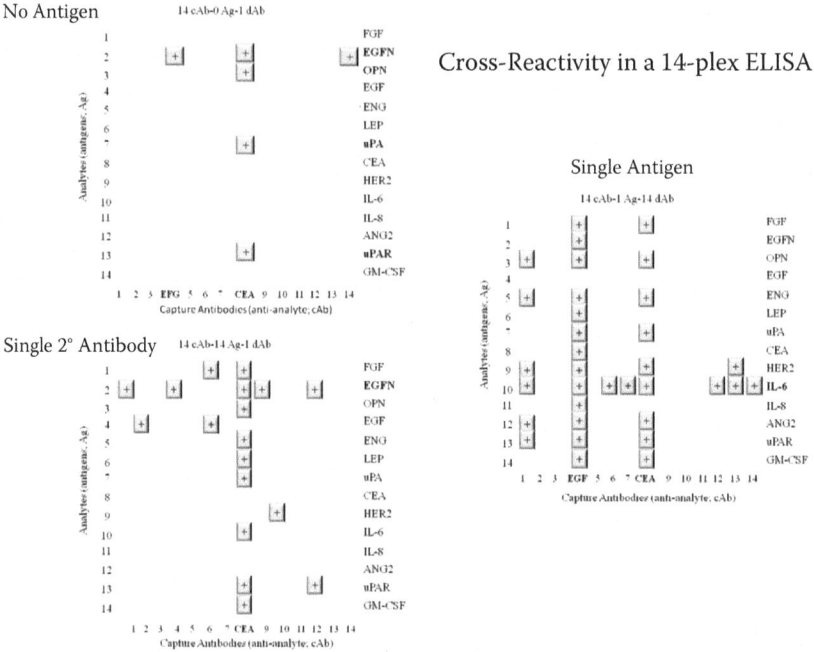

FIGURE 6.35 Cross-reactivity in a 14-plex ELISA. (From Pla-Roca M, Leulmi RF, Tourekhanova S, et al. Antibody colocalization microarray: A scalable technology for multiplex protein analysis in complex samples. *Mol. Cell. Proteomics* 11: 10.1074/mcp. M111.011460, 1–12, 2012.)

presence of interfering substances, including intrinsic or contaminating analytes and heterophilic antibodies.

This paper is an excellent testament to the criticality of understanding the cross-reactivity issues in multiplex immunoassays. The concepts of "liability pairs" and "vulnerability" offer a rationale to explain the shortcomings of multiplex assays (MSAs), but the antibody colocoalization microarray, ACM, approach described falls short of solving the problem. True, cross-reactivity among detection antibodies can be a major contributing factor and that was adequately addressed in this paper. Yet, this is not always the case. It remains that extensive optimization of all components in a multiplex assay are required to avoid a number of pitfalls, whether due to cross-reacting detection antibodies, the leaching of capture antibodies from the solid phase, heterophilic antibody interference, nonspecific adsorption of labeled conjugate, or the occurrence of unwanted surface-bound aggregates due to buffer, salt, pH, or sample matrix.

The ACM approach does have merit for use when highly complex multiplex assays are required in a sandwich ELISA format. For example, demonstration by these researchers of a 50-plex sandwich ELISA using ACM is impressive. However, the cost and availability of larger numbers of matched antibody pairs remain a current limitation for higher plex-level assay development on planar as well as bead-based platforms.

In summary, antibody arrays remain the predominant platform for both life science and clinical-based research. However, antigen microarrays have gained importance for use in allergy testing and auto-antibody screening.

REFERENCES

Angenendt P, Glokler J, Murphy D, Lehrach H, Cahill DJ. Toward optimized antibody microarrays: A comparison of current microarray support materials. *Anal. Biochem.* 309: 253–260, 2002.

Avseenko NV, Morozova T Ya, Ataullakhanov FI, Morozov VN. Immobilization of proteins in immunochemical microarrays fabricated by electrospray deposition. *Anal. Chem.* 73: 6047–6052, 2001.

Ball G, Mian S, Holding F, et al. An integrated approach utilizing neural networks and SELDI mass spectrometry for classification of human tumours and rapid identification of potential biomarkers. *Bioinformatics* 18(3): 395–404, 2002.

Batorfi J, Ye B, Mok SC, Cseh I, Berkowitz RS, Fulop V. Protein profiling of complete mole and normal placenta using ProteinChip analysis of laser capture microdissected cells. *Gynecol. Oncol.* 88: 424–428, 2003.

Bornhorst JA, Falke JJ. Purification of proteins using polyhistidine affinity tags. *Methods Enzymol.* 326: 245–254, 2000.

Boyle MDP, Romer TG, Meeker AK, Sledjeski DD. Use of surface-enhanced laser desorption ionization protein chip system to analyze streptococcal exotoxin B activity secreted by *Streptococcus pyogenes. J. Microbiol. Methods* 46: 87–97, 2001.

Brody EN, Gold L. Aptamers as therapeutic and diagnostic agents. *Rev. Mol. Biotechnol.* 74: 5–13, 2000.

Brody EN, Willis MC, Smith JD, Jayasena S, Zichi D, Gold L. The use of aptamers in large arrays for molecular diagnostics. *Mol. Diagn.* 4(4): 381–388, 1999.

Cox JC, Hayhurst A, Hesselberth J, Bayer TS, Georgiou G, Ellington AD. Automated selection of aptamers against protein targets translated in vitro: From gene to aptamers. *NAR* 30(20) e108: 1–14, 2002.

De Wildt RMT, Mundy CR, Gorick BD, Tomlinson IM. Antibody arrays for high-throughput screening of antibody-antigen interactions. *Nat. Biotechnol.* 18: 989–994, 2000.

Delehanty JB, Ligier FS. A microarray immunoassay for simultaneuos detection of proteins and bacteria. *Anal. Chem.* 74: 5681–5687, 2002.

Ekins RP. Ligand assays: From electrophoresis to minaturized microarrays. *Clin. Chem.* 44(9): 2015–2030, 1998.

Ekins R, Chu F. Ultrasensitive microarray-based analytical methods: Principles and practice. *Int'l. Bus. Commun.—Protein Microarray Technol.*, 2001.

Ekins RP. Multi-analyte immunoassay. *J. Pharm. & Biomed. Analysis* 7(2): 155–168, 1989.

Ekins R, Chu F, Biggart E. Fluorescence spectroscopy and its application to a new generation of high sensitivity, multi-microspot, multianalyte, immunoassay. *Clin. Chim. Acta* 194: 91–114, 1990.

Ellington AD, Szostak JW. In vitro selection of RNA molecules that bind specific ligands. *Nature* 346: 818–822, 1990.

Ezan E, Grassi J. Chapter 7: Optimization in *Immunoassays: A Practical Approach,* Gosling, JP, ed., Oxford University Press, Oxford, UK, 2000, pp. 187–210.

Fountoulakis M, Juranville J-F, Jian L, et al. Depletion of the high-abundance plasma proteins. *Amino Acids* 27: 249–259, 2004.

Fung ET, Thulasiraman V, Weinberger SR, Dalmasso EA. Protein biochips for differential profiling. *Curr. Opin. Biotechnol.* 12: 65–69, 2001.

Ge H. UPA, a universal protein array system for quantitative detection of protein-protein, protein-DNA, protein-RNA and protein-ligand interactions. *NAR* 28(2): i–vii, 2000.

Golden MC, Collins BD, Willis MC, Koch TH. Diagnostic potential of PhotoSELEX-evolved ssDNA aptamers. *J. Biotech.* 81: 167–178, 2000.

Haab BB, Dunham MJ, Brown PO. Protein microarrays for highly parallel detection and quantitation of specific proteins and antibodies in complex solutions. *Genome Biol.* 2(2): research0004.1–0004.13, 2001.

He M, Taussig MJ. Single step generation of protein arrays from DNA by cell-free expression and in situ immobilization (PISA method). *NAR* 29(15) e73: 1–6, 2001.

He M, Taussig MJ. DiscernArray technology: A cell-free method for the generation of protein arrays from PCR DNA. *J. Immunol. Methods* 274: 265–270, 2003.

Huang R-P. Detection of multiple proteins in an antibody-based protein microarray system. *J. Immunol. Methods* 255: 1–13, 2001a.

Huang R-P, Huang R, Fan Y, Lin Y. Simultaneous detection of multiple cytokines from conditioned media and patient's sera by an antibody-based protein array system. *Anal. Biochem.* 294(1): 55–62, 2001b.

Jayasena SD. Aptamers: An emerging class of molecules that rival antibodies in diagnostics. *Clin. Chem.* 45(9): 1628–1650, 1999.

Joos TO, Schrenk M, Hopfl P, et al. A microarray enzyme-linked immunosorbent assay for autoimmune diagnostics. *Electrophoresis* 21: 2641–2650, 2000.

LaBrie S. Protein analysis in an array format using a novel, highly sensitive detection system. *Int'l. Bus. Commun.—Drug Discovery Technol.*, Session F1, 2001.

Lizardi PM, Huang H, Zhu Z, Bray-Ward P, Thomas DC, Ward DC. Mutation detection and single-molecule counting using isothermal rolling-circle amplification. *Nat. Genet.* 19(3): 225–232, 1998.

Lueking A, Horn M, Eickhoff H, Bussow K, Lehrach H, Walter G. Protein microarrays for gene expression and antibody screening. *Anal. Biochem.* 270: 103–111, 1999.

Luo W, Pla-Roca M, Juncker D. Taguchi design-based optimization of sandwich immunoassay microarrays for detecting breast cancer biomarkers. *Anal. Chem.* 83: 5767–5774, 2011.

MacBeath G, Schreiber SL. Printing proteins as microarrays for high-throughput function determination. *Science* 289: 1760–1763, 2000.

Matson RS, Little MC. Strategy for the immobilization of monoclonal antibodies on solid-phase supports. *J. Chromatogr.* 458: 67–77, 1988.

Matson RS, Milton RC, Cress MC, Rampal JB. Microarray-based cytokine immunosorbent assay. *Oak Ridge Conference*, Poster No. 20, 2001.

Matson RS, Milton RC, Rampal JB, Chan TS, Cress MC. Overprint immunoassay using protein A microarrays. *Methods Mol. Biol.* 382: 273–286, 2007.

McBride JD, Gabriel FG, Fordham J, et al. Screening autoantibody profiles in systemic rheumatic disease with a diagnostic protein microarray that uses a filtration-assisted nanodot array luminometric immunoassay (NALIA). *Clin. Chem.* 54(5): 883–890, 2008.

Melnyk O, Duburcq X, Olivier C, Urbes F, Auriault C, Gras-Masse H. Peptide arrays for highly sensitive and specific antibody-binding fluorescence assays. *Bioconjug. Chem.* 13(4): 713–720, 2002.

Mendoza LG, McQary P, Mongan A, Gangadharan R, Brignac S, Eggers M. High-throughput microarray-based enzyme-linked immunosorbent assay (ELISA). *BioTechiques* 24(4): 778–788, 1999.

Mezzasoma L, Bacarese-Hamilton T, Di Cristina M, Rossi R, Bistoni R, Crisanti A. Antigen microarrays for serodiagnosis of infectious diseases. *Clin. Chem.* 48(1): 121–130, 2002.

Moody MD, Van Arsdell SW, Murphy KP, Orencole SF, Burns C. Array-based ELISAs for high-throughput analysis of human cytokines. *BioTechniques* 31(1): 1–7, 2001.

Nagele E, Han M, Demarshall C, Belinka B, Nagele R. Diagnosis of Alzheimer's disease based on disease specific autoantibody profiles in human sera. *PLOS ONE* 6(8): e23112.

Nock S, Wilson DS. Recent developments in protein microarray technology. *Angew. Chem. Int. Ed.*, 42(5): 494–500, 2003.

Pawlak M, Schick E, Bopp MA, Schneider MJ, Oroszlan P, Ehrat M. Zeptosens' protein microarrays: A novel high performance microarray platform for low abundance protein analysis. *Proteomics* 2: 383–393, 2002.

Peluso P, Wilson DS, Do D, et al. Optimizing antibody immobilization strategies for the construction of protein microarrays. *Anal. Biochem.* 312: 113–124, 2003.

Phizicky E, Bastiaens PIH, Zhu H, Snyder M, Fields S. Protein analysis on a proteomic scale. *Nature* 422: 208–215, 2003.

Pla-Roca M, Leulmi RF, Tourekhanova S, et al. Antibody colocalization microarray: A scalable technology for multiplex protein analysis in complex samples. *Mol. Cell. Proteomics* 11: 10.1074/mcp.M111.011460, 1–12, 2012.

Poon TCW, Yip T-T, Chan ATC, et al. Comprehensive proteomic profiling identifies serum proteomic signatures for detection of hepatocellualr carcinoma and its subtypes. *Clin. Chem.* 49(5) 752–760, 2003.

Ruiz-Taylor LA, Martin TL, Zaugg FG, Indermuhle P, Nock S, Wagner P. Monolayers of derivatized poly(L-lysine)-grafted poly(ethylene glycol) on metal oxides as a class of biomolecular interfaces. *PNAS* 98(3): 852–857, 2001.

Sapsford KE, Liron Z, Shubin YS, Ligler FS. Kinetics of antigen binding to arrays of antibodies in different sized spots. *Anal. Chem.* 73: 5518–5524, 2001.

Schena M, Shalon D, Davis RW, Brown PO. Quantitative monitoring of gene expression patterns with a complementary DNA microarray. *Science* 270: 467–470, 1995.

Schroder C, Jacob A, Tonack S, et al. Dual-color proteomic profiling of complex samples with a microarray of 810 cancer-related antibodies. *Mol. Cell. Proteomics* 9: 1271–1280, 2010.

Schweitzer B, Roberts S, Grimwade B, et al. Multiplexed protein profiling on microarrays by rolling-circle amplification. *Nat. Biotechnol.* 20: 359–365, 2002.

Schweitzer B, Wiltshire S, Lamber J, et al. Immunoassays with rolling circle DNA amplification: A versatile platform for ultrasensitive antigen detection. *PNAS* 97(18): 10113–10119, 2000.

Seong S-Y. Microimmunoassay using a protein chip: Optimizing conditions for protein immobilization. *Clin. Diagn. Lab. Immunol.* 9(4): 927–930, 2002.

Silzel JW, Cerecek B, Dodson C, Tsong T, Obremski RJ. Mass-sensing, multianalyte microarray immunoassay with imaging detection. *Clin. Chem.* 44(9): 2036–2043, 1998.

Stempfer R, Syed P, Vierlinger K, et al. Tumour auto-antibody screening: Performance of protein microarrays using SEREX derived antigens. *BMC Cancer* 10: 627, 2010.

Thulasiraman V, McCutchen-Maloney SL, Motin VL, Garcia E. Detection and identification of virulence factors in *Yersinia pestis* using SELDI ProteinChip System. *BioTechniques* 30(2): 428–432, 2001.

Tuerk C, Gold L. Systematic evolution of ligands by exponential enrichment: RNA ligands to bacteriophage T4 DNA polymerase. *Science* 249: 505–510, 1990.

Valle RPC, Jendoubi M. Antibody-based technologies for target discovery. *Curr. Opin. Drug Discovery Dev.* 6(2): 197–203, 2003.

Wagner P. Protein biochips as powerful new tools in proteomics. *Int'l. Bus. Commun.—Protein Microarrays Conference*, 2002.

Wellmann A, Wollscheid V, Lu H, et al. Analysis of microdissected prostate tissue with ProteinChip arrays—A way to new insights into carcinogenesis and to diagnostic tools. *Int'l. J. Mol. Med.* 9: 341–347, 2002.

Weng S, Gu K, Hammond PW, et al. Generating addressable protein microarrays with PROfusion covalent mRNA-protein fusion technology. *Proteomics* 2: 48–57, 2002.

Wiese R, Belosludtsev Y, Powdrill T, Thompson P, Hogan M. Simultaneous multianalyte ELISA performed on a microarray platform. *Clin. Chem.* 47(8): 1451–1457, 2001.

Willis MC, LeCuyer KA, Meisenheimer KM, Uhlenbeck OC, Koch TH. An RNA-protein contact determination by 5-bromouridine substitution, photocrosslinking and sequencing. *NAR* 22(23): 4947–4952, 1994.

Woodbury RL, Varnum SM, Zanger RC. Elevated HGF levels in serum from breast cancer patients detected using a protein microarray ELISA. *J. Proteome Res.* 1: 233–237, 2002.

Yguerabide J, Yguerabide EE. Light-scattering submicroscopic particles as highly fluorescent analogs and their use as tracer labels in clinical and biological applications. *Anal. Biochem.* 262(2): 137–156, 157–176, 1998.

Yguerabide J, Yguerabide EE. Resonance light scattering particles as ultrasensitive labels for detection of analytes in a wide range of applications. *J. Cell. Biochem.* Supp. 37: 71–81, 2001.

Zhong L, Zhang W, Zer C, Ge K, Gao X, Kernstine KH. Protein microarray: Sensitive and effective immunodetection for drug residues. *BMC Biotechnol.* 10: 12, 2010.

Zhu H, Klemic JF, Chang S, et al. Analysis of yeast protein kinases using protein chips. *Nat. Genet.* 26: 283–289, 2000.

Zichi D, Koga T, Greef C, Ostroff R, Petach H. Photoaptamer technology: Development of multiplexed microarray protein assays. *Clin. Chem.* 48(10): 1865–1868, 2002.

7 Multiplex Assays

DNA microarrays allow scientists to correlate gene expression signatures with disease progression, to screen for disease-specific mutations, and to treat patients according to their individual genetic profiles; however, the real key is proteins and their manifold functions.

Yu, Schneiderhan-Marra, and Joos, 2010

INTRODUCTION

In February 2012 the Chinese government approved Affymetrix's GeneChip DNA microarray system for IVD use, thereby opening the floodgate to personalized medicine into a country steeped in ancient apothecary. In March 2012 Protagen (Dortmund, Germany), a company at the forefront of high-throughput protein expression, established a collaborative effort with researchers at the University of Tubingen to develop multiplex protein assays for autoimmune and cancer diagnostics. The "omics" paradigm that was first embraced in the mid-1990s (largely through the commercialization of DNA microarray platforms) has begun its transition into clinical diagnostics. Multiplex analyses of biomarkers has gained worldwide acceptance, albeit, there is still work to be done on standardization and the establishment of guidelines for governing validation.

BENEFIT

There are several important benefits associated with multiplexing (Table 7.1). Sample and reagent are both conserved. In most cases, there is also a significant drop in labor costs, especially if the test can be automated. Overall, the cost per test can be dramatically reduced depending upon the degree of multiplexing achieved. It is also well recognized now that with many diseases, such as cancer, the etiology is very complex. Multiplexing of biomarkers into disease-associated analyte panels has been shown to provide greater predictive value for a more accurate prognosis and diagnosis of the disease.

DEFINITION

For the purpose of our discussion, we define the multiplex assay as *the introduction of more than one unique analyte-specific sensor into a single sample compartment that results in the detection of one or more analytes within the sample.*

The key to our definition is that we are undertaking multiple analyses simultaneously within the confine of a single container (Figure 7.1). The container could be a test tube, a microarray, a flow cell, and so forth. A simple example would be that of

TABLE 7.1
Benefits of the Multiplexed Assay

Feature	Benefit
Simultaneous measure of multiple analytes from a single sample in a single well	Creation of disease-associated analyte panels for a more accurate prognosis or diagnosis
	"Weight of evidence" when comparing different analytes and/or their ratios in the same reaction
Reduction in reagent consumption	Reduction in cost per test
Reduction in sample size	Save precious clinical research samples for other tests
SBS compliance for full automation	High-throughput/high-content assays

multiplex polymerase chain reaction (PCR) (Figure 7.2). A sample containing, for example, genomic DNA is placed in a single Eppendorf tube to which is added more than one primer pair to amplify different alleles. So, we have a single compartment, the tube, to which we have added allele "analyte" specific primers (the sensors).

In the first chapter, we discuss the importance of the omic's approach to biological investigation. Such methodologies rely upon global assessment of biological variation. For example, the most commonly applied tool for measuring gene expression is the DNA microarray. Here a gene population under study is compared with that of a control population. The sample genomic DNA is labeled with a red dye, the control genomic DNA is labeled with a green dye, and the two are mixed together and applied to the microarray. Variation (gene expression) is indirectly discovered upon interpreting the dye ratio of the two populations. Thus, the gene expression analysis is based upon using a multitude of analyte sensors (the microarray's capture probe library) to detect differences between two gene populations, which also satisfies our definition (Figure 7.3).

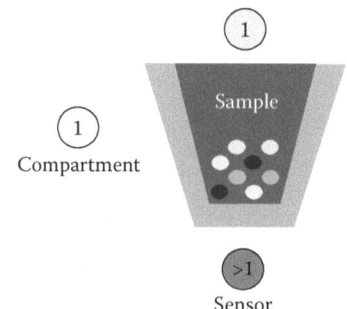

Multiplexing

Introduction of more than one unique analyte-specific **sensor** into a **single sample compartment** that results in the detection of one or more analytes within the sample.

FIGURE 7.1   (See color insert.) Multiplex assay defined.

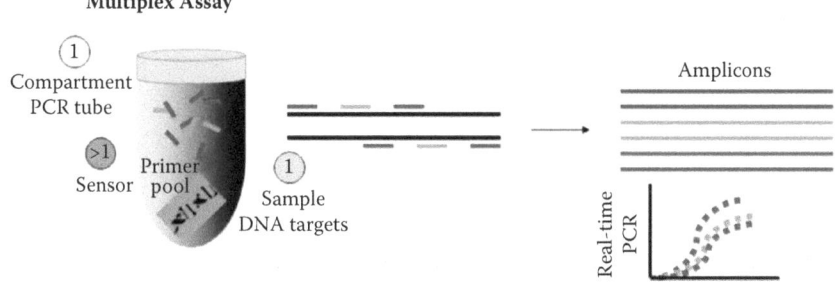

FIGURE 7.2 **(See color insert.)** Multiplex polymerase chain reaction (PCR) assay.

Recently, lateral flow "dipsticks" containing more than one capture probe were introduced. For example, different capture antibodies can be printed down as strips on the membrane. This permits detection of multiple analytes in a sample (Figure 7.4).

More complex multiplex assays are represented by bead-based "fluidic" arrays such as the Luminex system. Here 100 or more unique bead sensors can be introduced into a single sample to detect 100 or more different analytes. In this case, the sample compartment could be the well of a microplate allowing measurement of analytes in multiple wells of the microplate (Figure 7.5).

Other fluidic arrays can be prepared from microfabricated bar-coded particles allowing multiplexing of thousands of probes. A related technology based upon planar microarrays utilizes an "array of arrays" approach where multiple capture

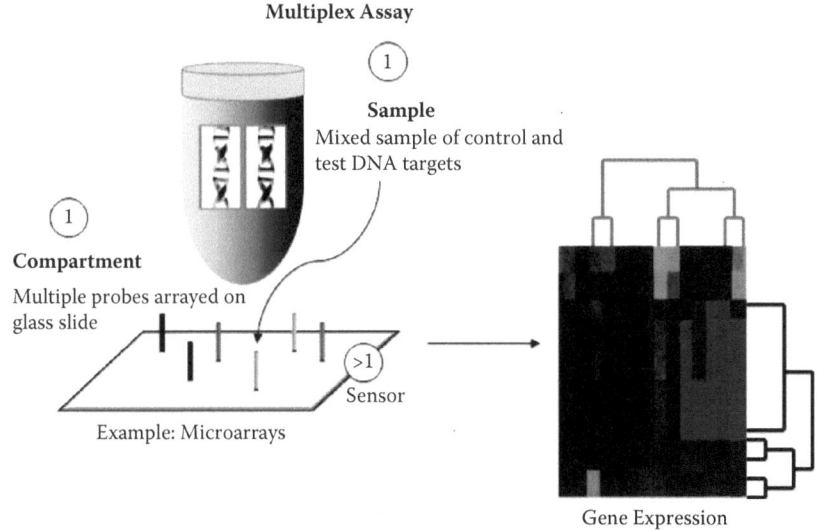

FIGURE 7.3 Gene expression microarray.

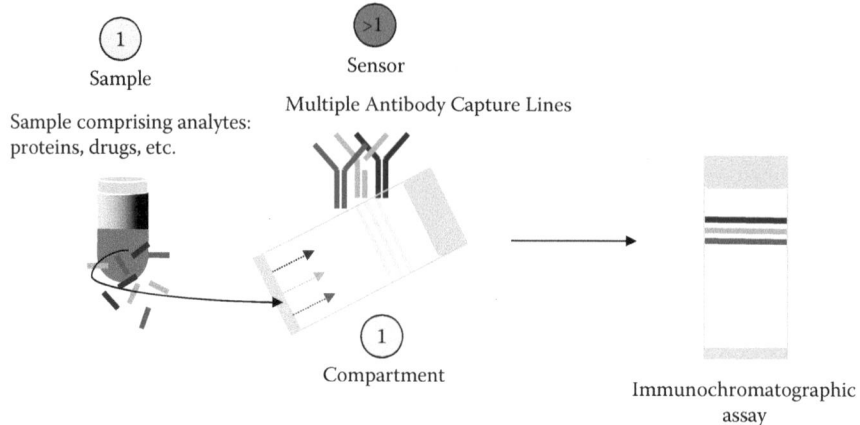

FIGURE 7.4   (See color insert.) Multiplex lateral flow (LF) assay.

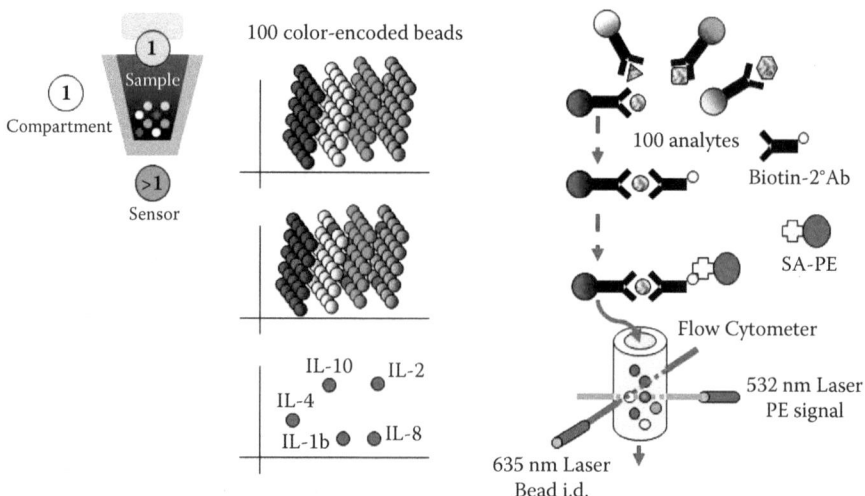

FIGURE 7.5   Multiplex bead-based assay.

probes are arrayed in the bottom of the wells of a conventional microtiter plate (Figure 7.6).

So, we now have an idea of what we mean by multiplex assays and some of the different approaches used. But, how useful are these? For what purpose then do multiplex assays best serve science and medicine? What are their strengths and limitations? Let's look at some recent examples from the literature. We will begin by examining the utility of different multiplexing technologies in similar analyses of pathogens.

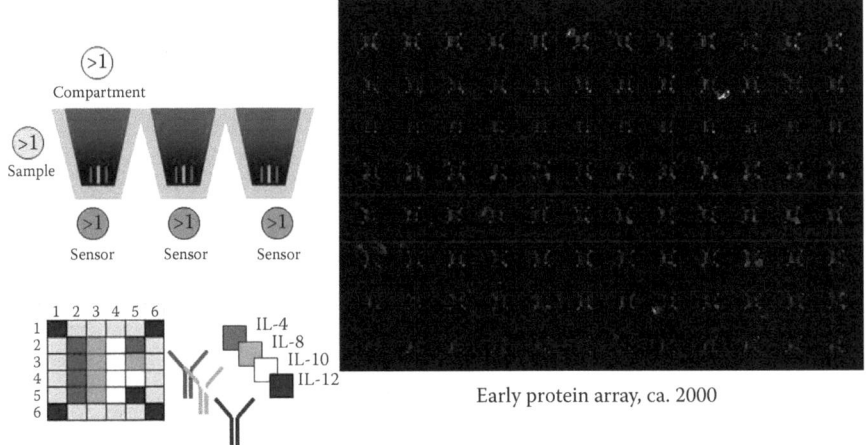

Example: Microplate-based "array of arrays" format

Early protein array, ca. 2000

FIGURE 7.6 (**See color insert.**) Multiplex microtiter plate (MTP) assay.

MULTIPLEX POLYMERASE CHAIN REACTION (PCR)

Pui and collaborators (2011) have developed a multiplex PCR assay to distinguish between three Salmonella species that can cause food poisoning and typhoid fever. The availability of a rapid and sensitive test that would identity these food-borne pathogens is of great health benefit in the United States and abroad, and especially during times of natural disaster when potable water supplies and sanitation conditions are lacking. On a side note, the authors remind us that while PCR is such a powerful diagnostic tool, it does not distinguish between live (infective) cells and dead cells, so follow-up culturing is still required to monitor treatment.

Key to a successful PCR assay is careful optimization. Consider the fundamental elements of a PCR reaction: primer pairs, dNTPs, and the Taq polymerase. They all have to work together, in concert, to attain sensitivity and specificity. Pui et al. have done a good job at working these conditions out in their study. It is important to note that these researchers proceeded in an incremental or stepwise manner. Such an approach is advisable when dealing with multiparameter processes, especially in multiplexing.

Single primer pairs were first optimized for each pathogen, then primer sets for detection of two pathogens, and finally for the complete 3-plex amplification. What they found is that in the case of a multiplex it was necessary to increase the primer length over that used in monoplex in order to avoid self-annealing. That is, increasing the T_m for annealing reduced the likelihood of nonspecific priming that often leads to false positives. Similarly, it was important to balance out the $MgCl_2$ concentration. Mg^{+2} is required for Taq enzyme activity. However, too much Mg^{+2} can reduce the enzyme's specificity (fidelity) causing nonspecific amplification. Adjusting the dNTP concentration is also essential because chelation of Mg^{+2} (sequestering) reduces the availability of Mg^{+2} required to maintain Taq enzymatic activity, thereby inhibiting the PCR. As a result, they found that 1.5 mM $MgCl_2$ produced no PCR, while 3 mM resulted in a nonspecific amplification smear on gels, and 2.5 mM concentration was just right.

In conclusion, the multiplex PCR is useful in providing a rapid, highly specific, and presumably sensitive test for a limited number of analytes. Unfortunately, Pui et al. (2011) failed to discuss anything regarding sensitivity.

MULTIPLEX PCR COMPARED TO DNA MICROARRAY

Peterson et al. (2010) undertook the development of multiplex assays in an effort to increase their testing coverage for *Salmonella enterica*. The pathogenic enteric include over 1500 subspecies or serovars that cause severe food poisoning in humans which can be fatal. Testing by conventional serology is costly. The authors sought to produce an oligonucleotide microarray that could screen for the serovars and provide a high-throughput and less costly approach. The DNA microarray was compared to a co-developed multiplex PCR with validation against serotype (agglutination).

A multiplex PCR including five primer sets was used to identify 42 enteric serovars. Two separate multiplex PCR assays were required for coverage. Following PCR, the resulting amplicons were analyzed by gel electrophoresis to determine molecular weight and serovar identity.

Probes for inclusion on the microarray were selected on the basis of those identified in previous work leading to differentiation of known serovars by microarrays and validation of new probes found to associate with the multiplex PCR. Also included were positive control hybridization probes: EUB (16S conserved region in eubacteria) and rpoB (conserved region, β subunit, bacterial RNA polymerase); negative control probe: nonhomologous, 25mer oligonucleotide. The microarray included 37 unique, 70mer oligonucleotides printed in two fields on the Ultra Gap slide (Corning, Inc., Life Sciences, Tewksbury, MA).

The authors undertook a multiplex blind test. For the multiplex PCR a total of 111 Salmonella culture samples (requiring DNA extraction) and 31 bacterial DNA samples were analyzed. In the case of the microarray study, the number of cultures was reduced to 20, while 36 bacterial DNA samples were analyzed. The concordance with serology was 100% for multiplex PCR at a sensitivity and specificity of 95% (135/142 serovars) and 93% (52/56 serovars) for the microarray in correctly identifying serotypes. Although the reduction in sample size for the microarray study may have some effect on the outcome, both methodologies performed very well.

MULTIPLEX LATERAL FLOW

Immuno-chromatographic or lateral flow (LF) strips are widely used for rapid testing throughout the world, in particular in developing nations. They are generally regarded as a low-cost qualitative or at best semi-quantitative measure. A typical design for an LF strip is described (Figure 7.7) and the process explained as follows.

Essentially, the sample is placed upon the sample pad. Capillary lateral flow causes the sample to flow along the strip. Sample moves onto the conjugate-release pad where it mixes with a reporter antibody immobilized on a bead such as colloidal gold. The reporter antibody binds the specific analyte in the sample and carries it

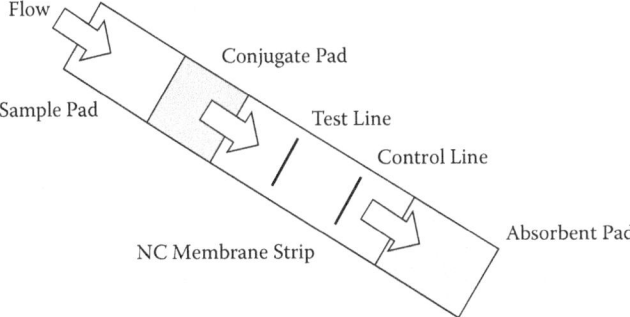

FIGURE 7.7   Lateral flow strip diagram.

onto a nitrocellulose (NC) membrane. The NC membrane has analyte-specific capture antibody adsorbed at a defined location, usually referred to as a test line. As the
analyte-bead complex moves along the NC membrane it is captured by an analyte-
specific antibody. In the case where a colloidal gold bead conjugate is utilized, the
capture line will turn red as the gold bead becomes sequestered on the line. A second
capture line is used as a control. Here reporter antibody-specific capture antibody
binds up residual beads that passed by the test line. The fluid continues to flow onto
the absorbent pad to complete the assay. The test line and control lines are then read.
This is done largely by visual inspection, although strip reader instrumentation has
recently entered the marketplace permitting a more quantitative result.

In the paper by Park et al. (2010), antibodies to four food- or water-borne pathogens (*Salmonella typhimurium, Staphylococcus aureus, Legionella pneumophila,*
and *Escherichia coli*) were spotted onto a single NC membrane at defined locations.
HRP-antibody conjugates of the corresponding reporter antibodies were placed on
the conjugate-release pad. Samples containing various levels of the pathogens were
applied, and the lateral flow strip was processed. Following completion of the wicking process, an enzyme substrate (TMB) loading pad was placed on the NC membrane. TMB substrate was placed on the substrate loading pad to generate the visible
blue color signal.

Unfortunately, the developed strips showed considerable background and nonspecific signals. This has limited the utility of multiplex LF for quantification over a
larger dynamic range. The colorimetric LF assay reported by Park et al. could detect
between 10^4 and 10^7 CFU/mL of the selected pathogens within 20 minutes.

Few examples of multiplex lateral flow immunoassay (LFIA) testing have been
documented in the primary literature. Those that are described are generally limited
to the simultaneous analysis of two to three analytes. The reasons for the apparent
lack of adoption of lateral flow for multiplexing are understandable. First is the matter of sensitivity. LF strips are largely developed for colorimetric analysis in which
the detection limits (either by eye or reflectance reader) are relatively high. For gold
nanoparticles detection of 1 µg/mL is achieved. In contrast, many protein analytes
require measurements in the nanogram to picogram per milliliter range (i.e., requiring 10- to 100-fold greater sensitivity). Second, proteins are present over a wide
dynamic range, perhaps as much as a five log distribution in concentration. This is

problematic for many multiplexing technologies. How do you measure one protein present at 1 pg/mL and another that is found in the same sample at 1 ng/mL (10^3 pg/mL) or 1 µg/mL (10^6 pg/mL)? For example, serum myoglobin protein occurs in the ng/mL range, while serum troponin is present in the pg/mL level in adult humans. Both are important in the monitoring of a heart attack (acute myocardial infarction [AMI]) but are difficult or impossible to detect using conventional LFIA.

Zhu et al. (2011) have devised a rather clever approach to simultaneously measure cardiac troponin I and myoglobin by lateral flow immunoassay. The key was to alter the "detectability" of the two analytes by using different-sized gold nanoparticles. A fast moving (13 nm diameter) gold conjugate was used for detection of troponin, while a larger (41 nm diameter), slower moving gold particle was used to detect myoglobin.

As the old idiom goes, "the devil is in the details." Zhu and co-workers prepared one batch of fast particle coated with anti-troponin. The bound antibody was subsequently conjugated with a 48mer oligonucleotide that terminated 3′ with biotin. Another batch was coated with anti-myoglobin but not oligonucleotide. The slow particle was conjugated with streptavidin. On the LFIA strip test lines were striped containing capture antibody for troponin and myoglobin, respectively, to form a 2-plex assay. In addition, two conjugate pads were added. The first pad held strep-tavidin-gold, while the second pad contained a mixture of the 13 nm gold-antibody conjugates for troponin and myoglobin.

When sample is added to the LFIA strip, the faster moving particles bind up their respective analytes (troponin or myoglobin) and move off the pad, eventually to be sequestered at the test lines. At this point equivalent levels of detection are presented to each test line. Myoglobin being present at a higher concentration would be easily detected. Troponin, on the other hand, would present a faint line. However, the slower moving streptavidin gold particle eventually crosses the test lines and becomes bound to the troponin conjugate containing the biotinylated oligonucleotide. The troponin test line is then amplified by the addition of the second gold particle.

Although the authors characterized their study as a "proof of principle" test, there was concordance with a standard clinical chemistry analyzer (Hitachi 7600). For troponin (cTnI), $r = 0.96$; and myoglobin, $r = 0.98$ with serum samples obtained following AMI. Troponin samples ranged from 9.2 pg/mL to 13,900 pg/mL. In contrast, myoglobin was determined over the range of 5.9 to 2771 ng/mL (approximately 6000 to 2.8 million pg/mL).

NUCLEIC ACID LATERAL FLOW IMMUNOASSAY (NALFIA)

Noguera and co-workers (2011) validated a nucleic acid lateral flow immunoassay (NALFIA) for detection of *E. coli* pathogenic strains commonly associated with food poisoning. As implied, the NALFIA methodology incorporates a PCR amplification step in front of the lateral flow immunoassay (LFIA). The strategy therefore is to increase the likelihood of early pathogen detection by lateral flow through mass amplification of virulence genes, and then capture the resulting amplicons on specific antibody test lines. In this case, carbon nanoparticles were used for visible line readout. Performance was compared against q-PCR.

Shiga toxin-producing *E. coli* (STEC) strains are transmitted from cattle to humans largely due to the occurrence of unsanitary conditions, such as the eating of undercooked meat or vegetables contaminated by improper handling or the presence of fecal material. The pathogen's virulence factors that are the causative agents for infection in humans are well known. PCR primer-pairs are available for amplifying these genes for detection by a number of methods.

Lateral flow (LF) offers several advantages for field testing. LF is relatively inexpensive, easily performed, and is a rapid test. It suffers the disadvantage of low sensitivity. These authors sought to overcome this issue by offering a multiplex approach with quantitative performance. In order to accomplish multiplex analysis of virulence factors from *E. coli* strains, the PCR primer sets used were tagged with dyes commonly used in labeling of the amplified DNA (amplicons). Antibodies raised to these dyes were also available and could be immobilized in lines on the lateral flow nitrocellulose strip for specific capture of the amplicons. For example, the virulence gene, vt2, was amplified using the forward primer tagged with FITC dye, while the reverse primer for all cases was labeled with biotin. The anti-FITC capture antibody line would thus bind the dual-labeled FITC-biotin vt2 DNA. Carbon nanoparticles with adsorbed neutravidin would localize on test lines that had biotin-labeled amplicons to permit a visible detection and thus determine the presence or absence of a virulence gene associated with a particular pathogenic *E. coli* strain.

For the multiplex NALFIA, optimization of the capture antibody test lines was essential. In this case, the concentrations were selected based upon achieving acceptable levels of line sensitivity at the lowest antibody load and lowest PCR sample volume. Keeping the PCR volume at 1 μL, the antibody loadings were determined to be for the anti-dye species: Texas Red (100 μg antibody/mL), DIG (100), DNP (125), FITC (250), and Cy5 (300). As a case in point, different antibodies possess different binding affinities. In the design and preparation of any multiplex immunoassay such differences need to be taken into consideration.

As mentioned, the NALFIA was compared against q-PCR as the reference method. Not surprisingly, q-PCR was found to be much more sensitive. For example, detection limits (LOD) for NALFIA ranged from a low of 6.7×10^4 cfu/mL (cfu = colony forming units) to 4.1×10^5 cfu/mL across the five virulence factors determined, while q-PCR ranged from 8.4×10^1 to 5.9×10^5 cfu/mL. From the stated ranges it is apparent that at least one NALFIA determination was more sensitive than for q-PCR. For the virulence factor, vt2, the LOD for NALFIA = 6.7×10^4 cfu/mL (0.6 ng/mL), while the LOD for q-PCR = 5.9×10^5 cfu/mL (5.4 ng/mL). In the testing of real samples from cattle, the NALFIA method yielded both false-positive and false-negative results relative to q-PCR. These were attributable to differences in sensitivity. For example, NALFIA was unable to detect a low positive in four cases (reported as false negative), and q-PCR failed to detect a low positive in two cases (NFLFIA reports as a false positive). However, the overall performance of NALFIA was quite comparable with q-PCR in terms of quality of fit. The kappa coefficients for NALFIA and q-PCR ranged from 0.80 to 1.00, meaning that the results using the two methods were in agreement.

Are there advantages in using a **multiplex NALFIA** over that of q-PCR? For the NALFIA a 30 minute standard PCR needs to be done prior to the 10 minute lateral flow step, while q-PCR can be accomplished within the same time period. In a standard laboratory with the availability of the required thermo-cycler, and taking into consideration the higher sensitivity achievable with q-PCR, the advantage would be to q-PCR. Albeit, as the authors point out, the multiplex NALFIA does offer some cost savings over that of q-PCR, at least for low-throughput applications. Comparing a duplex q-PCR with a two-line NALFIA the authors calculate that the NALFIA can be run at about one-third the cost. However, such savings may be overstated when one takes into consideration the labor savings associated with the higher throughput achievable by automated q-PCR systems.

NALFIA would be advantageous in **point-of-use testing** provided that an inexpensive, portable thermo-cycler could be used in the field. Multiplex NALFIA lowers test costs by a reduction in reagent and materials consumption. There are also no transport costs or post-contamination issues because the test would be performed at the site. Finally, the results of the test would be available immediately to the client.

MULTIPLE TEST LINES

In the study by Corstjens et al. (2007), a three-step, multiplex lateral flow immunoassay for the detection of HIV, HCV, and *Mycobacterium tuberculosis* (TB) is described. The flow steps involve the consecutive additions (CF, consecutive flow) of first, diluted sample, followed by a wash step, and concluding with the addition of protein A-UPT (up-converting phosphor technology) particles for detection. The authors refer to this as the CF-UPT format for lateral flow. A semi-automated microfluidic cassette device for use with CF-UPT is also described. Protein A from *S. aureus* binds to the Fc region of many but not all antibodies. There are both IgG subtype and species differences in its relative affinity. For more detailed information on UPT see the review by Achatz et al. (2011).

Of particular concern here is a problem encountered by the authors during their systematic development of the multiplex, which they refer to as "preceding capture line interference." They observed that the location of the antibody test line relative to the distance from the sample application pad was important. In the case of an HCV-HIV multiplex LFIA, the HCV test line signal strength increased as the line was moved closer to the sample pad. In developing a multiplex LFIA test, pay careful attention to line widths, antibody loading, and the relationship between the number of lines and their relative distance.

MULTIPLEX BEAD-BASED ASSAYS

Bead-based technology has been developed for multiplex assays. These are sometimes referred to as "fluidic" arrays. Automated high-throughput systems such as xMAP (Luminex, Inc., Austin, TX) utilized fluorescence-coded dyed beads (microspheres) with flow cytometry for detection and analysis. Essentially, varying the dye ratio for two dyes across different bead populations establishes the level of multiplex. Dyed particles are identified as the particles streamed through the sheath flow using laser excitation and a photomultiplier to capture the emission spectra. Likewise, a particular bead population can be coated with specific capture antibodies and multiple bead types mixed together to provide a "fluidic" array to determine 100 to 500 different analytes when using a sandwich immunoassay. A different laser is used to interrogate the beads for analyte-specific capture. Recently, Luminex introduced the MagPlex microsphere that is dye encoded and super-paramagnetic. This would enable a more efficient means to address sample preparation.

Kim et al. (2010) from the Naval Research Laboratory sought to evaluate the utility of the MagPlex-based flow cytometry assay for the detection of pathogens in foods. The advantage of such an approach would be to provide a rapid test that could displace bacterial culturing, thereby reducing the analysis time from days to a few hours. Magnetic beads offer a more efficient means over filter plates or centrifugation in bead processing prior to analysis. This is because food particles are difficult to separate from the analytical microspheres by these methods. Kim and co-workers introduced pathogens including heat-killed Salmonella, Campylobacter, E. coli, Listeria, and staphylococcal enterotoxin B and spiked known levels into various food extracts. Recovery levels were determined relative to the pathogens spiked in a simple buffer. Unfortunately, working with the MagPlex beads in complex food matrices turned out to be problematic, leading to significant reduction in sensitivity and high backgrounds. Such assay interferences, attributed to the nonspecific binding of proteins, carbohydrates, and small molecules such as polyphenols released in food extracts, resulted in a dramatic decrease in sensitivity. For example, the limit of detection (LOD) for Listeria was suppressed by two to five orders of magnitude—that is, detection of about 100 colony forming units (cfu/mL) in buffer shifting upwards (less sensitive) to greater than 10 million units in lettuce extracts. This is about the same level of sensitivity achieved by Park et al. (2010) using a simple lateral flow strip test.

MULTIPLEX BEAD-BASED ASSAY COMPARED TO THE ELISA

Elshal and McCoy (2006) reviewed various bead-based assay commercial platforms in comparison to standard ELISA. The performances of the Cytometric Bead Array (CBA, BD Biosciences), Multi-Analyte Profiling (xMAP, Luminex) and Coupled Particle Light Scattering (Copalis, Diasorin) were cited. The question the authors wished to address is whether or not there is good concordance between the data obtained by these platforms relative to other methods. The authors examined the measurement of human cytokines from serum by others. The problem faced is that there is invariably a lack of full disclosure in such reports. For example, were the same antibody pairs used? It is difficult to compare results for cases in which the multiplex and ELISA did not utilize identical antibody pairs. What diluents were required to overcome matrix effects? Multiplexing of standards, especially from recombinant sources, can be problematic and require careful design of diluents. What correlation method(s) were used? The statistical treatment or assumptions were applied to obtain the correlation should be stated.

Here an attempt is made to provide evidence to support the notion that bead-based multiplexing platforms in most instances compare favorably with the results obtained from ELISA. The authors conclude that by a "careful consideration of (these) variables, it should be possible to utilize multiplex bead array assays in lieu of ELISAs" (Elshal and McCoy, 2006, p. 317). It is perhaps a cautionary tale that we should subscribe. For example, in the cited study of DuPont et al. (2005), excellent correlation between Luminex and ELISA were obtained for seven cytokines, a fair correlation for one cytokine, and poor correlation for another. Another way of interpreting these results is that the multiplex assay was about 80% in agreement with the ELISA determinations.

Here are some findings:

1. "The degree of correlation has varied widely."
2. "Kits from different vendors yield different absolute concentration."
3. "Multiplex, on average, 2.36-fold higher than ELISA values."
4. "Assays from different vendors should not be considered interchangeable."

Siawaya et al. (2008) have undertaken an extensive comparison of analytical performance for Luminex cytokine assay kits from Bio-Rad and LINCO Research (now EMD Millipore) with those of R&D Systems Fluorokine Multianalyte Profiling MAP kit (bead-based cytokine kit) and Quantikine standard ELISA kits. This is a very detailed accounting of immunoassay performance for each platform across 13 to 29 cytokines, depending upon vendor for serum and whole blood supernatants. All bead-based assays were also compared against the IFN-gamma ELISA for response in antigen stimulated whole blood culture supernatants.

The accuracy for each assay (recovery of spiked samples) was measured for each cytokine. The acceptable recovery from spiked samples was taken at 70 to 130% of the spiked input. I believe this is a generous recovery window. Others suggest 80 to 120% recovery is more in keeping with analytical standards for immunoassays. In the case of positive readings from whole blood supernatants, the Bio-Rad 17-plex reported 21% within the recovery range; LINCO 29-plex, 78% within range; and R&D MAP 13-plex, 67%. Correlation with the IFN-gamma ELISA using the Pearson correlation coefficient (r) was poor for the Bio-Rad at −0.01; LINCO, 0.84; and R&D MAP, 0.99 (Figure 7.8). The outcome of these experiments led these authors to conclude that the Luminex technology is most suitable as a screening tool from which candidates may be selected and then validated by more accurate and reliable means.

Breen et al. (2011) undertook an extensive multisite study of Luminex bead-based platforms from Bio-Rad, BioSource, and LINCO, as well as the Meso Scale Discovery electro-chemiluminescent detection system. Using cohort materials linked to HIV infectivity, a total of 13 cytokines were analyzed using the commercial platforms in six different laboratories. While all platforms were designed to detect cytokines with high sensitivity, the study revealed considerable lot-to-lot assay variation. As the authors report, "no single multiplex panel detected all cytokines, and there were highly significant differences ($P < 0.001$) between laboratories and/or lots with all kits" (Breen et al., 2011, p. 1229). In essence, these platforms failed to deliver the level of inter-laboratory precision required for such comprehensive evaluations (e.g., the level of performance that would be required for clinical diagnostic assays).

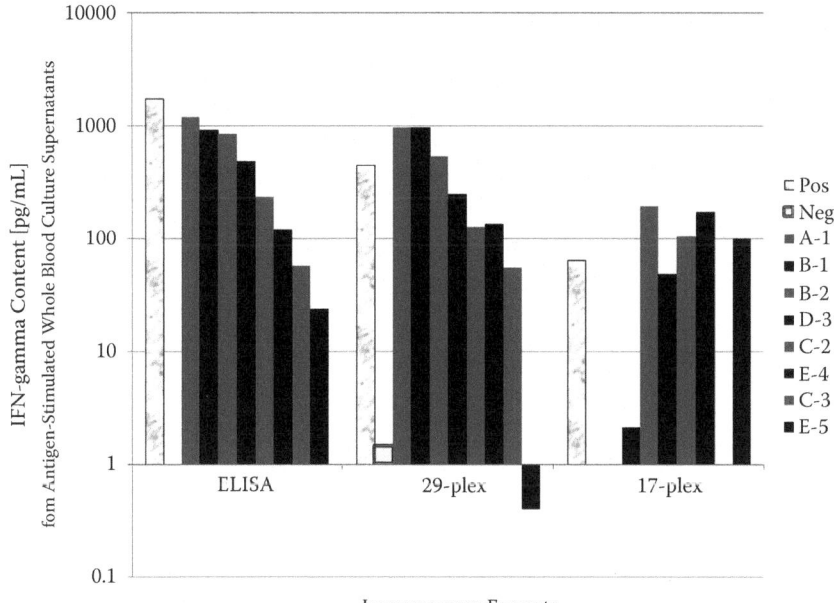

FIGURE 7.8 A comparison between commercial bead-based multiplex ELISA kits and standard ELISA in the determination of IFN-gamma levels in antigen-stimulated blood culture. (From Siawaya JFD, Roberts T, Babb C, et al. An evaluation of commercial fluorescent bead-based Luminex cytokine assays. *PLoS ONE* 3(7): e2535. doi:10.1371/journal.pone.0002535, 2008. With permission.) Positive control, phytohemagglutinin (PHA)-stimulated blood cells; negative control, unstimulated supernatant from healthy donor; Mtb (*Mycobacterium tuberculosis*) derived antigens (1 to 5) IFN-gamma responses in blood culture supernatants (A to E).

I believe the nemesis surrounding multiplexing technology (bead-based, arrays, etc.) remains the lack of a complete and proper assay validation process with standardized reference materials and protocols in place.

Multiplex Bead-Based Herpes Simplex Virus (HSV) Assay
Compares Favorably with Serological Methods

Martins and co-workers (2009) compared Alere's U.S. Food and Drug Administration (FDA)–approved AtheNA Multi-Lyte bead-based assay for the discrimination of Herpes simplex virus (HSV) species with that of ELISA and serological methods. HSV is a sexually transmitted pathogen that is responsible for orofacial (e.g., eye infections) as well as genital infections in adults. At great risk are infants who have been infected by transmittal from the mother at birth which can lead to neurological damage of the neonate. There is a high mortality with those who remain untreated due to missed diagnosis.

HSV can be serologically classified as type 1 (HSV-1) and type 2 (HSV-2). HSV-2 is largely associated with neonatal infection. HSV-1 is associated with genital infectivity. Both types are reported to be increasing in their worldwide prevalence.

The availability of recombinant forms of type-specific glycoprotein, gG-1 (HSV-1) and gG-2 (HSV-2), have allowed for the development of ELISA and immunoblot tests. These have displaced the more traditional Western blot methodology that was originally used in the diagnosis. Multiplex tests for sexually transmitted viral diseases, such as the ToRCH (*Toxoplasma gondii, Rubella virus, Cytomegalovirus, and Herpes Simplex Virus Types 1 and 2*) assay have also been introduced primarily for prenatal screening in order to begin early treatment and reduce further health risk of the neonate.

The **Multi-Lyte** bead assay (run on the Luminex flow platform) test includes a mixture of beads (in multiplex jargon, a 7-plex) containing gG-1/gG-2 antigens, a nonspecific control, as well as a set of four calibrator beads. The **ELISAs** (Focus Diagnostics) measure either HSV-1 or HSV-2 in wells coated with recombinant gG-1 or gG-2 antigens. The **immunoblots**, HerpeSelect 1 and 2 (Focus Diagnostics) are made up of nitrocellulose strips with four antigen bands: anti-human serum; HSV band (mix of HSV-1,2 antigens), gG-1 band, and gG-2 band. The **Western blot** was developed from HSV-1,2 infected human fibroblasts and the viral antigens recovered onto electrophoresis gels. These were subsequently transferred onto nitrocellulose membrane.

In the Martins et al. (2009) study 332 patient samples were analyzed for HSV. Of these, 140 were HSV-1 positive, 34 HSV-2 positive, 56 both HSV-1 and HSV-2 positive, 88 HSV negative, and 14 indeterminate by immunoblot. Sensitivity and specificity for all methods were high (>90%) with agreement. In this example, concordance of the bead-based multiplex with the other methods was excellent (Table 7.2).

The author's conclusion that the multiplex assay was "statistically equal" to the other methods is well supported. Furthermore, in consideration of reagent usage and labor, the multiplex assay offered significant cost savings over the ELISA (20% cost reduction) and immunoblot (60% cost reduction). Keep in mind that we have with

---

TABLE 7.2

Concordance between Multi-Lyte Bead Assay and Other Methods in Discrimination for Herpes Simplex Virus

Multiplex versus	Concordance (%)		
	ELISA	**Immunoblot**	**Western Blot**
HSV-1	94.8	95.8	95.8
HSV-2	96.9	97.9	98.8

Source: Martins TB, Welch RJ, Hill HR, Liwin CM. Comparison of a multiplexed Herpes Simplex virus type-specific immunoglobulin g serology assay to immunoblot, Western blot, and enzyme-linked immunosorbent assays. *Clin. Vaccine Immunol.* 16(1): 55–60, 2009. With permission.

this study a rather low complexity example (essentially a 2-plex) from which to glean the virtues of the bead-based assay designed to assess 50 to 100 biomarkers.

FALSE READINGS

Binnicker et al. (2010) evaluated the BioPlex 2200 (Bio-Rad Laboratories) bead-based multiplex assay for ToRC (Toxoplasma, Rubella, CMV) IgG and IgM relative to conventional plate-based immunoassays. Assessment of 600 serum samples by BioPlex and enzyme-linked immunoassays revealed excellent concordance for ToRC IgG but rather poor agreement for that of ToRC IgM. The BioPlex ToRC IgM showed false positive on 39 samples.

False readings have both health and economic consequences. In the case of IgM, its presence is often associated with a recent or current infection. Thus, confirmatory testing must be conducted on the mother and infant. As the authors point out by raising the positive cutoff value, the false-positive rate would drop dramatically but not completely. While the sample throughput for the bead-based assay would be over threefold that of the immunoassay (the authors report, "530 samples versus 260 samples per 9 hours"), any re-testing would reduce the advantage.

OBSERVED DIFFERENCES BETWEEN ELISA AND BEAD-BASED MULTIPLEX

De Jager et al. (2003) prepared and evaluated a 15-plex bead-based assay using the Bio-Plex system (Bio-Rad) and compared its performance relative to standard ELISA from various vendors. Assessment of cytokines in antigen-stimulated PBMC (peripheral blood mononuclear cells) cell culture media was undertaken. Cells were obtained from healthy individuals as well as patients with rheumatoid arthritis or juvenile idiopathic arthritis.

During the process the authors noted that a number of commercial antibodies used for ELISA failed to work in the bead-based assay. This could be the consequence of covalent attachment of the antibodies to beads, perhaps by alteration of the antigen binding site. Another possibility is poor response to recombinant antigens.

Dynamic ranges (pg/mL) for ELISA and Bio-Plex were compared for the 15-plex cytokine panel. Results were comparable at the lower end of the range, while the Bio-Plex had an extended upper level over that of the ELISA in all cases. There were some differences in performance. For example, IL-2 by ELISA was 6-fold less sensitive than Bio-Plex, while IL-4 was about 40-fold and IL-8 was about 5-fold more sensitive by ELISA than with the Bio-Plex.

In the measurement of PBMC cytokines in culture media, a good correlation (r^2, 0.750 to 0.991) between the two techniques was observed. Serum matrix effects (presumed to be due to presence of heterophilic antibodies) were observed with IL-13 by both techniques. Heterophilic antibodies are anti-animal (e.g., human anti-mouse) immunoglobulin (all classes) found in human serum of individuals who have had contact with an animal species. This can occur as a natural immune response to a foreign "animal" antigen, by vaccination, or by other means. For a good review on the etiology see the article by Kricka (1999). The problem is often encountered during use of a monoclonal (mouse hybridoma) based sandwich ELISA using human serum. In this case, interfering agents completely blocked development of the assay. Replacement of the serum diluents with buffer elevated the problem.

As a side note, we have also observed the presence of low but detectable levels of analytes in a number of serums. When developing multiplex assays it is very important to thoroughly screen serum-based diluents for interference as well as intrinsic or contaminant analyte species.

COMBINED MULTIPLEX PCR AND MULTIPLEX BEAD-BASED ASSAY

Bøving et al. (2009) sought out an alternative amplification method for the determination of bacterial meningitis from cerebrospinal fluid (CSF). Although 16S rRNA gene amplification by PCR may be used, the authors expressed concern with potential contaminations by other bacterial species' DNA in handling or trace presence in buffers and enzymes used in PCR. As a result, species-specific multiplex PCR coupled with Luminex bead-based detection was elected for this study. The advantage for this approach cited by the authors was that detection of multiple targets by real-time PCR was limited by current instrumentation, while the Luminex 100 system could easily handle the detection of multiple PCR products. So, an 8-plex PCR amplification of bacterial and viral pathogens causing meningitis was undertaken with bead-based capture of the resulting biotinylated amplicons through hybridization. Following sample processing, the Luminex system was used to detect and quantify.

The Danish group analyzed 1187 CSF samples over a period of 1 year. Of these, 55 were scored positive by the PCR-Luminex method. However, 25/55 (45%) were determined to be false-positives. There were six bacterial pathogens examined. A significant number of false positives were scored in five of six species. For *S. aureus* a high false-negative rate was also reported leading to a sensitivity of 33% by the method, and that of *S. pneumonia*, sensitivity of 90.5%. For the viral pathogens (HSV-1,2 and VZV) significant rates of false positives were scored as well. Overall specificity for the 8-plex ranged from 99.1 to 100%; sensitivity excluding the above two cases was 100%.

The authors conclude that the PCR-Luminex method is useful for screening purposes to identify pathogen species "from patients with suspected meningitis" (Bøving et al., 2009, p. 908). While reported analytical sensitivity and specificity are reasonable, the clinical metric matters most here. False positives and false negatives can have an impact on a clinical outcome. False positives can lead to unnecessary surgery or therapies (Kricka, 2000), while a missed diagnosis may lead to delayed treatment (Tate & Ward, 2004).

What did this study accomplish? First, the authors demonstrated the utility of combining two powerful multiplexing technologies: multiplex PCR for the mass amplification of multiple pathogens, and the Luminex bead suspension array system providing the capability for sensitive multiplex detection. The PCR-Luminex method was successfully validated (albeit, with reservation regarding the false positive-negative rates) in a clinical setting, offering the advantage of a rapid and higher-throughput screening of patients.

MULTIPLEX MICROARRAYS

In 2008, Claudon et al. reported on studies measuring biomarkers for osteoporosis using Roche's IMPACT chip (Immunological Multi-Parameter Chip Technology). IMPACT employs a streptavidin coated polystyrene substrate to which biotinylated analyte capture antibodies are spotted. The secondary antibody, conjugated with digoxin, is added to serum to capture the analyte. After formation of the sandwich ELISA, anti-digoxin antibodies immobilized on fluorescent latex particles are used to generate signal with detection using a CCD camera. The fully automated platform processes 41 samples in duplicate within 1 hour.

The following bone turnover biomarkers were used in the multiplex immunoassay for the evaluation of osteoporosis: CTX-1 (C-terminal cross-linked telopeptide of type 1 collagen), PINP (N-terminal propeptide of type 1 collagen), OC (osteocalcin), and PTH (parathyroid hormone). These biomarkers are important in monitoring treatment and risk assessment of potential bone fracture. In this study, the Elecys 2010 automated analyzer (Roche Diagnostics, GmbH) was used to measure individual biomarkers for comparison to the multiplex platform. Samples for evaluation were obtained from three groups: 157 healthy premenopausal women, 56 postmenopausal women with osteoporosis evaluated before and after treatment with ibandronate (bone resorption drug), and 74 healthy men.

. There has been some concern in the immunoassay community that the multiplex immunoassay suffers from reduced sensitivity relative to singlet assays due to increased cross-reactivity between antibodies. This is certainly not the case in the reported study where limits of detection (LLOD) and quantification (LLOQ) were lower for the multiplex (Figure 7.9). Except for PINP, the intra- and inter-assay analytical imprecision (% CV) were comparable or lower for the multiplex format. The PINP intra-assay CV, singlet versus multiplex (1 to 2.1% versus 6 to 6.8%); and inter-assay CV, singlet versus multiplex (2 to 4.4% versus 4.7 to 9.2%) are still acceptable levels of imprecision. In concordance of singlet assay and multiplex assay across patient samples, $r = 0.93$ to 0.97 ($P < 0.0001$) for the four biomarkers.

In a more recent study, Chandra et al. (2011) utilized the IMPACT platform to stratify patients with early-stage rheumatoid arthritis (RA). Even though the etiology of rheumatoid arthritis is not well understood, a number of biomarkers have been implicated. Candidate biomarkers include RF (rheumatoid factor), CRP (C-reactive protein), and CCPs (cyclic citrullinated peptides). The problem is that individually they all lack sufficient specificity or sensitivity, and no single biomarker has been identified that is prognostic of this complex disease. Hence, the creation of a multiplex panel of RA biomarkers could potentially provide the opportunity to identify and differentiate various RA disease states.

Chandra et al. (2011) prepared a set of seven different multiplex biomarker IMPACT chips (a total of 41 biomarkers including 18 synovial-related antigens, 11 citrullinated peptides, 12 capture antibodies) for use in their study of a subset of patient serum samples from the ARAMIS (Arthritis, Rheumatism and Aging Medical Information System, Stanford University) cohort. This included patients with various forms of the disease: RA (120 patients), PsA (psoriatic arthritis, 28 patients), AS (ankylosing spondylitis, 27 patients), and 25 healthy individuals. The

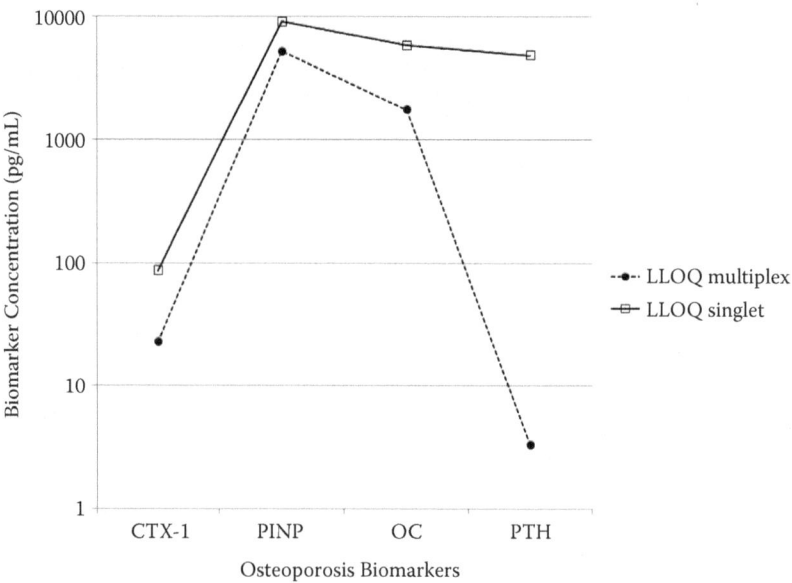

FIGURE 7.9 Analytical sensitivity in the assessment of bone biomarkers by IMPACT multiplex immunoassay and ELISA. (CTX-1 [C-terminal cross-linked telopeptide, type 1 collagen]; PINP [N-terminal propeptide, type 1 collagen]; OC [osteocalcin]; PTH [parathyroid hormone]; LLOQ [lower limit of quantification, the lowest concentration measurable at <20% CV].) (From Claudon A, Vergnaud P, Valverde C, Mayr A, Klause U, Garnero P. New automated multiplex assay for bone turnover markers in osteoporosis. *Clin. Chem.* 54(9): 1554–1563, 2008. With permission.)

Luminex platform was also employed to measure the levels of 13 pro-inflammatory cytokines that are often associated with RA.

Following normalization of the biomarker against healthy control samples, the results were subjected to hierarchical clustering and displayed to reveal any differentiation between disease states on the basis of autoantibody reactivity or cytokine levels. In essence, the graphical display permits the selection and development of RA biomarker signatures that may stratify RA patients by disease subtype. In this manner, the resulting hierarchical profile obtained from the cohort of 120 patients for 54 biomarkers and cytokines identified six candidate biomarkers. These six biomarkers when combined in various panel groupings provided increased sensitivity and specificity toward the early-stage RA diagnosis.

Multiplex Microarray Plate Assay Compared to the ELISA

Kang et al. (2012) created an ELISA array for the detection of five viruses associated with encephalitis and compared the multiplex assay with that of a conventional ELISA. Twenty-three monoclonal antibodies were prepared that varied in specificity toward the virus antigens. Of these, six were selected as capture antibodies, while three were biotinylated for use as detection antibodies. Capture antibodies were spotted directly onto ELISA plates in an array format to create the ELISA array. The same antibodies were coated in wells to produce a conventional ELISA plate format. Following optimization studies the two formats were evaluated in terms of cross-reactivity, sensitivity, and specificity in the detection of the pathogens.

The encephalitis viruses—dengue virus (DV-2; DV-4), Japanese encephalitis virus (JEV), tick-borne encephalitis virus (TBE), sindbis virus (SV), and eastern equine encephalitis virus (EEEV)—were cultured in host cells. The titers were estimated using the 50% tissue culture infectious dose ($TCID_{50}$) method. Here virus is serially diluted and inoculated into the host cell tissue culture. The viral load that kills 50% of the infected cells is calculated. The titers were evaluated by ELISA array and conventional ELISA for these pathogens. Cross-reactivity was determined by challenge using other viruses (Yellow fever, West Nile, Western equine). Neither assay format showed cross-reactivity. No false-positive or false-negative results were obtained based upon examination of virus combinations with the biotinylated detection antibodies.

Sensitivity for the two assay formats was compared based upon titer determination from the serially diluted viral cultures. The $TCID_{50}$ for the ELISA array was lower in five of six viral titers in comparison to that of the standard ELISA. The authors concluded that the ELISA array provided a higher sensitivity (Figure 7.10). They also noted that the microarray format used 60-fold less antibody, requiring

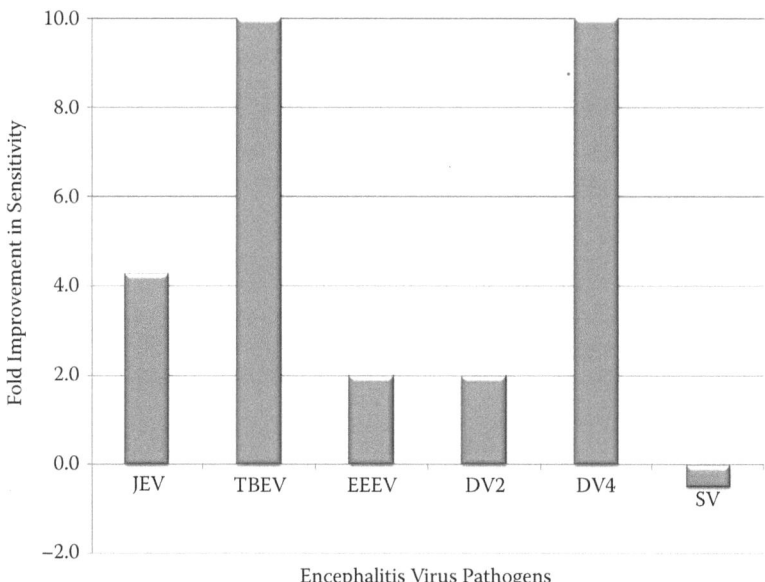

FIGURE 7.10 Increased sensitivity by multiplex ELISA in the detection of encephalitis viral species. Fold improvement in detection limit ($TCID_{50}$/mL) by ELISA array relative to standard ELISA format. (From Kang X, Li Y, Fan L, et al. Development of an ELISA-array for simultaneous detection of five encephalitis viruses. *Virol. J.* 9(56): 1–8, 2012. With permission.)

1.5 ng of monoclonal antibody per spot, while optimal conditions for the standard ELISA used 100 ng of each antibody for well coating. Finally, the multiplex format was significantly faster to perform. The conventional ELISA took 6 hours to complete, while the ELISA array required only 4 hours.

A LACK OF STANDARDIZATION

Toedter and co-workers (2008) compared two commercially available plate-based multiplex ELISA formats with a standard ELISA (R&D Systems, Minneapolis, MN) for eight cytokines present in human serum. The Pierce SearchLight system (currently offered by Aushon Biosystems, Billerica, MA) is based upon the spotting of capture antibodies onto nitrocellulose membrane in 96-well microplates. A CCD camera system imaging of chemiluminescent signal is used for detection. Meso Scale Discovery's platform utilizes electrochemiluminescent detection using a CCD camera. Capture antibodies are spotted onto carbon-coated electrode pad arrays formed within the microplate well. The R&D System's ELISA for single analytes also utilizes chemiluminescent substrates for detection. The MSD assay dynamic range is very broad covering the analysis of cytokine from about 1 pg/mL to 40,000 pg/mL, while the SearchLight product dynamic range is 0.4 pg/mL to 5000 pg/mL. The MSD system uses undiluted serum samples, and the assay is complete within 4.5 hours. The SearchLight platform requires 1:5 v/v dilution of serum and about 2.5 hours for a complete assay. The R&D System ELISA for the cytokines tested covers a range similar to that of SearchLight.

Endogenous levels of various cytokines from healthy individuals were measured using these platforms. Although the reported concentration (pg/mL) of certain cytokines such as MCP-1 and VEGF were in agreement across platforms, others such as IL-6 and IL-12p40 varied dramatically (Figure 7.11). Correlation of the MSD platform with that of the Pierce SearchLight and R&D Systems ELISA in the determination of cytokine levels present in various disease states revealed even greater differences (Figure 7.12). The author's evaluation of the eight serum cytokine levels measured by these platforms resulted in the following conclusion: "The consistency of cytokine detection across all three platforms varied depending upon the cytokine being tested and the disease state."

This is to be expected. All assays were found to exhibit low variability within their respective system (intra-assay), but the lack of standardization across platforms is self-evident. A comparison of the various multiplex assay formats is provided in Table 7.3.

LABEL FREE DETECTION OF A MULTIPLEX SMALL MOLECULE MICROARRAY

Fei et al. (2010) created a small molecule compound microarray of more than 10,000 elements to screen protein ligands. Measurement of compound-protein ligand binding kinetics was conducted using label free detection by scanning microscopy. As the authors discuss, one of the paramount challenges in producing a microarray-based compound library is how to immobilize small molecules without bias or destruction of functionality. Moreover, the compound screening of protein ligands can be

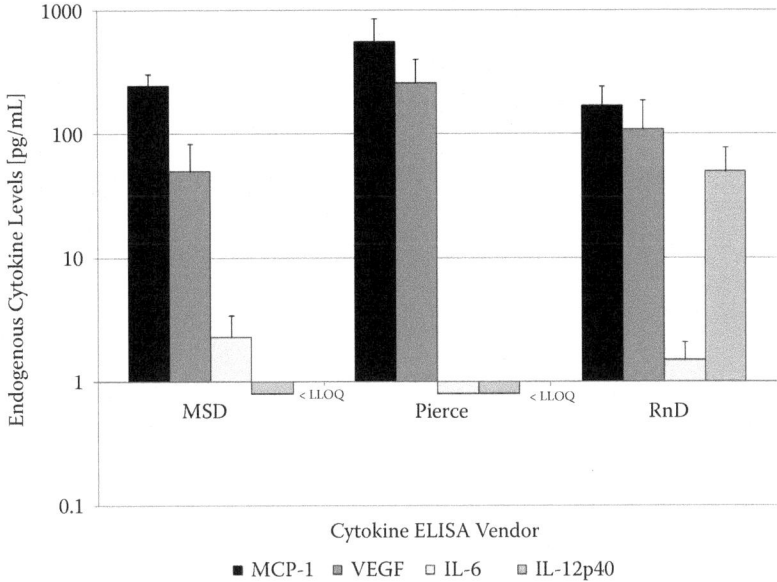

FIGURE 7.11   Correlation (coefficient) between commercial cytokine assay kits in measuring serum cytokines in various disease state samples—ulcerative colitis, rheumatoid arthritis (RA), and Crohn's disease. (From Toedter G, Hayden K, Wagner C, Brodmerkel C. Simultaneous detection of eight analytes in human serum by two commercially available platforms for multiplex cytokine analysis. *Clin. Vaccine Immunol.* 15(1): 42–48, 2008. With permission.)

compromised by modification of the protein during labeling, thereby leading to an alteration in the native binding site. Thus, it would be advantageous to use label free, native proteins.

For label free work, a special scanning microscope based upon detection of changes in light polarization upon ligand interaction at a surface was developed. The OI-RD (oblique-incidence reflectivity difference) scanning microscope measures the difference in reflectance between *p*-polarized light in the plane of the substrate (*rp*) and that of *s*-polarized light that is perpendicular to the substrate surface (*rs*) as delta Δ *rp/rs*, where *r* is the reflectance coefficient. The OI-RD can scan a microscope slide array (2 cm × 4 cm) of 10,000 elements in 1.5 hours.

As a demonstration, Fei et al. (2010) constructed a 20-plex library of biotinylated synthetic test compounds by printing down onto a streptavidin coated glass slide (ArrayIt, Sunnyvale, CA) using an OmniGrid 100 contact arrayer (Digilab, Holliston, MA). Each slide contained two microarrays of 100 elements each, printed with 10 biotin compounds present in duplicate at five different loading concentrations from 1 mM to 0.0625 mM, presumably by 1:2 (v/v) serial dilution (i.e., 10 × 2 × 5 = 100 spots). A Jurkat cell lysate was applied and the binding events measured in real time by OI-RD at 60 second intervals between microarray elements. Jurkat cells are immortalized T lymphocytes commonly used in drug screenings.

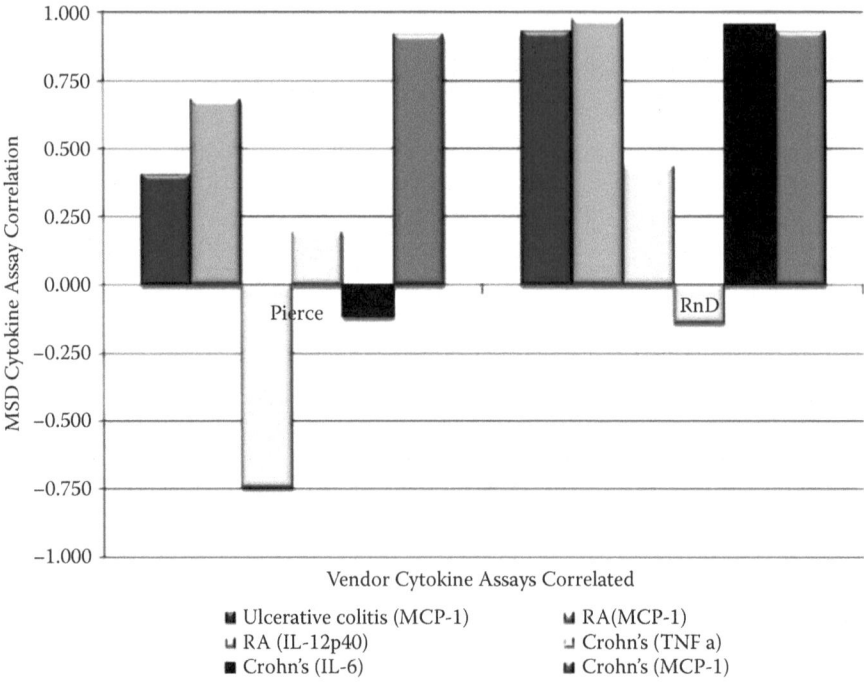

FIGURE 7.12   Comparison between plate-based multiplex cytokine assays in the measurement of endogenous levels of cytokines in human serum from healthy patients. RnD (standard chemiluminescent ELISA; R&D Systems); Pierce (Pierce Endogen SearchLight—multiplex chemiluminescent assay); MSD (Meso Scale Discovery—multiplex electro-chemiluminescence assay). (From Toedter G, Hayden K, Wagner C, Brodmerkel C. Simultaneous detection of eight analytes in human serum by two commercially available platforms for multiplex cytokine analysis. *Clin. Vaccine Immunol.* 15(1): 42–48, 2008. With permission.)

Only a single compound was found not to bind to the lysate proteins. This appeared to be due to a missed printing of that compound. As a higher-throughput example, the authors printed down 4608 biotinylated small molecule compounds in duplicate along with controls to achieve a microarray at a spot density of 10,804 targets. The microarray was then exposed to mouse anti-biotin IgG (Fab fragment) at a concentration of 13 μg/mL (260 nM) for 1 hour. The microarray was then scanned by OI-RD and the end-point signal determined. As reported, variation in signal across replicates was observed that is most likely attributable to the variation in spot density due to printing or immobilization. Nevertheless, the label free detection of a small molecule microarray was accomplished using the OI-RD microscope system.

ADOPTION OF MULTIPLEX ASSAYS

As with any technology there are pro and con arguments concerning acceptance (Table 7.4). And, as we have seen, only a few multiplex assay platforms for in vitro diagnostics are now commercially available. Multiplex panels may improve

TABLE 7.3
Comparison of Multiplex Assay Formats: Summary of Assay Formats and Platform Characteristics

Multiplex Format	Plex-Size (Features)	Analyte	Matrix	Samples	First Answer	Sensitivity (units/mL)	Detection	Dynamic Range (linear)	Imprecision (% CV)
Lateral flow	<10	Protein/DNA	Blood; serum	1	10–15 minutes	10E4-10E7 cfu ng protein	Colorimetric fluorescence	2–3 logs	>10%
Polymerase chain reaction (PCR)	<10	DNA	Blood; serum	96/384/1536	Several hours DNA prep 2–3 hours assay	10–20 genomes 10E2 cfu	Fluorescence		
Bead	2 to 100	DNA/protein	Serum; media	96	3 hours	pg protein 30 genomes 10E5 cfu	Fluorescence	3–3.5 logs	Inter-Assay 5-85% / Inter-Assay 5-15%
Microarray slide									
Oligonucleotides	100s to 1000s	DNA	Serum; media	1 to 16	2–18 hours	10E4 cfu	Fluorescence		5-15%
Antibody	10s to 100s	Proteins	Serum; media	1 to 16	1–3 hours	pg protein	Fluorescence		
Microarray plate	2 to 20	Proteins	Serum; media	96/384	2–3 hours	pg protein	Fluorescence chemiluminescence	2.5–3 logs	Inter-5-15% / Intra-5-10%
Electrochemical chip	4 to 10	Proteins	Serum; media	96	4 hours	pg protein	Electro-chemiluminescence	4–5 logs	Inter-10 to 100% / Intra-5-15%
Microarray plate									

Sources: See Wong H-L, Pfeiffer RM, Fears TR, Vermeulen R, Ji S, Rabkin CS. Reproducibility and correlations of multiplex cytokine levels in asymptomatic persons. *Cancer Epidemiol. Biomarkers Prev.* 17(2): 3450–3456, 2008; Montague J. Investigating assay precision and validation on different multiplex immunoassay platforms, *Decision Biomarkers*, August 18, 2009; Bastarche JA, Koyama T, Wickersham NE, Mitchell DB, Mernaugh RL, Ware LB. Accuracy and reproducibility of a multiplex immunoassay platform: A validation study. *J. Immunol. Methods* 367(1–2): 33–39, 2011; and Anderson M, David J, Pearlman S, Schmidt, J. Evaluation of multiplex immunoassay results. *Genet. Eng. Biotechnol. News,* April 15, 2011. With permission.

TABLE 7.4

Adoption of Multiplexed Assays for Diagnostics

Pro	Con
• Certain diagnostic panels run as singlets already exist	• Installed instrumentation based upon singlet menu expansion
• Multiplexed panels provide increase in reimbursement money	• QA/QC/Manufacturability
• Attractive for POC/POL where a single sample draw is desired	• Small batch sizes
• Encoded beads offer potential for random access	• Robustness
	• Complexity
	• Multi-step
	• Many reagents
	• Lack of experience:
	– Image analysis
	– Multi-variant analysis
	• Key content may need to be licensed (royalty stack)

prognosis and diagnosis, but the economic incentives (reimbursement) must be in place. There also is a continuing need to educate. The technology is complex. In order for the multiplexing assay to gain further acceptance as a diagnostic tool, it must be well understood by users and regulators alike (Boja et al., 2011).

TRENDS

As we have reviewed, there is a significant level of interest and support in the advocacy and adoption of multiplex assays in the scientific community. DNA microarrays have successfully bridged the technology gap. We have seen how nucleic acid–based microarrays have progressed from research beginning in the late 1980s, commercialization in the mid-1990s, and the first FDA cleared test based upon genetic analysis by a microarray (GeneChip by Affymetrix for Roche's AmpliChip Cytochrome P450 Genotyping Test) approved in December 2004. Vermillion's OVA1 test became the first FDA cleared protein biomarker multiplex assay (5-plex) in 2009 for ovarian cancer based upon multivariant analysis scoring (Smith, 2011).

It is unclear what formats and platforms will be most utilized in the future within the biopharmaceutical arena and in clinical settings. It is apparent that the complexity of multiplex testing, at least in the case of biomarkers, can be reduced to a manageable level (i.e., panels including 3 to 20 biomarkers are likely sufficient) (Figure 7.13). As companion diagnostics evolve and reimbursement strategies are clarified, multiplex assays are well positioned for adoption. There are, of course, technologies that have emerged such as next-generation sequencing (NGS) that will undoubtedly impact the future utility of DNA microarrays. However, next-generation sequencing is still finding its way and is by no means a mature, validated, and standardized commercial platform. It is plausible that such tools in the future will continue to be refined into useful and complementary platforms (Teng

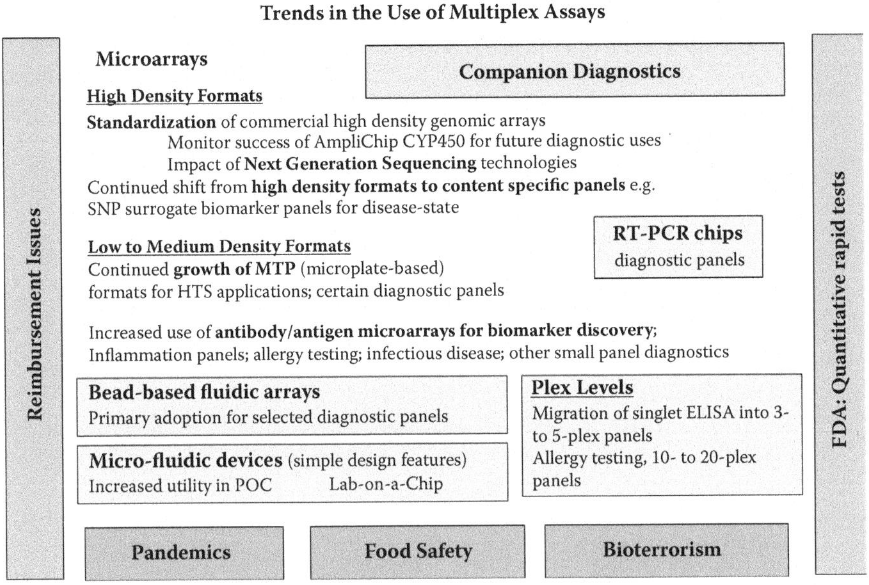

Trends in the Use of Multiplex Assays

FIGURE 7.13   Future trends for multiplex assay use.

and Xiao, 2009). Finally, in the face of catastrophic events such as pandemics (e.g., H1N1), issues of food safety (e.g., U.S. Food Safety Modernization Act), and the threat from bio-terrorism, there is a growing need for more quantitative and rapid multiplex tests.

REFERENCES

Achatz DE, Ali R, Wolfbeis OS. Luminescent chemical sensing, biosensing and screening using upconverting nanoparticles. *Top. Curr. Chem.* 300: 29–50, 2011. doi:10.1007/128-2010-98.

Anderson M, David J, Pearlman S, Schmidt, J. Evaluation of multiplex immunoassay results. *Genet. Eng. Biotechnol. News*, April 15, 2011.

Bastarche JA, Koyama T, Wickersham NE, Mitchell DB, Mernaugh RL, Ware LB. Accuracy and reproducibility of a multiplex immunoassay platform: A validation study. *J. Immunol. Methods* 367(1–2): 33–39, 2011.

Binnicker MJ, Jespersen DJ, Harring JA. Multiplex detection of IgM and IgG class antibodies to *Toxoplasma gondii*, Rubella virus and Cytomegalovirus using a novel multiplex flow immunoassay. *Clin. Vaccine Immunol.* 17(11): 1734–1738, 2010.

Boja ES, Jortani SA, Ritichie J, et al. The journey to regulation of protein-based multiplex quantitative assays. *Clin. Chem.* 57(4): 560–567, 2011.

Bøving MK, Pedersen LN, Moller JK. Eight-plex PCR and liquid-array detection of bacterial and viral pathogens in cerebrospinal fluid from patients with suspected meningitis. *J. Clin. Microbiol.* 47(4): 908–913, 2009.

Breen EC, Reynolds SM, Cox C, et al. Multisite comparison of high-sensitivity multiplex cytokine assays. *Clin. Vaccine Immunol.* 18(8): 1229–1241, 2011.

Chandra PE, Sokolove J, Hipp BG, et al. Novel multiplex technology for diagnostic characterization of rheumatoid arthritis. *Arthritis Res. Ther.* 13: R102, 2–13, 2011.

Claudon A, Vergnaud P, Valverde C, Mayr A, Klause U, Garnero P. New automated multiplex assay for bone turnover markers in osteoporosis. *Clin. Chem.* 54(9): 1554–1563, 2008.

Corstjens PLM, Chen Z, Zuiderwijk M, et al. Rapid assay format for multiplex detection of humoral immune responses to infectious disease pathogens (HIV, HCV, and TB). *Ann. N.Y. Acad. Sci.* 1098: 437–445, 2007.

De Jager W, te Velthruis H, Prakken BJ, Kuis W, Rijkers GT. Simultaneous detection of 15 human cytokines in a single sample of stimulated peripheral blood mononuclear cells. *Clin. Diagn. Lab. Immunol.* 10(1): 133–139, 2003.

Dupont NC, Wang K, Wadhwa PD, Culhane JF, Nelson EL. Validation and comparison of Luminex multiplex cytokine analysis kits with ELISA: Determinations of a panel of nine cytokines in clinical samples of culture supernatants. *Cancer Epidemiol. Biomarkers Prev.* 11(11): 175–191, 2005.

Elshal MF, McCoy JM. Multiplex bead array assays: Performance evaluation and comparison of sensitivity to ELISA. *Methods.* 38(4): 317–323, 2006.

Fei Y, Landry JP, Sun Y, Zhu X. Screening small-molecule compound microarrays for protein ligands without fluorescence labeling with a high-throughput scanning microscope. *J. Biomed. Optics* 15(1): 016018-1–016018-8, 2010.

Kang X, Li Y, Fan L, et al. Development of an ELISA-array for simultaneous detection of five encephalitis viruses. *Virol. J.* 9(56): 1–8, 2012.

Kim JS, Taitt CR, Ligler FS, Anderson GP. Multiplexed magnetic microsphere immunoassays for detection of pathogens in foods. *Sens. Instrum. Food Qual. Saf.* 4(2): 73–82. doi: 10.1007/s11694-010-9097-x, 2010.

Kricka LJ. Human anti-animal antibody interferences in immunological assays. *Clin. Chem.* 45(7), 942–956, 1999.

Kricka LJ. Interferences in immunoassay—Still a threat. *Clin. Chem.* 46(8): 1037, 2000.

Martins TB, Welch RJ, Hill HR, Liwin CM. Comparison of a multiplexed herpes simplex virus type-specific immunoglobulin g serology assay to immunoblot, Western blot, and enzyme-linked immunosorbent assays. *Clin. Vaccine Immunol.* 16(1): 55–60, 2009.

Montague J. Investigating assay precision and validation on different multiplex immunoassay platforms. *Decision Biomarkers*, August 18, 2009.

Noguera P, Posthuma-Trumpie GA, van Tuil M., et al. Carbon nanoparticles in lateral flow methods to detect genes encoding virulence factors of Shiga toxin-producing *Escherichia coli. Anal. Bioanal. Chem.* 399, 831–838, 2011.

Park J, Park S, Kim Y-K. Multiplex detection of pathogens using an immunochromatographic assay strip. *BioChip J.* 4(4): 305–312, 2010.

Peterson G, Gerdes B, Berges J, et al. Development of microarray and multiplex polymerase reaction assays for identification of serovars and virulence genes in *Salmonella enteric* of human or animal origin. *J. Vet Diagn. Invest.* 22: 559–569, 2010.

Pui CF, Wong WC, Chai LC, et al. Multiplex PCR for the concurrent detection and differentiation of *Salmonella* spp., *Salmonella typhi* and *Salmonella typhimurium. Tropical Med. Health* 39(1): 9–15, 2011.

Siawaya JFD, Roberts T, Babb C, et al. An evaluation of commercial fluorescent bead-based Luminex cytokine assays. *PLoS ONE* 3(7): e2535. doi:10.1371/journal.pone.0002535, 2008.

Smith KM. Exploring FDA-approved IVDMIAs. *IVD Technol.* 17(3): May/June, 2011.

Tate J, Ward G. Interferences in immunoassay. *Clin. BioChem. Rev.* 25: 105–120, 2004.

Teng X, Xiao H. Perspectives on DNA microarray and next-generation DNA sequencing technologies. *Sci. China C-Life Sci.* 52(1): 7–18, 2009.

Toedter G, Hayden K, Wagner C, Brodmerkel C. Simultaneous detection of eight analytes in human serum by two commercially available platforms for multiplex cytokine analysis. *Clin. Vaccine Immunol.* 15(1): 42–48, 2008.

Wong H-L, Pfeiffer RM, Fears TR, Vermeulen R, Ji S, Rabkin CS. Reproducibility and correlations of multiplex cytokine levels in asymptomatic persons. *Cancer Epidemiol. Biomarkers Prev.* 17(2): 3450–3456, 2008.

Yu A, Schneiderhan-Marra N, Joos TO. Protein microarrays for personalized medicine. *Clin. Chem.* 56(3): 376–387, 2010.

Zhu J, Zou N, Zhu D, et al. Simultaneous detection of high-sensitivity cardiac troponin I and myoglobin by modified sandwich lateral flow immunoassay: Proof of principle. *Clin. Chem.* 57(12): 1732–1738, 2011.

Index